Test Bank

to accompany

CHEMISTRY

The Molecular Nature of Matter and Change

Second Edition

Martin S. Silberberg

Prepared by
S. Walter Orchard
University of Puget Sound

Boston Burr Ridge, IL Dubuque, IA Madison, WI New York San Francisco St. Louis
Bangkok Bogotá Caracas Lisbon London Madrid
Mexico City Milan New Delhi Seoul Singapore Sydney Taipei Toronto

McGraw-Hill Higher Education
A Division of The ***McGraw-Hill*** *Companies*

Test Bank to accompany
CHEMISTRY: THE MOLECULAR NATURE OF MATTER AND CHANGE, SECOND EDITION

Printed in the United States of America.

This book is printed on acid-free paper.

1 2 3 4 5 6 7 8 9 0 QPD QPD 9 0 3 2 1 0 9

ISBN 0-697-39602-9

www.mhhe.com

CONTENTS

PREFACE

This revised and expanded Test Bank contains over 2,000 questions. Almost all of the questions in the previous Test Bank, written by Dennis Flentge, have been retained, and more than 800 new questions have been added. The Test Bank now has three categories of questions: Multiple Choice (including over 200 new ones), Short Answer, and True/False. Each question has been assigned a difficulty rating: E (easy), M (moderate) or H (hard).

In preparing the new Test Bank, I have attempted to add more variety to the questions in terms of style, content and level of difficulty. Most of the new questions have been field tested in my classes at the University of Puget Sound. Answers to all questions are given at the end of each chapter. The answers to the Short Answer questions reflect my own particular emphasis, and should be treated accordingly.

In combining the old questions with the new, some inconsistency in the fonts and styles of certain symbols has crept in. This should not affect the usability of the Test Bank in any way.

I would like to thank my colleagues at the University of Puget Sound, particularly Steven Neshyba, who helped in various ways; my students; and Margaret Horn at McGraw-Hill for calmly keeping me on track. Finally, I thank my wife, Christine, and daughters, Sam and Gogga, for their interest, encouragement and inspiration.

Walter Orchard
University of Puget Sound
Tacoma, WA

Keys to the Study of Chemistry
Chapter 1

Multiple Choice Questions

1. Which one of the following is a "substance" in the sense of the word as used in your text book?
E

a. sea water
b. tap water
c. air
d. silver
e. wood

2. Select the best statement.
E

a. Physical changes may be reversed by changing the temperature.
b. Physical changes alter the composition of the substances involved.
c. Physical properties are not valid characteristics for identifying a substance.
d. Physical properties are mostly extensive in nature.

3. Select the best statement.
E

a. Chemical changes provide the only valid basis for identification of a substance.
b. Chemical changes are easily reversed by altering the temperature of the system.
c. Chemical changes always produce substances different from the starting materials.
d. Chemical changes are associated primarily with extensive properties.

4. Which of the following is a chemical change?
E

a. vaporizing water
b. melting wax
c. broiling a steak on a grill
d. condensing water vapor into rainfall

5. During the swing of a frictionless pendulum, what energy form(s) remain constant?
H

a. kinetic energy only
b. potential energy only
c. both kinetic energy and potential energy
d. kinetic plus potential energy

6. The most significant contribution to modern science made by alchemists was
M

a. their fundamental work in the transmutation of the elements
b. their widespread acceptance of observation and experimentation
c. their systematic method of naming substances
d. their understanding of the nature of chemical reactions

7. Select the best statement about chemistry before 1800.
M

a. Alchemy focused on objective experimentation rather than mystical explanations of processes.
b. The phlogiston theory laid a valuable theoretical basis for modern chemistry.
c. Lavoisier's quantitative work on the role of oxygen in combustion was the beginning of modern chemistry.
d. The interpretation of data by alchemists was not biased by their overall view of life.

8. The distance between carbon atoms in ethylene is 134 picometers. Which of the following
E expresses that distance in meters?

a. 1.34×10^{-13} m
b. 1.34×10^{-10} m
c. 1.34×10^{-7} m
d. 1.34×10^{-6} m

9. The average distance from Earth to the Sun is 150 megameters. What is that distance in meters?
E

a. 1.5×10^{8} m
b. 1.5×10^{6} m
c. 1.5×10^{5} m
d. 1.5×10^{3} m

10. The mass of a sample is 550 milligrams. Which of the following expresses that mass in kilograms?
E

a. 5.50×10^{8} kg
b. 5.50×10^{5} kg
c. 5.50×10^{-4} kg
d. 5.50×10^{-1} kg

11. A dose of medication was prescribed to be 35 microliters. Which of the following expresses that
E volume in centiliters?

a. 3.5×10^{5} cL
b. 3.5×10^{4} cL
c. 3.5×10^{-4} cL
d. 3.5×10^{-3} cL

12. Which of the following represents the largest volume?
M

a. 10,000 μL
b. 1000 pL
c. 100 mL
d. 10 nL

13. You prepare 1000 mL of tea and transfer it to a 1.00 quart pitcher for storage. Which of the
M following statements is true?

a. The pitcher will be filled to 100% of its capacity with no tea spilled.
b. The pitcher will be filled to about 95% of its capacity.
c. The pitcher will be filled to about 50% of its capacity.
d. The pitcher will be completely filled and tea will spill over onto the counter.

14. M Sulfuric acid is consistently at the top of the list of chemicals produced in the United States and is widely used in the production of fertilizers. In 1994, 89.20 billion pounds of sulfuric acid were produced. How many kilograms was this?

a. 4.045×10^{13} kg
b. 1.967×10^{11} kg
c. 4.045×10^{10} kg
d. 1.967×10^{8} kg

15. M In an average year the American chemical industry produces more than 9.5 million metric tons of sodium carbonate. Over half of this is used in the manufacture of glass while another third is used in the production of detergents and other chemicals. How many pounds of sodium carbonate are produced annually?

a. 2.1×10^{10} lb
b. 4.3×10^{9} lb
c. 1.1×10^{7} lb
d. 2.2×10^{6} lb

16. M A large pizza has a diameter of 15 inches. Express this diameter in centimeters.

a. 38 cm
b. 24 cm
c. 9.3 cm
d. 5.9 cm

17. E The average distance between the Earth and the Moon is 240,000 miles. Express this distance in kilometers.

a. 6.1×10^{5} km
b. 3.9×10^{5} km
c. 1.5×10^{5} km
d. 9.4×10^{4} km

18. M The area of a 15-inch pizza is 176.7 in^2. Express this area in square centimeters.

a. 1140. cm^2
b. 448.8 cm^2
c. 69.57 cm^2
d. 27.39 cm^2

19. M Acetone, which is used as a solvent and as a reactant in the manufacture of Plexiglas®, boils at 56.1°C. What is the boiling point in degrees Fahrenheit?

a. 159°F
b. 133°F
c. 69.0°F
d. 43.4°F

20. E
Isopropyl alcohol, commonly known as rubbing alcohol, boils at 82.4°C. What is the boiling point in kelvins?

a. 387.6 K
b. 355.6 K
c. 323.6 K
d. 190.8 K

21. M
Acetic acid boils at 244.2°F. What is its boiling point in degrees Celsius?

a. 167.7°C
b. 153.4°C
c. 117.9°C
d. 103.7°C

22. M
The speed needed to escape the pull of Earth's gravity is 11.3 km/s. What is this speed in mi/h?

a. 65,500 mi/h
b. 25,300 mi/h
c. 18,200 mi/h
d. 1,090 mi/h

23. M
The density of mercury, the only metal to exist as a liquid at room temperature, is 13.6 g/cm^3. What is that density in pounds per cubic inch?

a. 849 lb/in^3
b. 376 lb/in^3
c. 0.491 lb/in^3
d. 1.83×10^{-3} lb/in^3

24. M
Given that 1 inch = 2.54 cm, 1 cm^3 is equal to

a. 16.4 in^3
b. 6.45 in^3
c. 0.394 in^3
d. 0.155 in^3
e. 0.0610 in^3

25. M
At a pressure of one billionth (10^{-9}) of atmospheric pressure, there are about 2.7×10^{10} molecules in one cubic centimeter of a gas. How many molecules is this per cubic meter?

a. 2.7×10^{16}
b. 2.7×10^{14}
c. 2.7×10^{12}
d. 2.7×10^{8}
e. 2.7×10^{4}

26. M
If the price of gold at the morning fixing in London was $4730 per lb, what would a kilogram of gold have cost in £ (pounds)? (Assume an exchange rate of $1.00 = £ 0.611)

a. 1310
b. 3510
c. 6370
d. 10400
e. 17100

27. The S.I. unit of mass is
E

a. mg
b. g
c. kg
d. metric ton
e. lb

28. The S.I. unit of volume is
M

a. mL
b. cm^3
c. L
d. m^3
e. fluid ounce

29. The S.I. unit of speed (velocity) is
E

a. km/h
b. km/s
c. m/h
d. m/s
e. none of the above

30. Which of the following is an extensive property of oxygen?
E

a. boiling point
b. temperature
c. mass
d. density

31. A flask has a mass of 78.23 g when empty and 593.63 g when filled with water. When the same
M flask is filled with concentrated sulfuric acid, H_2SO_4, its mass is 1026.57 g. What is the density of concentrated sulfuric acid? (Assume water has a density of 1.00 g/cm^3 at the temperature of the measurement.)

a. 1.992 g/cm^3
b. 1.840 g/cm^3
c. 1.729 g/cm^3
d. 1.598 g/cm^3

32. Talc is a mineral that has low conductivity for heat and electricity and that is not attacked by
H acid. It is used as talcum powder and face powder. A sample of talc weighs 35.97 g in air and 13.65 g in mineral oil (d = 1.75 g/cm^3). What is the density of talc?

a. 2.82 g/cm^3
b. 2.63 g/cm^3
c. 2.44 g/cm^3
d. 1.61 g/cm^3

33. M Select the answer that expresses the result of this calculation with the correct number of significant figures.

$$\frac{13.602 \times 1.90 \times 3.06}{4.2 \times 1.4097} =$$

a. 13.35678
b. 13.357
c. 13.4
d. 13

34. E Select the answer that expresses the result of this calculation with the correct number of significant figures.

$$25.234\ \text{g} \div 1.35\ \text{cm} \div 4.662\ \text{cm} \div 3.20\ \text{cm} =$$

a. 1.2529 g/cm^3
b. 1.253 g/cm^3
c. 1.25 g/cm^3
d. 1.2 g/cm^3

35. M Select the answer that expresses the result of this calculation with the correct number of significant figures and with correct units.

$$16.18\ \text{cm} \times 9.6114\ \text{g} \div 1.4783\ \text{cm}^2 =$$

a. 105.2 g/cm^2
b. 105.2 g/cm
c. 72.13 g/cm^2
d. 72.13 g/cm

36. E Which measurement is expressed to 4 significant figures.

a. 0.00423 kg
b. 24.049 cm
c. 62.40 g
d. 82,360 m

37. E Express 96,342 m using 2 significant figures.

a. 9.60×10^4 m
b. 9.6×10^4 m
c. 9.60×10^{-4} m
d. 9.6×10^{-4} m

38. M Select the answer with the correct number of decimal places for the following sum:

$$13.914\ \text{cm} + 243.1\ \text{cm} + 12.0046\ \text{cm} =$$

a. 269.0186 cm
b. 269.019 cm
c. 269.02 cm
d. 269.0 cm

39. Select the answer with the correct number of decimal places for the following sum:
M

$$4.65 \text{ g} + 2.0 \text{ g} + 19.6442 \text{ g} =$$

a. 26.2942 g
b. 26.29 g
c. 26.3 g
d. 26 g

40. The appropriate number of significant figures in the result of 15.234×15.208 is
M

a. 2
b. 3
c. 4
d. 5
e. 6

41. The appropriate number of significant figures in the result of 15.234 - 15.208 is
M

a. 1
b. 2
c. 3
d. 4
e. 5

42. The result of $(3.8621 \times 1.5630) - 5.98$ is properly written as
M

a. 0.06
b. 0.056
c. 0.0565
d. 0.05646
e. 0.056462

43. As chief chemist at Superior Analytical Products (SAP) you must design an experiment to
M determine the density of an unknown liquid to three (3) significant figures. The density is of the order of 1 g/cm^3. You have approximately 7 mL of the liquid and only graduated cylinders and balances are available for your use. Which of the following combinations of equipment will allow you to meet but not exceed your goal?

a. Graduated cylinder with ±0. 1 mL uncertainty; balance with ±0. 1 g uncertainty
b. Graduated cylinder with ±0.01 mL uncertainty; balance with ±0. 1 g uncertainty
c. Graduated cylinder with ±0.01 mL uncertainty; balance with ±0.01 g uncertainty
d. Graduated cylinder with ±0.001 mL uncertainty; balance with ±0.001 g uncertainty

44. Bud N. Chemist must determine the density of a mineral sample. His four trials yield densities of
M 4.77 g/cm^3, 4.67 g/cm^3, 4.69 g/cm^3, and 4.81 g/cm^3. Independent studies found the correct density to be 4.75 g/cm^3. Which of the following statements represents the best analysis of the data?

a. Bud's results have much greater accuracy than precision
b. Bud's results have much greater precision than accuracy
c. Bud's results have high accuracy and high precision
d. Bud's results have low accuracy and low precision

45. As part of an experiment to determine the density of a new plastic developed in her laboratory, Sara
M Ann Dippity measures the volume of a solid sample. Her four trials yield volumes of 12.37 cm^3, 12.41 cm^3, 12.39 cm^3, and 12.38 cm^3. Measurements of other scientists in the lab give an average volume of 12.49 cm^3. Which of the following statements represents the best analysis of the data?

a. Sara's results have low precision and high accuracy
b. Sara's results have high precision and high accuracy
c. Sara's results have greater precision than accuracy
d. Sara's results have greater accuracy than precision

46. Which of the following correctly expresses 52,030.2 m in scientific notation?
E

a. 5.20302×10^4 m
b. 5.203×10^4 m
c. 5.20×10^4 m
d. 5.2×10^4 m

47. Which of the following correctly expresses 0.000007913 g in scientific notation?
E

a. 7.913×10^6 g
b. 7.913×10^5 g
c. 7.913×10^{-5} g
d. 7.913×10^{-6} g

Short Answer Questions

48. Give an example of a physical property and a chemical property of
E

a. oxygen gas
b. octane
c. copper

49. Briefly explain the relationship between hypothesis and experiment in the scientific method.
H

50. Calculate the numerical part of the conversion factors needed to carry out the following unit
M conversions:

a. density in g/cm^3 to kg/m^3
b. speed in mi/h to ft/s
c. area in km^2 to mi^2
d. area in km^2 to cm^2
e. mass/area of aluminum foil in mg/cm^2 to g/m^2
f. number of gas molecules per unit volume from $/m^3$ to $/ft^3$
g. number of bacteria per unit area on a microscope slide from $/mm^2$ to $/in^2$

51. The S.I. unit of energy is the joule, J. 1 J = 1 $kg.m^2/s^2$. Another energy unit, the erg, was once in
M widespread use. 1 erg = 1 $g.cm^2/s^2$. Calculate the number of ergs in 1 J, showing all your working.

52. Classify the following properties of hydrogen gas as either intensive or extensive.
M

a. the mass of the gas sample
b. the average speed of a molecule in the sample
c. temperature
d. density
e. number of molecules present

53. An evacuated 276 mL glass bulb weighs 129.6375 g. Filled with an unknown gas, the bulb weighs
M 130.0318 g. Calculate the gas density in g/L, and express it with an appropriate number of significant figures.

54. Use the relationship between temperatures in Celsius and Fahrenheit to calculate the temperature at which
M

a. the numerical value is the same on both scales
b. the Fahrenheit number is exactly twice the Celsius number

55. Write the following numbers and results in scientific notation, with appropriate significant figures.
E

a. 654
b. 1234560
c. 0.000000673
d. 0.002590
e. 200.4
f. 260.0
g. πr^2, where $r = 8.7$ cm
h. $23.24 + 18.6 - 5$

56. Write the following numbers and results in standard notation, with appropriate significant figures.
E

a. 7.85×10^{-3}
b. 7.85×10^{4}
c. 5.920×10^{3}
d. $7.85 \times 10^{12} \div 10^{10}$
e. 7.00×10^{-5}
f. circumference of a circle, $2\pi r$, where r = 8.7 cm
g. $$\frac{6.626 \times 10^{-34} \times 6.02214 \times 10^{23} \times 2.9979 \times 10^{8}}{5.23 \times 10^{-6}}$$

57. In each of the sets below, choose the one quantity or number which is exact.
M

a. i. the human population
ii. the distance in light years from the sun to Alpha Centauri, a nearby star
iii. the winning time for the 100 m dash in the Olympic Games

b. i. the weight of a particular one cent coin in g
ii. the boiling point of lead, in °C
iii. the number of cm in 1 yd

c. i. the measured value of the speed of light ($2.998... \times 10^{8}$ m/s)
ii. π (3.141...)
iii. the volume of milk in a 1-gallon jug

True/False Questions

58. The ripening of fruit, once picked, is an example of physical change.
E

59. The potential energy of a car moving on a level road does not depend on its speed.
M

60. When a wooden match burns in air, chemical potential energy is converted to kinetic energy.
E

61. When applying the scientific method, it is important to avoid any form of hypothesis.
M

62. When applying the scientific method, a model or theory should be based on experimental data.
E

63. The numerical value of any temperature expressed in Celsius is always different from the numerical value of the same temperature in Fahrenheit.
E

64. The numerical value of any temperature expressed in Celsius is always different from the numerical value of the same temperature in kelvin.
E

65. The number 6.0448, rounded to 3 decimal places, becomes 6.045
E

66. The number 6.0448, rounded to 2 decimal places, becomes 6.05.
M

67. The weight of a coin measured as 1.96235 g on one balance is definitely more accurate than a weight measurement of 1.95 g on another balance.
M

Keys to the Study of Chemistry
Chapter 1
Answer Key

1.	d	17.	b	33.	d
2.	a	18.	a	34.	c
3.	c	19.	b	35.	b
4.	c	20.	b	36.	c
5.	d	21.	c	37.	b
6.	b	22.	b	38.	d
7.	c	23.	c	39.	c
8.	b	24.	e	40.	d
9.	a	25.	a	41.	b
10.	c	26.	c	42.	a
11.	d	27.	c	43.	c
12.	c	28.	d	44.	a
13.	d	29.	d	45.	c
14.	c	30.	c	46.	a
15.	a	31.	b	47.	d
16.	a	32.	a		

48. Answers could all be the same, but some possibilities are
    a. boiling point, reaction with sodium
    b. boiling point, reaction with oxygen
    c. electrical conductivity, reaction with nitric acid

49. A hypothesis should be capable of leading to a prediction which is testable by experiment. If the experimental result differs from the prediction, the hypothesis should be modified.

50. a. 10^3
    b. 1.4666
    c. 0.38
    d. 10^{10}
    e. 10
    f. 0.028
    g. 645

51. $1\ J = 10^7$ erg

52. a. E
    b. I
    c. I
    d. I
    e. E

53. 1.43 g/L

54. a. 40.°F = -40.°C
    b. 320.°F = 160.°C

55. a. 6.54×10^2
b. 1.23456×10^6
c. 6.73×10^{-7}
d. 2.590×10^{-3}
e. 2.004×10^2
f. 2.600×10^2
g. 2.4×10^2 cm^2
h. 3.7×10^1

56. a. 0.00785
b. 78500
c. 5920.
d. 785
e. 0.0000700
f. 55 cm
g. 22900

57. a. i
b. iii
c. ii

58. F

59. T

60. T

61. F

62. T

63. F

64. T

65. T

66. F

67. F

The Components of Matter
Chapter 2

Multiple Choice Questions

1. Kaolinite, a clay mineral with the formula $Al_4Si_4O_{10}(OH)_8$, is used as a filler in slick-paper for
M magazines and as a raw material for ceramics. Analysis shows that 14.35 g of kaolinite contains 8.009 g of oxygen. Calculate the mass percent of oxygen in kaolinite.

a. 24.80 mass %
b. 30.81 mass %
c. 34.12 mass %
d. 55.81 mass %

2. Gay-Lussac studied the formation of water from hydrogen and oxygen. Which of the following
M statements about his work is correct?

a. It provided support for Dalton's atomic theory and Dalton strongly supported Gay-Lussac's conclusions.
b. It was the basis of Avogadro's proposal that hydrogen gas consisted of particles which contained two hydrogen atoms each.
c. It provided the basis for Sir William Crookes's study of the atom.
d. It reaffirmed Lavosier's assignment of a relative atomic mass of 8 to oxygen.

3. J. J. Thomson studied cathode ray particles (electrons) and was able to measure the mass/charge ratio.
E His results showed that

a. the mass/charge ratio varied with as the cathode material was changed.
b. the charge was always a whole-number multiple of some minimum charge.
c. matter included particles much smaller than the atom.
d. atoms contained dense areas of positive charge.

4. Who is credited with measuring the mass/charge ratio of the electron?
E

a. Dalton
b. Gay-Lussac
c. Thomson
d. Millikan
e. Rutherford

5. Who is credited with first measuring the charge of the electron?
E

a. Dalton
b. Gay-Lussac
c. Thomson
d. Millikan
e. Rutherford

6. Millikan's oil-drop experiment
E

a. established the charge on an electron.
b. showed that all oil drops carried the same charge.
c. provided support for the nuclear model of the atom.
d. suggested the presence of a neutral particle in the atom.

7. In a Millikan oil-drop experiment, the charges on several different oil drops were as follows: -5.92; -4.44; -2.96; -8.88. The units are arbitrary. What is the likely value of the electronic charge in these arbitrary units?
M

a. -1.11
b. -1.48
c. -2.22
d. -2.96
e. -5.55

8. Who is credited with discovering the atomic nucleus?
E

a. Dalton
b. Gay-Lussac
c. Thomson
d. Millikan
e. Rutherford

9. Rutherford bombarded gold foil with alpha (α) particles and found that a small percentage of the particles were deflected. Which of the following was not accounted for by the model he proposed for the structure of atoms?
M

a. the small size of the nucleus
b. the charge on the nucleus
c. the total mass of the atom
d. the existence of protons

10. Which one of the following statements about atoms and subatomic particles is correct?
M

a. Rutherford discovered the atomic nucleus by bombarding gold foil with electrons
b. The proton and the neutron have identical masses
c. The neutron's mass is equal to that of a proton plus an electron
d. A neutral atom contains equal numbers of protons and electrons
e. An atomic nucleus contains equal numbers of protons and neutrons

11. Bromine is the only nonmetal that is a liquid at room temperature. Consider the isotope bromine- 81, $^{81}_{35}Br$ Select the combination which lists the correct atomic number, neutron number, and mass number, respectively.
E

a. 35, 46, 81
b. 35, 81, 46
c. 81, 46, 35
d. 46, 81, 35

12. Atoms X, Y, Z, and R have the following nuclear compositions:
E

$$^{410}_{186}X \quad ^{410}_{183}Y \quad ^{412}_{186}Z \quad ^{412}_{185}R$$

Which two are isotopes?

a. X & Y
b. X & Z
c. Y & R
d. Z & R

13. Lithium forms compounds which are used in dry cells and storage batteries and in high-temperature
H lubricants. It has two naturally occurring isotopes, ^{6}Li (isotopic mass = 6.015121 amu) and ^{7}Li (isotopic mass = 7.016003 amu). Lithium has an atomic mass of 6.9409 amu. What is the percent abundance of lithium-6?

a. 86.66%
b. 46.16%
c. 7.503%
d. 6.080%

14. Silicon, which makes up about 25% of Earth's crust by mass, is used widely in the modern electronics
M industry. It has three naturally occurring isotopes, ^{28}Si, ^{29}Si, and ^{30}Si. Calculate the atomic mass of silicon.

Isotope	Isotopic Mass (amu)	Abundance %
^{28}Si	27.976927	92.23
^{29}Si	28.976495	4.67
^{30}Si	29.973770	3.10

a. 29.2252 amu
b. 28.9757 amu
c. 28.7260 amu
d. 28.0855 amu

15. Which of the following elements are largely unreactive substances?
E

a. alkali metals
b. noble gases
c. halogens
d. alkaline earth metals

16. Which of the following is a non-metal?
E

a. lithium, Li, $Z = 3$
b. bromine, Br, $Z = 35$
c. mercury, Hg, $Z = 80$
d. bismuth, Bi, $Z = 83$

17. Which of the following is a metal?
M

a. nitrogen, N, $Z = 7$
b. phosphorus, P, $Z = 15$
c. arsenic, $Z = 33$
d. thallium, Tl, $Z = 81$

18. Which of the following is a metalloid?
M

a. carbon, C, $Z = 6$
b. sulfur, S, $Z = 16$
c. germanium, Ge, $Z = 32$
d. iridium, $Z = 77$

19. A column of the periodic table is called a
E

a. group
b. period
c. isotopic mixture
d. pillar

20. A row of the periodic table is called a
E

a. group
b. period
c. isotopic mixture
d. family

21. Which of the following compounds is ionic?
M

a. PF_3
b. CS_2
c. $MgCl_2$
d. SO_2

22. Which of the following ions forms commonly from its element?
E

a. N^{3+}
b. S^{6+}
c. O^{2-}
d. Ca^{+}

23. Which of the following ions forms commonly from its element?
E

a. P^{3+}
b. Br^{7+}
c. O^{6+}
d. Ca^{2+}

24. Which of the following compounds is covalent?
M

a. $CaCl_2$
b. MgO
c. PCl_3
d. Cs_2S

25. Which of the following is the empirical formula for hexane, C_6H_{14}?
E

a. $C_{12}H_{28}$
b. C_6H_{14}
c. C_3H_7
d. $CH_{2\,3}$

26. Sodium oxide combines violently with water. Which of the following gives the formula and
E the bonding for sodium oxide?

a. NaO, ionic compound
b. NaO, covalent compound
c. Na_2O, ionic compound
d. Na_2O, covalent compound

27. Barium fluoride is used in embalming and in glass manufacturing. Which of the following gives the
E formula and bonding for barium fluoride?

a. BaF_2, ionic compound
b. BaF_2, covalent compound
c. BaF, ionic compound
d. BaF, covalent compound

28. The colorless substance, MgF_2, is used in the ceramics and glass industry. What is its name?
M

a. magnesium difluoride
b. magnesium fluoride
c. magnesium(II) fluoride
d. monomagnesium difluoride

29. The compound, BaO, absorbs water and carbon dioxide readily and is used to dry gases and organic
M solvents. What is its name?

a. barium oxide
b. barium(II) oxide
c. barium monoxide
d. baric oxide

30. What is the name of CsBr?
M

a. cesium bromide
b. cesium(I) bromide
c. cesium monobromide
d. cesious bromide

31. What is the name of BeI_2?
M

a. beryllium diiodide
b. beryllium(I) iodide
c. beryllium(II) iodide
d. beryllium iodide

32. The substance, CaSe, is used in materials which are electron emitters. What is its name?
M

a. calcium monoselenide
b. calcium(II) selenide
c. calcium selenide
d. calcium(I) selenide

33. M The substance, $CoCl_2$, is useful as a humidity indicator because it changes from pale blue to pink as it gains water from moist air. What is its name?

a. cobalt dichloride
b. cobalt(II) chloride
c. cobalt chloride
d. cobaltic chloride

34. H Which one of the following combinations of names and formulas of ions is incorrect?

a. O_2^- oxide
b. Al^{3+} aluminum
c. NO_3^- nitrate
d. PO_4^{3-} phosphate
e. CrO_4^{2-} chromate

35. H Which one of the following combinations of names and formulas of ions is incorrect?

a. O^{2-} oxide
b. Cd^{2+} cadmium
c. ClO_3^- chlorate
d. HCO_3^- hydrogen carbonate
e. NO_2^- nitrate

36. M Which one of the following combinations of names and formulas of ions is incorrect?

a. Ba^{2+} barium
b. S^{2-} sulfate
c. CN^- cyanide
d. ClO_4^- perchlorate
e. HCO_3^- bicarbonate

37. H Which one of the following combinations of names and formulas of ions is incorrect?

a. NH_4^+ ammonium
b. S^{2-} sulfide
c. CN^- cyanide
d. $S_2O_3^{2-}$ thiosulfate
e. ClO_3^- perchlorate

38. M A red glaze on porcelain can be produced by using $MnSO_4$. What is its name?

a. manganese disulfate
b. manganese(II) sulfate
c. manganese(IV) sulfate
d. manganese sulfate

39. M The compound, $(NH_4)_2S$, can be used in analysis for trace amounts of metals present in a sample. What is its name?

a. ammonium sulfide
b. diammonium sulfide
c. ammonium sulfite
d. ammonia(I) sulfite

40. The substance, $KClO_3$, is a strong oxidizer used in explosives, fireworks, and matches. What is its name?
M

a. potassium chlorite
b. potassium chlorate
c. potassium(I) chlorite
d. potassium(I) chlorate

41. The compound, NaH_2PO_4, is present in many baking powders. What is its name?
M

a. sodium biphosphate
b. sodium hydrogen phosphate
c. sodium dihydrogen phosphate
d. sodium hydrophosphate

42. Zinc acetate is used in preserving wood and in manufacturing glazes for porcelain. What is its formula?
M

a. $ZnAc_2$
b. $ZnCH_3COO$
c. $Zn(CH_3COO)_2$
d. Zn_2CH_3COO

43. Silver chloride is used in photographic emulsions. What is its formula?
E

a. $AgCl$
b. $AgCl_2$
c. Ag_2Cl
d. Ag_2Cl_3

44. Barium sulfate is used in manufacturing photographic paper. What is its formula?
M

a. $BaSO_4$
b. $Ba(SO_4)_2$
c. Ba_2SO_4
d. $Ba_2(SO_4)_3$

45. Sodium peroxide is an oxidizer used to bleach animal and vegetable fibers. What is its formula?
M

a. NaO
b. NaO_2
c. Na_2O_2
d. Na_2O

46. What is the formula for magnesium sulfide?
E

a. MgS
b. MgS_2
c. Mg_2S
d. Mg_2S_3

47. Ferric oxide is used as a pigment in metal polishing. Which of the following is its formula?
E

a. FeO
b. Fe_2O
c. Fe_2O_3
d. Fe_2O_5

48. What is the formula for lead (II) oxide?
E

a. PbO
b. PbO_2
c. Pb_2O
d. PbO_4

49. Potassium permanganate is a strong oxidizer that reacts explosively with easily oxidized materials.
M What is its formula?

a. $KMnO_3$
b. $KMnO_4$
c. K_2MnO_4
d. $K(MnO_4)_2$

50. Calcium hydroxide is used in mortar, plaster and cement. What is its formula?
E

a. $CaOH$
b. $CaOH_2$
c. Ca_2OH
d. $Ca(OH)_2$

51. What is the formula for lithium nitrite?
E

a. $LiNO_2$
b. Li_2NO_2
c. $LiNO_3$
d. Li_2NO_3

52. Iron (III) chloride hexahydrate is used as a coagulant for sewage and industrial wastes. What is its formula?
M

a. $FeCl_3 \cdot 6H_2O$
b. $Fe_3Cl \cdot 6H_2O$
c. $FeCl_3(H_2O)_6$
d. $Fe_3Cl(H_2O)_6$

53. Which one of the following formulas of ionic compounds is the least likely to be correct?
M

a. NH_4Cl
b. $Ba(OH)_2$
c. Na_2SO_4
d. Ca_2NO_3
e. $Cu(CN)_2$

54. Which one of the following formulas of ionic compounds is the least likely to be correct?
M

a. $CaCl_2$
b. $NaSO_4$
c. $MgCO_3$
d. KF
e. $Cu(NO_3)_2$

55. What is the name of the acid formed when H_2S gas is dissolved in water?
H

a. sulfuric acid
b. sulfurous acid
c. hydrosulfuric acid
d. hydrosulfurous acid

56. What is the name of the acid formed when HBr gas is dissolved in water?
M

a. bromic acid
b. bromous acid
c. hydrobromic acid
d. hydrobromous acid

57. What is the name of the acid formed when $HClO_4$ liquid is dissolved in water?
M

a. hydrochloric acid
b. perchloric acid
c. chloric acid
d. chlorous acid

58. What is the name of the acid formed when HCN gas is dissolved in water?
M

a. cyanic acid
b. hydrocyanic acid
c. cyanous acid
d. hydrocyanous acid

59. Which one of the following combinations of names and formulas is incorrect?
M

a.	H_3PO_4	phosphoric acid
b.	HNO_3	nitric acid
c.	$NaHCO_3$	sodium carbonate
d.	H_2CO_3	carbonic acid
e.	KOH	potassium hydroxide

60. What is the name of PCl_3?
E

a. phosphorus chloride
b. phosphoric chloride
c. phosphorus trichloride
d. trichlorophosphide

61. The compound, P_4S_{10}, is used in the manufacture of safety matches. What is its name?
M

a. phosphorus sulfide
b. phosphoric sulfide
c. phosphorus decasulfide
d. tetraphosphorus decasulfide

62. What is the name of BBr_3?
M

a. boron bromide
b. boric bromide
c. boron tribromide
d. tribromoboride

63. What is the name of IF_7?
M

a. iodine fluoride
b. iodic fluoride
c. iodine heptafluoride
d. heptafluoroiodide

64. What is the name of P_4Se_3
M

a. phosphorus selenide
b. phosphorus triselenide
c. tetraphosphorus selenide
d. tetraphosphorus triselenide

65. Diiodine pentaoxide is used as an oxidizing agent that converts carbon monoxide to carbon dioxide. What is its chemical formula?
E

a. I_2O_5
b. IO_5
c. $2IO_5$
d. I_5O_2

66. Tetrasulfur dinitride decomposes explosively when heated. What is its formula?
E

a. S_2N_4
b. S_4N_2
c. $4SN_2$
d. S_4N

67. Chlorine dioxide is a strong oxidizer that is used for bleaching flour and textiles and for purification of water. What is its formula?
E

a. ClO_2
b. Cl_2O
c. Cl_2O_2
d. Cl_2O_4

68. Ammonium sulfate, $(NH_4)_2SO_4$, is a fertilizer widely used as a source of nitrogen. Calculate its molecular mass.
E

a. 63.07 amu
b. 114.10 amu
c. 118.13 amu
d. 132.13 amu

69. Sodium chromate is used to protect iron from corrosion and rusting. Determine its molecular mass.
M

a. 261.97 amu
b. 238.98 amu
c. 161.97 amu
d. 138.98 amu

70. Iodine pentafluoride reacts slowly with glass and violently with water. Determine its molecular mass.
E

a. 653.52 amu
b. 259.89 amu
c. 221.90 amu
d. 202.90 amu

71. Determine the molecular mass of iron (III) bromide hexahydrate, a substance used as a catalyst in organic reactions.
M

a. 403.65 amu
b. 355.54 amu
c. 317.61 amu
d. 313.57 amu

Short Answer Questions

72. Name the three important "laws" that were accounted for by Dalton's atomic theory.
M

73. Dalton's atomic theory has required some modifications in the light of subsequent discoveries. For any three appropriate postulates of Dalton's atomic theory
M

a. state the postulate in its original form
b. in one sentence, describe why the postulate has needed modification.

74. Fill in the blank spaces and write out all the symbols in the left hand column in full, in the form $^{A}_{Z}X$ (i.e. include the appropriate values of Z and A as well as the correct symbol X).
M

Symbol	# protons	# neutrons	# electrons
	17	18	
Au		118	
		20	20

75. The following charges on individual oil droplets were obtained during an experiment similar to Millikan's. Use them to determine a charge for the electron in coulombs (C), showing all your working.
M

Charges (C): -3.184×10^{-19}; -4.776×10^{-19}; -7.960×10^{-19}

76. State the two important experimental results (and the names of the responsible scientists) which enabled the mass of the electron to be determined.
M

77. For each of the following elements, indicate whether it is a metal, a non-metal or a metalloid:
E

a. S
b. Ge
c. Hg
d. H
e. I
f. Si

78. Give the common name of the group in the periodic table to which each of the following elements belongs:
E

a. Rb
b. Br
c. Ba
d. Ar

79. a. Give the names of the following ions:
M
(i) NH_4^+
(ii) SO_3^{2-}
b. Write down the formulas of the following ions:
(i) aluminum
(ii) carbonate

80. a. Give the names of the following ions:
M
(i) O_2^{2-}
(ii) SO_4^{2-}
b. Write down the formulas of the following ions:
(i) ammonium
(ii) nitrate

81. For each of the following names, write down the corresponding formula, including charge where appropriate (atomic numbers and mass numbers are not required):
M

a. zinc ion
b. nitrite ion
c. carbonic acid
d. cyanide ion

82. Calculate the molecular masses of the following:
E

a. Cl_2
b. H_2O_2
c. $(NH_4)_2SO_4$
d. $Ba(NO_3)_2$

True/False Questions

83. In nature, some elements exist as molecules, while others do not.
E

84. Modern studies have shown that the Law of Multiple Proportions is not valid.
M

85. Atoms of one element cannot be converted to another element by any known method.
E

86. The mass of a neutron is equal to the mass of a proton plus the mass of an electron.
E

87. All neutral atoms of tin have 50 protons and 50 electrons.
E

88. Copper (Cu) is a transition metal.
E

89. Lead (Pb) is a main-group element.
E

90. Ionic compounds may carry a net positive or negative charge.
E

91. When an alkali metal combines with a non-metal, a covalent bond is normally formed.
E

92. The formula C_9H_{20} is an empirical formula.
E

The Components of Matter
Chapter 2
Answer Key

1.	d	26.	c	51.	a
2.	b	27.	a	52.	a
3.	c	28.	b	53.	d
4.	c	29.	a	54.	b
5.	d	30.	a	55.	c
6.	a	31.	d	56.	c
7.	b	32.	c	57.	b
8.	e	33.	b	58.	b
9.	c	34.	a	59.	c
10.	d	35.	e	60.	c
11.	a	36.	b	61.	d
12.	b	37.	e	62.	c
13.	c	38.	b	63.	c
14.	d	39.	a	64.	d
15.	b	40.	b	65.	a
16.	b	41.	c	66.	b
17.	d	42.	c	67.	a
18.	c	43.	a	68.	d
19.	a	44.	a	69.	c
20.	b	45.	c	70.	c
21.	c	46.	a	71.	a
22.	c	47.	c		
23.	d	48.	a		
24.	c	49.	b		
25.	c	50.	d		

72. Laws of conservation of mass; definite composition; multiple proportions

73. Matter consists of atoms which are indivisible, cannot be created or destroyed. But, atoms are divisible, as the existence of subatomic particles shows.
Atoms of one element cannot be converted into atoms of another element. They can be converted in various nuclear reactions, including radioactive decay.
Atoms of an element are identical in mass and other properties. Isotopes of an element differ in their masses and other properties.

74.

Symbol	# protons	# neutrons	# electrons
$^{35}_{17}Cl$	17	18	17
$^{197}_{79}Au$	79	118	79
$^{40}_{20}Ca$	20	20	20

75. -1.59×10^{-19} C

76. Thomson measured *m/e*, the mass-to-charge ratio. Millikan measured *e*, the charge. Thus, the mass *m* could be calculated.

77. a. nonmetal
b. metalloid
c. metal
d. nonmetal
e. nonmetal
f. metalloid

78. a. alkali metals
b. halogens
c. alkaline earth metals
d. noble gases

79. a. (i) ammonium
(ii) sulfite
b. (i) Al^{3+}
(ii) CO_3^{2-}

80. a. (i) peroxide
(ii) sulfate
b. (i) NH_4^+
(ii) NO_3^-

81. a. Zn^{2+}
b. NO_2^-
c. H_2CO_3
d. CN^-

82. a. 70.90 amu
b. 34.02 amu
c. 132.2 amu
d. 261.3 amu

83. T

84. F

85. F

86. F

87. T

88. T

89. T

90. F

91. F

92. T

Stoichiometry: Mole-Mass-Number Relationships in Chemical Systems
Chapter 3

Multiple Choice Questions

1. E — Calcium fluoride, CaF_2, is a source of fluorine and is used to fluoridate drinking water. Calculate its molar mass.

a. 118.15 g/mol
b. 99.15 g/mol
c. 78.07 g/mol
d. 59.08 g/mol

2. E — Calculate the molar mass of tetraphosphorus decaoxide, P_4O_{10}, a corrosive substance which can be used as a drying agent.

a. 469.73 g/mol
b. 283.89 g/mol
c. 190.97 g/mol
d. 94.97 g/mol

3. E — Calculate the molar mass of rubidium carbonate, Rb_2CO_3.

a. 340.43 g/mol
b. 255.00 g/mol
c. 230.94 g/mol
d. 113.48 g/mol

4. E — Calculate the molar mass of $(NH_4)_3AsO_4$.

a. 417.80 g/mol
b. 193.03 g/mol
c. 165.02 g/mol
d. 156.96 g/mol

5. E — Aluminum sulfate, $A1_2(SO_4)_3$, is used in tanning leather, purifying water, and manufacture of antiperspirants. Calculate its molar mass.

a. 450.06 g/mol
b. 342.15 g/mol
c. 315.15 g/mol
d. 278.02 g/mol

6. M — Calculate the molar mass of $Ca(BO_2)_2{\cdot}6H_2O$.

a. 273.87 g/mol
b. 233.79 g/mol
c. 183.79 g/mol
d. 143.71 g/mol

7. E — Calculate the mass in grams of 2.35 mol of sodium chloride, or common table salt.

a. 221 g
b. 137 g
c. 93.9 g
d. 58.4 g

8. E Magnesium fluoride is used in the ceramics and glass industry. What is the mass of 1.72 mol of magnesium fluoride?

a. 107 g
b. 74.5 g
c. 62.3 g
d. 43.3 g

9. H Sodium bromate is used in a mixture which dissolves gold from its ores. Calculate the mass in grams of 4.68 mol of sodium bromate.

a. 706 g
b. 482 g
c. 32.2 g
d. 0.0310 g

10. M What is the mass in grams of 0.250 mol of the common antacid calcium carbonate?

a. 4.00×10^2 g
b. 25.0 g
c. 4.00×10^{-2} g
d. 2.50×10^{-3} g

11. E Calculate the number of moles in 17.8 g of the antacid magnesium hydroxide, $Mg(OH)_2$.

a. 3.28 mol
b. 2.32 mol
c. 0.431 mol
d. 0.305 mol

12. E Phosphorus pentachloride, PCl_5, a white solid that has a pungent, unpleasant odor, is used as a catalyst for certain organic reactions. Calculate the number of moles in 38.7 g of PCl_5.

a. 5.38 mol
b. 3.55 mol
c. 0.282 mol
d. 0.186 mol

13. E Aluminum oxide, Al_2O_3, is used as a filler for paints and varnishes as well as in the manufacture of electrical insulators. Calculate the number of moles in 47.51 g of Al_2O_3.

a. 2.377 mol
b. 2.146 mol
c. 0.4660 mol
d. 0.4207 mol

14. H Calculate the number of oxygen atoms in 29.34 g of sodium sulfate, Na_2SO_4.

a. 1.244×10^{23} O atoms
b. 4.976×10^{23} O atoms
c. 2.915×10^{24} O atoms
d. 1.166×10^{25} O atoms

15. H Potassium dichromate, $K_2Cr_2O_7$, is used in tanning leather, decorating porcelain and water proofing fabrics. Calculate the number of chromium atoms in 78.82 g of $K_2Cr_2O_7$.

a. 2.248×10^{24} Cr atoms
b. 1.124×10^{24} Cr atoms
c. 3.227×10^{23} Cr atoms
d. 1.613×10^{23} Cr atoms

16. Copper(II) sulfate pentahydrate, $CuSO_4 \cdot 5H_2O$, is used as a fungicide and algicide. Calculate the mass of oxygen in 1.000 mol of $CuSO_4 \cdot 5H_2O$.
M

a. 249.7 g
b. 144.0 g
c. 80.00 g
d. 64.00 g

17. Lead (II) nitrate is a poisonous substance which has been used in the manufacture of special explosives and as a sensitizer in photography. Calculate the mass of lead in 139 g of $Pb(NO_3)_2$.
M

a. 107 g
b. 90.8 g
c. 87.0 g
d. 83.4 g

18. Household sugar, sucrose, has the molecular formula $C_{12}H_{22}O_{11}$. What is the % of carbon in sucrose, by mass?
M

a. 26.7
b. 33.3
c. 41.4
d. 42.1
e. 52.8

19. Sulfur trioxide can react with atmospheric water vapor to form sulfuric acid that falls as acid rain. Calculate the mass in grams of 3.65×10^{20} molecules of SO_3.
M

a. 6.06×10^{-4} g
b. 4.85×10^{-2} g
c. 20.6 g
d. 1650 g

20. Calculate the mass in grams of 8.35×10^{22} molecules of CBr_4.
M

a. 46.0 g
b. 7.21 g
c. 0.139 g
d. 0.0217 g

21. The number of hydrogen atoms in 0.050 mol of $C_3H_8O_3$ is
E

a. 3.0×10^{22}
b. 1.2×10^{23}
c. 2.4×10^{23}
d. 4.8×10^{23}
e. none of the above

22. Gadolinium oxide, a colorless powder which absorbs carbon dioxide from the air, contains 86.76 mass % Gd. Determine its empirical formula.
M

a. Gd_2O_3
b. Gd_3O_2
c. Gd_3O_4
d. Gd_4O_3

23. M Hydroxylamine nitrate contains 29.17 mass % N, 4.20 mass % H, and 66.63 mass % O. Determine its empirical formula.

a. HNO
b. H_2NO_2
c. HN_6O_{16}
d. $HN_{16}O_7$

24. H Hydroxylamine nitrate contains 29.17 mass % N, 4.20 mass % H, and 66.63 mass O. If its molar mass is between 94 and 98 g/mol, what is its molecular formula?

a. NH_2O_5
b. $N_2H_4O_4$
c. $N_3H_3O_3$
d. $N_4H_8O_2$

25. M A compound of bromine and fluorine is used to make UF_6, which is an important chemical in processing and reprocessing of nuclear fuel. The compound contains 58.37 mass percent bromine. Determine its empirical formula.

a. BrF
b. BrF_3
c. Br_2F_3
d. Br_3F

26. M A compound containing chromium and silicon contains 73.52 mass percent chromium. Determine its empirical formula.

a. $CrSi_3$
b. Cr_2Si_3
c. Cr_3Si
d. Cr_3Si_2

27. M Alkanes are compounds of carbon and hydrogen with the general formula C_nH_{2n+2}. An alkane component of gasoline has a molar mass of between 125 and 130 g/mol. What is the value of n for this alkane?

a. 4
b. 9
c. 10
d. 13
e. 14

28. H Terephthalic acid, used in the production of polyester fibers and films, is composed of carbon, hydrogen, and oxygen. When 0.6943 g of terephthalic acid was subjected to combustion analysis it produced 1.471 g CO_2 and 0.226 g H_2O. What is its empirical formula?

a. $C_2H_3O_4$
b. $C_3H_4O_2$
c. $C_4H_3O_2$
d. $C_5H_{12}O_4$

29. H Terephthalic acid, used in the production of polyester fibers and films, is composed of carbon, hydrogen, and oxygen. When 0.6943 g of terephthalic acid was subjected to combustion analysis it produced 1.471 g CO_2 and 0.226 g H_2O. If its molar mass is between 158 and 167 g/mol, what is its molecular formula?

a. $C_4H_6O_7$
b. $C_6H_8O_5$
c. $C_7H_{12}O_4$
d. $C_8H_6O_4$

30. M Hydroxylamine hydrochloride is a powerful reducing agent which is used as a polymerization catalyst. It contains 5.80 mass % H, 20.16 mass % N, 23.02 mass % O, and 51.02 mass % Cl. What is its empirical formula?

a. $H_2N_7O_8Cl_{18}$
b. $H_2N_2O_2Cl$
c. $HN_3O_4Cl_9$
d. H_4NOCl

31. M In the combustion analysis of 0.1127 g of glucose ($C_6H_{12}O_6$), what mass, in grams, of CO_2 would be produced?

a. 0.0451
b. 0.0825
c. 0.1652
d. 0.4132
e. 1.466

32. E Balance the following equation:

$B_2O_3(s) + HF(l) \rightarrow BF_3(g) + H_2O(l)$

a. $B_2O_3(s) + 6HF(l) \rightarrow 2BF_3(g) + 3H_2O(l)$
b. $B_2O_3(s) + H_6F_6(l) \rightarrow B_2F_6(g) + H_6O_3(l)$
c. $B_2O_3(s) + 2HF(l) \rightarrow 2BF_3(g) + H_2O(l)$
d. $B_2O_3(s) + 3HF(l) \rightarrow 2BF_3(g) + 3H_2O(l)$

33. E Balance the following equation:

$UO_2(s) + HF(l) \rightarrow UF_4(s) + H_2O(l)$

a. $UO_2(s) + 2HF(l) \rightarrow UF_4(s) + H_2O(l)$
b. $UO_2(s) + 4HF(l) \rightarrow UF_4(s) + 2H_2O(l)$
c. $UO_2(s) + H_4F_4(l) \rightarrow UF_4(s) + H_4O_2(l)$
d. $UO_2(s) + 4HF(l) \rightarrow UF_4(s) + 4H_2O(l)$

34. E Balance the following equation for the combustion of benzene:

$C_6H_6(l) + O_2(g) \rightarrow H_2O(g) + CO_2(g)$

a. $C_6H_6(l) + 9O_2(g) \rightarrow 3H_2O(g) + 6CO_2(g)$
b. $C_6H_6(l) + 9O_2(g) \rightarrow 6H_2O(g) + 6CO_2(g)$
c. $2C_6H_6(l) + 15O_2(g) \rightarrow 6H_2O(g) + 12CO_2(g)$
d. $C_6H_6(l) + 15O_2(g) \rightarrow 3H_2O(g) + 6CO_2(g)$

35. M Balance the following equation:

$C_8H_{18}O_3(l) + O_2(g) \rightarrow H_2O(g) + CO_2(g)$

a. $C_8H_{18}O_3(l) + 8O_2(g) \rightarrow 9H_2O(g) + 8CO_2(g)$
b. $C_8H_{18}O_3(l) + 11O_2(g) \rightarrow 9H_2O(g) + 8CO_2(g)$
c. $2C_8H_{18}O_3(l) + 22O_2(g) \rightarrow 9H_2O(g) + 16CO_2(g)$
d. $C_8H_{18}O_3(l) + 13O_2(g) \rightarrow 18H_2O(g) + 8CO_2(g)$

36. M

Balance the following equation:

$Ca_3(PO_4)_2(s)$ + $SiO_2(s)$ + $C(s)$ → $CaSiO_3(s)$ + $CO(g)$ + $P_4(s)$

a. $Ca_3(PO_4)_2(s)$ + $3SiO_2(s)$ + $8C(s)$ → $3CaSiO_3(s)$ + $8CO(g)$ + $P_4(s)$
b. $Ca_3(PO_4)_2(s)$ + $3SiO_2(s)$ + $14C(s)$ → $3CaSiO_3(s)$ + $14CO(g)$ + $P_4(s)$
c. $Ca_3(PO_4)_2(s)$ + $3SiO_2(s)$ + $8C(s)$ → $3CaSiO_3(s)$ + $8CO(g)$ + $2P_4(s)$
d. $2Ca_3(PO_4)_2(s)$ + $6SiO_2(s)$ + $10C(s)$ → $6CaSiO_3(s)$ + $10CO(g)$ + $P_4(s)$

37. E

Balance the following equation:

$B_5H_9(l)$ + $O_2(g)$ → $B_2O_3(s)$ + $H_2O(g)$

a. $B_5H_9(l)$ + $12O_2(g)$ → $5B_2O_3(s)$ + $9H_2O(g)$
b. $2B_5H_9(l)$ + $12O_2(g)$ → $5B_2O_3(s)$ + $9H_2O(g)$
c. $B_5H_9(l)$ + $9O_2(g)$ → $3B_2O_3(s)$ + $9H_2O(g)$
d. $2B_5H_9(l)$ + $16O_2(g)$ → $5B_2O_3(s)$ + $18H_2O(g)$

38. E

Sulfur dioxide reacts with chlorine to produce thionyl chloride (used as a drying agent for inorganic halides) and dichlorine oxide (used as a bleach for wood, pulp and textiles).

$SO_2(g)$ + $2Cl_2(g)$ → $SOCl_2(g)$ + $Cl_2O(g)$

If 0.400 mol of Cl_2 reacts with excess SO_2, how many moles of Cl_2O are formed?

a. 0.800 mol
b. 0.400 mol
c. 0.200 mol
d. 0.100 mol

39. E

Aluminum will react with bromine to form aluminum bromide (used as an acid catalyst in organic synthesis).

$Al(s)$ + $Br_2(l)$ → $Al_2Br_6(s)$ [unbalanced]

How many moles of Al are needed to form 2.43 mol of Al_2Br_6?

a. 7.29 mol
b. 4.86 mol
c. 2.43 mol
d. 1.62 mol

40. E

Ammonia will react with fluorine to produce dinitrogen tetrafluoride and hydrogen fluoride (used in production of aluminum, in uranium processing, and in frosting of light bulbs).

$2NH_3(g)$ + $5F_2(g)$ → $N_2F_4(g)$ + $6HF(g)$

How many moles of NH_3 are needed to react completely with 13.6 mol of F_2?

a. 34.0 mol
b. 6.80 mol
c. 5.44 mol
d. 2.27 mol

41. E Titanium(IV) oxide (extensively used as a white pigment) reacts with bromine trifluoride to form titanium(IV) fluoride, bromine, and oxygen.

$$3TiO_2(s) + 4BrF_3(l) \rightarrow 3TiF_4(s) + 2Br_2(l) + 3O_2(g)$$

How many moles of bromine trifluoride are needed to react completely with 9.68 mol of titanium(IV) oxide?

a. 12.9 mol
b. 9.68 mol
c. 7.26 mol
d. 3.23 mol

42. M Ammonia, an important source of fixed nitrogen that can be metabolized by plants, is produced using the Haber process in which nitrogen and hydrogen combine.

$$N_2(g) + 3H_2(g) \rightarrow 2NH_3(g)$$

How many grams of nitrogen are needed to produce 325 grams of ammonia?

a. 1070 g
b. 535 g
c. 267 g
d. 178 g

43. M How many grams of sodium fluoride (used in water fluoridation and manufacture of insecticides) are needed to form 485 g of sulfur tetrafluoride?

$$3SCl_2(l) + 4NaF(s) \rightarrow SF_4(g) + S_2Cl_2(l) + 4NaCl(s)$$

a. 1510 g
b. 754 g
c. 205 g
d. 51.3 g

44. M How many grams of oxygen are needed to react completely with 200.0 g of ammonia, NH_3?

$$4NH_3(g) + 5O_2(g) \rightarrow 4NO(g) + 6H_2O(g)$$

a. 469.7 g
b. 300.6 g
c. 3.406 g
d. 2.180 g

45. M How many grams of IO_2F are needed to react completely with 150.0 g of BrF_3?

$$3IO_2F(s) + 4BrF_3(l) \rightarrow 3IF_5(l) + 2Br_2(l) + 3O_2(g)$$

a. 260.0 g
b. 146.2 g
c. 121.8 g
d. 64.98 g

46. Phosphine, an extremely poisonous and highly reactive gas, will react with oxygen to form
M tetraphosphorus decaoxide and water.

$PH_3(g) + O_2(g) \rightarrow P_4O_{10}(s) + H_2O(g)$ [unbalanced]

Calculate the mass of $P_4O_{10}(s)$ formed when 225 g of PH_3 reacts with excess oxygen.

a. 1880 g
b. 940 g
c. 470 g
d. 56.3 g

47. Calculate the mass of chromium(III) chloride formed when 125 g of chlorine gas (used as bleach
M and as a disinfectant for water supplies) reacts with excess chromium.

$Cr(s) + Cl_2(g) \rightarrow CrCl_3(s)$ [unbalanced]

a. 419 g
b. 186 g
c. 135 g
d. 59.9 g

48. Potassium chlorate (used in fireworks, flares and safety matches) forms oxygen and potassium
M chloride when heated.

$KClO_3(s) \rightarrow KCl(s) + O_2(g)$ [unbalanced]

How many grams of oxygen are formed when 26.4 g of potassium chlorate is heated?

a. 223 g
b. 99.1 g
c. 10.3 g
d. 4.60 g

49. Aluminum metal reacts with chlorine gas to form solid aluminum trichloride, $AlCl_3$. What mass of
M chlorine gas is needed to react completely with 163 g of aluminum?

a. 643 g
b. 321 g
c. 245 g
d. 214 g

50. Barium hydroxide (used in corrosion inhibitors and lubricants) reacts with chloric acid to form barium
H chlorate and water. What mass of water is formed when 138 g of barium hydroxide reacts with excess chloric acid?

a. 32.5 g
b. 29.0 g
c. 16.2 g
d. 7.31 g

51. Lead(II) sulfide was once used in glazing earthenware. It will also react with hydrogen peroxide to form
H lead(II) sulfate and water. How many grams of hydrogen peroxide are needed to react completely with 265 g of lead(II) sulfide?

a. 151 g
b. 123 g
c. 37.7 g
d. 9.41 g

52. M An important reaction sequence in the industrial production of nitric acid is the following:

$$N_2(g) + 3H_2(g) \rightarrow 2NH_3(g)$$
$$4NH_3(g) + 5O_2(g) \rightarrow 4NO(g) + 6H_2O(l)$$

Starting from 20.0 mol of nitrogen gas in the first reaction, how many moles of oxygen gas are required in the second one?

a. 12.5
b. 20.0
c. 25.0
d. 50.0
e. 100.

53. H Aluminum oxide (used as an adsorbent or a catalyst for organic reactions) forms when aluminum reacts with oxygen.

$$4Al(s) + 3O_2(g) \rightarrow 2Al_2O_3(s)$$

A mixture of 82.49 g of aluminum ($\mathcal{M}$= 26.98 g/mol) and 117.65 g of oxygen ($\mathcal{M}$= 32.00 g/mol) is allowed to react. What mass of aluminum oxide ($\mathcal{M}$= 101.96 g/mol) can be formed?

a. 155.8 g
b. 249.9 g
c. 311.7 g
d. 374.9 g

54. H Aluminum reacts with oxygen to produce aluminum oxide which can be used as an adsorbent, desiccant or catalyst for organic reactions.

$$4Al(s) + 3O_2(g) \rightarrow 2Al_2O_3(s)$$

A mixture of 82.49 g of aluminum ($\mathcal{M}$ = 26.98 g/mol) and 117.65 g of oxygen ($\mathcal{M}$ = 32.00 g/mol) is allowed to react. Identify the limiting reactant and determine the mass of the excess reactant present m the vessel when the reaction is complete.

a. oxygen is the limiting reactant; 19.81 g of aluminum remain
b. oxygen is the limiting reactant; 35.16 g of aluminum remain
c. aluminum is the limiting reactant; 16.70 g of oxygen remain
d. aluminum is the limiting reactant; 44.24 g of oxygen remain

55. H Magnesium reacts with iron(III) chloride to form magnesium chloride (which can be used in fireproofing wood and in disinfectants) and iron.

$$3Mg(s) + 2FeCl_3(s) \rightarrow 3MgCl_2(s) + 2Fe(s)$$

A mixture of 41.0 g of magnesium ($\mathcal{M}$ = 24.31 g/mol) and 175 g of iron(III) chloride ($\mathcal{M}$ = 162.2 g/mol) is allowed to react. What mass of magnesium chloride = 95.21 g/mol) is formed?

a. 154 g $MgCl_2$
b. 107 g $MgCl_2$
c. 71.4 g $MgCl_2$
d. 68.5 g $MgCl_2$

56. H Magnesium (used in the manufacture of light alloys) reacts with iron(III) chloride to form magnesium chloride and iron.

$$3Mg(s) + 2FeCl_3(s) \rightarrow 3MgCl_2(s) + 2Fe(s)$$

A mixture of 41.0 g of magnesium (M = 24.31 g/mol) and 175 g of iron(III) chloride (M = 162.2 g/mol) is allowed to react. Identify the limiting reactant and determine the mass of the excess reactant present in the vessel when the reaction is complete.

a. Limiting reactant is Mg; 67 g of $FeCl_3$ remain
b. Limiting reactant is Mg; 104 g of $FeCl_3$ remain
c. Limiting reactant is $FeCl_3$; 2 g of Mg remain
d. Limiting reactant is $FeCl_3$; 87 g of Mg remain

57. M Potassium chloride is used as a substitute for sodium chloride for individuals with high blood pressure. Calculate the mass of potassium chloride formed when 7.00 g of chlorine gas is allowed to react with 5.00 g of potassium.

a. 14.7 g KCl
b. 9.53 g KCl
c. 7.36 g KCl
d. 4.77 g KCl

58. H Potassium chloride is used as a substitute for sodium chloride for individuals with high blood pressure. Identify the limiting reactant and determine the mass of the excess reactant remaining when 7.00 g of chlorine gas reacts with 5.00 g of potassium to form potassium chloride.

a. Potassium is the limiting reactant; 2.47 g of chlorine remain
b. Potassium is the limiting reactant; 7.23 g of chlorine remain
c. Chlorine is the limiting reactant; 4.64 g of potassium remain
d. Chlorine is the limiting reactant; 2.70 g of potassium remain

59. M Tetraphosphorus hexaoxide (M = 219.9 g/mol) is formed by the reaction of phosphorus with oxygen gas.

$$P_4(s) + 3O_2(g) \rightarrow P_4O_6(s)$$

If a mixture of 75.3 g of phosphorus and 38.7 g of oxygen produce 43.3 g of P_4O_6, what is the percent yield for the reaction?

a. 57.5%
b. 48.8%
c. 32.4%
d. 16.3%

60. M What is the percent yield for the reaction

$$PCl_3(g) + Cl_2(g) \rightarrow PCl_5(g)$$

if 119.3 g of PCl_5 (M = 208.2 g/mol) are formed when 61.3 g of Cl_2 (M = 70.91 g/mol) react with excess PCl_3?

a. 85.0%
b. 66.3%
c. 51.4%
d. 43.7%

61. M Methanol (CH_4O) is converted to bromomethane (CH_3Br) as follows:

$$CH_4O + HBr \rightarrow CH_3Br + H_2O$$

If 12.23 g of bromomethane are produced when 5.00 g of methanol is reacted with excess HBr, what is the % yield?

a. 40.9
b. 82.6
c. 100.
d. 121
e. 245

62. M A 0.150 $\mathcal{m}$ sodium chloride solution is referred to as a physiological saline solution because it has the same concentration of salts as normal human blood. Calculate the mass of solute needed to prepare 275.0 mL of a physiological saline solution.

a. 31.9 g
b. 16.1 g
c. 8.77 g
d. 2.41 g

63. M Sodium chlorate is used as an oxidizer in the manufacture of dyes, explosives and matches. Calculate the mass of solute needed to prepare 1.575 L of 0.00250 *M* $NaClO_3$ ($\mathcal{M}$= 106.45 g/mol).

a. 419 g
b. 169 g
c. 0.419 g
d. 0.169 g

64. M Lithium hydroxide is used in alkaline batteries. Calculate the molarity of a solution prepared by dissolving 35.8 g of LiOH ($\mathcal{M}$=23.95 g/mol) in enough water to give a final volume of 750.0 mL.

a. 1.99 *M*
b. 1.50 *M*
c. 1.12 *M*
d. 0.502 *M*

65. M Calculate the molarity of 1.25 L of solution which contains 238.7 g of cobalt (II) chloride, which is used in the manufacture of vitamin B_{12} and as a humidity indicator.

a. 2.30 *M*
b. 1.47 *M*
c. 0.435 *M*
d. 0.680 *M*

66. E Hydrochloric acid is widely used as a laboratory reagent, in refining ore for the production of tin and tantalum, and as a catalyst in organic reactions. Calculate the number of moles of HCl in 62.85 mL of 0.453 *M* hydrochloric acid.

a. 1.04 mol
b. 0.139 mol
c. 0.0285 mol
d. 0.00721 mol

67. Sodium hydroxide, also known as caustic soda, is used to neutralize acids and to treat cellulose in
E making of cellophane. Calculate the number of moles of solute in 1.875 L of 1.356 *M* NaOH solution.

a. 2.543 mol
b. 1.383 mol
c. 0.7232 mol
d. 0.3932 mol

68. Calculate the molarity of a 23.55-mL solution which contains 28.24 mg of sodium sulfate (used in
M dyeing and printing textiles, $\mathcal{M}$ = 139.04 g/mol).

a. 1.199 *M*
b. 0.8339 *M*
c. 0.2031 *M*
d. 0.008625 *M*

69. When 2.61 g of solid Na_2CO_3 is dissolved in sufficient water to make 250. mL of solution, the
E concentration of Na_2CO_3 is:

a. 0.0246 *M*
b. 10.4 *M*
c. 0.205 *M*
d. 0.0985 *M*
e. 0.141 *M*

70. What mass (in g) of solid $CuSO_4$ is needed to make up 250. mL of a 0.300 *M* solution of this salt?
M

a. 8.36
b. 12.0
c. 39.9
d. 47.9
e. 75.0

71. Calcium chloride is used to melt ice and snow on roads and sidewalks and to remove water from organic
M liquids. Calculate the molarity of a solution prepared by diluting 165 mL of 0.688 *M* calcium chloride to 925.0 mL.

a. 3.86 *M*
b. 0.222 *M*
c. 0.123 *M*
d. 0.114 *M*

72. What will be the final volume of a solution prepared by diluting 25 mL of 8.25 *M* sodium hydroxide
M to a concentration of 2.40 *M?*

a. 330 mL
b. 210 mL
c. 86 mL
d. 60 mL

73. What volume, in L, of 10.0 *M* HCl is needed to make 2.00 L of 2.00 *M* HCl solution by dilution
E with water?

a. 0.800
b. 0.400
c. 0.200
d. 0.100
e. none of the above

74. How many mL of concentrated nitric acid (HNO_3, 16.0 *M*) should be diluted with water in order to
E make 2.00 L of 2.00 *M* solution?

a. 32.0
b. 62.5
c. 125
d. 250.
e. 500.

75. Acetic acid is a weak acid which is used in printing calico and dyeing silk. A dilute solution is
H known as vinegar. An aqueous solution which contains 80.00% acetic acid ($\mathcal{M}$= 60.05 g/mol) by mass has a density of 1.0699 g/mL. Calculate the molarity of this solution.

a. 17.81 *M*
b. 14.25 *M*
c. 0.01781 *M*
d. 0.01425 *M*

76. Formic acid is the primary ingredient in the venom of ants. Concentrated formic acid (17.72 *M*) has
H a density of 1.161 g/mL. What is the mass percent of HCOOH ($\mathcal{M}$= 46.03 g/mol) in the solution?

a. 81.57%
b. 70.26%
c. 34.53%
d. 6.655%

77. How many milliliters of 1.58 *M* HCl are needed to react completely with 23.2 g of $NaHCO_3$
M ($\mathcal{M}$= 84.02 g/mol)?

$$HCl(aq) + NaHCO_3(s) \rightarrow NaCl(s) + H_2O(l) + CO_2(g)$$

a. 638 mL
b. 572 mL
c. 536 mL
d. 175 mL

78. Copper(II) sulfide, CuS, is used in the development of aniline black dye in textile printing. What is the
M maximum mass of CuS which can be formed when 38.0 mL of 0.500 *M* $CuCl_2$ are mixed with 42.0 mL of 0.600 *M* $(NH_4)_2S$? Aqueous ammonium chloride is the other product.

a. 2.41 g
b. 1.82 g
c. 1.21 g
d. 0.909 g

Short Answer Questions

79. Propane, C_3H_8, is commonly provided as a bottled gas for use as a fuel. In 0.200 mol of propane
M

a. what is the mass of propane?
b. what mass of carbon is present?
c. how many molecules of C_3H_8 are present?
d. how many hydrogen atoms are present?

80. For a sample consisting of 2.50 g of methane, CH_4, calculate
M

a. the number of moles of methane present
b. the total number of atoms present

81. In 0.20 mole of phosphoric acid, H_3PO_4

E

a. how many H atoms are there?
b. what is the total number of atoms?
c. how many moles of O atoms are there?

82. A compound consisting of C, H and O only, has a molar mass of 331.5 g/mol. Combustion of 0.1000 g of this compound caused a 0.2921 g increase in the mass of the CO_2 absorber and a 0.0951 g increase in the mass of the H_2O absorber. What is the empirical formula of the compound?

H

83. Analysis of a white solid produced in a reaction between chlorine and phosphorus showed that it contained 77.44% chlorine and 22.56% phosphorus. What is its empirical formula?

E

84. a. Balance the following equation for the combustion of butane, a hydrocarbon used in gas lighters:

M

$$C_4H_{10}(g) + O_2(g) \rightarrow CO_2(g) + H_2O(l)$$

85. Balance the equation

E

$$B_2O_3(s) + NaOH(aq) \rightarrow Na_3BO_3(aq) + H_2O(l)$$

86. Balance the following equation for partial oxidation of ammonia, an important reaction in the production of nitric acid:

M

$$NH_3(g) + O_2(g) \rightarrow NO(g) + H_2O(l)$$

87. Gaseous methanol (CH_4O) reacts with oxygen gas to produce carbon dioxide gas and liquid water. Write a balanced equation for this process.

M

88. Consider the balanced equation for the combustion of propane, C_3H_8

M

$$C_3H_8(g) + 5O_2(g) \rightarrow 3CO_2(g) + 4H_2O(l)$$

If propane reacts with oxygen as above

a. what is the limiting reagent in a mixture containing 5.00 g of C_3H_8 and 10.0 g of O_2?
b. what mass of CO_2 is formed when 1.00 g of C_3H_8 reacts completely?

89. Ammonia, NH_3, is produced industrially from nitrogen and hydrogen as follows:

M

$$N_2(g) + 3H_2(g) \rightarrow 2NH_3(g)$$

What mass, of which starting material, will remain when 30.0 g of N_2 and 10.0 g of H_2 react until the limiting reagent is completely consumed?

90. Consider the balanced equation:

M

$$Al_2S_3(s) + 6H_2O(l) \rightarrow 2Al(OH)_3(s) + 3H_2S(g)$$

If 15.0g of aluminum sulfide and 10.0g of water are allowed to react as above, and assuming a complete reaction

a. by calculation, find out which is the limiting reagent
b. calculate the maximum mass of H_2S which can be formed from these reagents
c. calculate the mass of excess reagent remaining after the reaction is complete

91. The insecticide DDT was formerly in widespread use, but now it is severely restricted owing to its
H adverse environmental effects. It is prepared as follows:

$$\underset{chloral}{C_2HCl_3O} + \underset{chlorobenzene}{2C_6H_5Cl} \rightarrow \underset{DDT}{C_{14}H_9Cl_5} + H_2O$$

If 10.00 g of chloral were reacted with 10.00 g of chlorobenzene

a. what is the maximum amount (mol) of DDT which could be formed?
b. what is the limiting reagent?
c. what is the % yield, if 12.15 g of DDT is produced?

92. a. You are provided with a 250 mL volumetric flask, deionized water and solid NaOH. How much NaOH
E should be weighed out in order to make 250. mL of 0.100 *M* solution?

93. A solution of methanol (CH_4O) in water has a concentration of 0.200 *M*. What mass of methanol,
E in grams, is present in 0.150 liters of this solution?

94. a. A solution of common salt, NaCl, in water has a concentration of 0.0921 *M*. Calculate the number of
M moles of HCl contained in 50.0 mL of this solution.
b. If, instead, an NaCl solution is prepared by dissolving 10.0 g of solid NaCl in enough water to make 250. mL of solution, what is the molarity?

95. Aluminum metal dissolved in hydrochloric acid as follows
M

$$2Al(s) + 6HCl(aq) \rightarrow 2AlCl_3(aq) + 3H_2(g)$$

a. What is the minimum volume of 6.0 *M* HCl(*aq*) needed to completely dissolve 3.20 g of aluminum in this reaction?
b. What mass of $AlCl_3$ would be produced by complete reaction of 3.20 g of aluminum?

True/False Questions

96. One mole of O_2 has a mass of 16.0 g.
E

97. One mole of methane (CH_4) contains a total of 3×10^{24} atoms.
E

98. The formula $CH_3O_{0.5}$ is an example of an empirical formula.
E

99. In combustion analysis, the carbon and hydrogen contents of a substance are determined from the
E CO_2 and H_2O, respectively, which are collected in the absorbers.

100. In combustion analysis, the oxygen content of a substance is equal to the total oxygen in the CO_2
M and H_2O collected in the absorbers.

101. Structural isomers have the same empirical formula but different molecular formulas.
M

102. Structural isomers have the same molecular formula but different structural formulas.
M

103. In a correctly balanced equation, the number of reactant molecules must equal the number of
E product molecules.

104. The correct method for preparing one liter of a 1.0 *M* solution of X is to dissolve exactly one mole of X in exactly one liter of water.
E

105. When a solution is diluted with water, the ratio of the initial to final volumes of solution is equal to the ratio of final to initial molarities.
E

Stoichiometry: Mole-Mass Relationships in Chemical Systems
Chapter 3
Answer Key

1.	c	27.	b	53.	a
2.	b	28.	c	54.	d
3.	c	29.	d	55.	a
4.	b	30.	d	56.	c
5.	b	31.	c	57.	b
6.	b	32.	a	58.	a
7.	b	33.	b	59.	b
8.	a	34.	c	60.	b
9.	a	35.	b	61.	b
10.	b	36.	d	62.	d
11.	d	37.	b	63.	c
12.	d	38.	c	64.	a
13.	c	39.	b	65.	b
14.	b	40.	c	66.	c
15.	c	41.	a	67.	a
16.	b	42.	c	68.	d
17.	c	43.	b	69.	d
18.	d	44.	a	70.	b
19.	b	45.	b	71.	c
20.	a	46.	c	72.	c
21.	c	47.	b	73.	b
22.	a	48.	c	74.	d
23.	b	49.	a	75.	b
24.	b	50.	b	76.	b
25.	b	51.	a	77.	d
26.	d	52.	d	78.	b

79. a. 8.82 g
    b. 7.21 g
    c. 1.20×10^{23} C_3H_8 molecules
    d. 9.64×10^{23} H atoms

80. a. 0.156 mol CH_4
    b. 4.69×10^{23} atoms

81. a. 3.61×10^{23} H atoms
    b. 9.64×10^{23} atoms
    c. 4.82×10^{23} O atoms

82. $C_{22}H_{35}O_2$

83. PCl_3

84. $2C_4H_{10}(g) + 13O_2(g) \rightarrow 8CO_2(g) + 10H_2O(l)$

85. $B_2O_3(s) + 6NaOH(aq) \rightarrow 2Na_3BO_3(aq) + 3H_2O(l)$

86. $4NH_3(g) + 5O_2(g) \rightarrow 4NO(g) + 6H_2O(l)$

87. $2CH_4O(g) + 3O_2(g) \rightarrow 2CO_2(g) + 4H_2O(l)$

88. a. Oxygen is the limiting reagent
    b. 2.99 g CO_2

89. 3.52 g of hydrogen gas remains

90. a. Water is the limiting reagent
    b. 9.50 g of H_2S
    c. 1.11 g Al_2S_3 remains

91. a. 0.0444 mol DDT
    b. Chlorobenzene is the limiting reagent
    c. 77.2%

92. 1.000 g

93. 0.961 g

94. a. 0.00461 g
    b. 0.684 *M*

95. a. 59.3 mL
    b. 15.8 g

96. F

97. T

98. F

99. T

100. F

101. F

102. T

103. F

104. F

105. T

The Major Classes of Chemical Reactions
Chapter 4

Multiple Choice Questions

1. E

Potassium chloride, KCl, sodium sulfate, K_2SO_4, glucose, $C_6H_{12}O_6$, and ammonium phosphate, $(NH_4)_3PO_4$, are soluble in water. Which one produces the largest number of dissolved particles per mole of dissolved solute?

a. KCl
b. K_2SO_4
c. $C_6H_{12}O_6$
d. $(NH_4)_3PO_4$

2. E

Potassium carbonate, K_2CO_3, sodium iodide, NaI, methyl alcohol, CH_3OH, and ammonium chloride, NH_4Cl, are soluble in water. Which produces the largest number of dissolved particles per mole of dissolved solute?

a. K_2CO_3
b. NaI
c. CH_3OH
d. NH_4Cl

3. E

How many moles of ions are released when 0.27 mol of cobalt(II) chloride, $CoCl_2$, is dissolved in water?

a. 0.81 mol
b. 0.54 mol
c. 0.27 mol
d. 0.090 mol

4. E

How many moles of ions are released when 1.6 mol of ammonium phosphate, $(NH_4)_3PO_4$, is dissolved in water?

a. 0.40 mol
b. 1.6 mol
c. 3.2 mol
d. 6.4 mol

5. E

How many moles of $H^+(aq)$ ions are present in 750 mL of 0.65 *M* hydrochloric acid?

a. 1.2 mol
b. 0.87 mol
c. 0.65 mol
d. 0.49 mol

6. E

How many moles of $H^+(aq)$ ions are present in 1.25 L of 0.75 *M* nitric acid?

a. 0.60 mol
b. 0.75 mol
c. 0.94 mol
d. 1.7 mol

7. M

How many sodium ions are present in 325 mL of 0.850 *M* Na_2SO_4?

a. 1.66×10^{23}
b. 3.33×10^{23}
c. 4.99×10^{23}
d. 6.20×10^{23}

8. Which of the following is most soluble in water?
E

a. benzene, C_6H_6
b. potassium nitrate, KNO_3
c. carbon tetrachloride, CCl_4
d. hexane, C_6H_{14}

9. Which of the following will be least soluble in water?
E

a. potassium sulfate, K_2SO_4
b. ammonium nitrate, NH_4NO_3
c. chloromethane, CH_3Cl
d. calcium chloride, $CaCl_2$

10. Which of the following solutions will be the poorest conductor of electrical current?
E

a. sucrose, $C_{12}H_{22}O_{11}(aq)$
b. sodium chloride, $NaCl(aq)$
c. potassium nitrate, $KNO_3(aq)$
d. lithium hydroxide, $LiOH(aq)$

11. Which of the following solutions will be the best conductor of electrical current?
E

a. methyl alcohol, $CH_3OH(aq)$
b. glucose, $C_6H_{12}O_6(aq)$
c. potassium chloride, $KCl(aq)$
d. bromine, $Br_2(aq)$

12. 1.0 *M* aqueous solutions of the following substances are prepared. Which one would you expect to
E have the lowest electrical conductivity?

a. $NaOH$
b. CH_3CH_2OH (ethanol)
c. KBr
d. CH_3COOH (acetic acid)
e. $HClO_4$

13. Which one of the following substances, when dissolved in water at equal molar concentrations,
E will give the solution with the lowest electrical conductivity?

a. $CaCl_2$
b. HNO_3
c. NH_3
d. $C_6H_{12}O_6$ (glucose)
e. CO_2

14. Which one of the following substances is the best electrolyte?
E

a. CO
b. CH_3Cl
c. H_2O
d. C_2H_5OH
e. $CsCl$

15. Select the spectator ions for the following reaction.
E

$Pb(NO_3)_2(aq)$ + $2NaCl(aq)$ → $PbCl_2(s)$ + $2NaNO_3(aq)$

a. $Pb^{2+}(aq)$, $Cl^-(aq)$
b. $Na^+(aq)$, $NO_3^-(aq)$
c. $Pb^{2+}(aq)$, $NO_3^-(aq)$
d. $Na^+(aq)$, $Cl^-(aq)$

16. Select the precipitate that forms when the following reactants are mixed.
E

$Na_2CO_3(aq)$ + $BaCl_2(aq)$ →

a. Ba_2CO_3
b. $BaCO_3$
c. NaCl
d. $NaCl_2$

17. Select the precipitate that forms when the following reactants are mixed.
E

$Mg(CH_3COO)_2(aq)$ + $LiOH(aq)$ →

a. $LiCH_3COO$
b. $Li(CH_3COO)_2$
c. MgOH
d. $Mg(OH)_2$

18. Select the precipitate that forms when aqueous ammonium sulfide reacts with aqueous copper(II) nitrate
M

a. CuS
b. Cu_2S
c. NH_4NO_3
d. $NH_4(NO_3)_2$

19. Select the precipitate that forms when aqueous lead(II) nitrate reacts with aqueous sodium sulfate.
M

a. $NaNO_3$
b. Na_2NO_3
c. $PbSO_4$
d. Pb_2SO_4

20. Select the correct name and chemical formula for the precipitate that forms when the following reactants are mixed.
M

$CuCl_2(aq)$ + $Na_2CO_3(aq)$ →

a. copper(I) carbonate, $Cu_2CO_3(s)$
b. copper(II) carbonate, $Cu_2CO_3(s)$
c. copper(I) carbonate, $CuCO_3(s)$
d. copper(II) carbonate, $CuCO_3(s)$

21. Select the correct name and chemical formula for the precipitate that forms when the following reactants are mixed.
M

$CoSO_4(aq)$ + $(NH_4)_3PO_4(aq)$ →

a. cobalt(II) phosphate, $Co_3(PO_4)_2$
b. cobalt(III) phosphate, $Co_3(PO_4)_2$
c. cobalt(II) phosphate, $CoPO_4$
d. cobalt(III) phosphate, $CoPO_4$

22. Select the net ionic equation for the reaction between sodium chloride and mercury(I) nitrate.
E

$$2NaCl(aq) + Hg_2(NO_3)_2(aq) \rightarrow NaNO_3(aq) + Hg_2Cl_2(s)$$

a. $Na^+(aq) + NO_3^-(aq) \rightarrow NaNO_3(aq)$
b. $Hg_2^{2+}(aq) + 2Cl^-(aq) \rightarrow Hg_2Cl_2(s)$
c. $NaCl(aq) \rightarrow Na^+(aq) + Cl^-(aq)$
d. $Hg_2(NO_3)_2(aq) \rightarrow Hg_2^{2+}(aq) + 2NO_3^-(aq)$

23. Which of the following is a weak acid?
M

a. H_2SO_4
b. HNO_3
c. HF
d. HBr

24. Which of the following is a strong acid?
E

a. H_3PO_4
b. HNO_3
c. HF
d. CH_3COOH

25. Which of the following is a strong base?
E

a. NH_3
b. $Ca(OH)_2$
c. $Al(OH)_3$
d. $B(OH)_3$

26. Which of the following is a weak base?
M

a. NH_3
b. $Ca(OH)_2$
c. $Ba(OH)_2$
d. NaOH

27. Which one of the following substances is a strong acid?
E

a. HNO_3
b. H_2CO_3
c. NH_3
d. CH_3COOH
e. H_3PO_4

28. Select the correct set of products for the following reaction.
E

$$Ba(OH)_2(aq) + HNO_3(aq) \rightarrow$$

a. $BaN_2(s) + H_2O(l)$
b. $Ba(NO_3)_2(aq) + H_2O(l)$
c. $Ba(s) + H_2(g) + NO_2(g)$
d. No reaction occurs

29. Select the net ionic equation for the reaction between lithium hydroxide and hydrobromic acid.
E

$$LiOH\,(aq) \;+\; HBr(aq) \;\rightarrow\; H_2O(l) \;+\; LiBr(aq)$$

a. $LiOH\,(aq) \rightarrow Li^+(aq) + OH^-(aq)$
b. $HBr(aq) \rightarrow H^+(aq) + Br^-(aq)$
c. $H^+(aq) + OH^-(aq) \rightarrow H_2O(l)$
d. $Li^+(aq) + Br^-(aq) \rightarrow LiBr(aq)$

30. A standard solution of 0.243 *M* NaOH was used to determine the concentration of a hydrochloric acid solution. If 46.33 mL of NaOH is needed to neutralize 10.00 mL of the acid, what is the molar concentration of the acid?
M

a. 0.0524 *M*
b. 0.888 *M*
c. 1.13 *M*
d. 2.43 *M*

31. Automobile batteries use 3.0 *M* H_2SO_4 as an electrolyte. How much 1.20 *M* NaOH will be needed to neutralize 225 mL of battery acid?
M

$$H_2SO_4(aq) \;+\; 2NaOH(aq) \;\rightarrow\; 2H_2O(l) \;+\; Na_2SO_4(aq)$$

a. 0.045 L
b. 0.28 L
c. 0.56 L
d. 1.1 L

32. Vinegar is a solution of acetic acid, CH_3COOH, dissolved in water. A 5.54-g sample of vinegar was neutralized by 30.10 mL of 0.100 *M* NaOH. What is the percent by weight of acetic acid in the vinegar?
M

a. 0.184%
b. 3.26%
c. 5.43%
d. 9.23%

33. A 0.00100 mol sample of $Ca(OH)_2$ requires 25.00 mL of aqueous HCl for neutralization according to the reaction below. What is the concentration of the HCl?
M

Equation: $Ca(OH)_2(s) \;+\; 2HCl(aq) \rightarrow CaCl_2(aq) \;+\; H_2O(l)$

a. 0.0200 *M*
b. 0.0400 *M*
c. 0.0800 *M*
d. 4.00×10^{-5} *M*
e. none of the above

34. Calculate the oxidation number of the chlorine in perchloric acid, $HClO_4$, a strong oxidizing agent.
E

a. -1
b. +4
c. +5
d. +7

35. Calculate the oxidation number of sulfur in sodium metabisulfite, $Na_2S_2O_5$.
E

a. -2
b. +2
c. +4
d. +5

36. Sodium tripolyphosphate is used in detergents to make them effective in hard water. Calculate the oxidation number of phosphorus in $Na_5P_3O_{10}$.
E

a. +3
b. +5
c. +10
d. +15

37. Calculate the oxidation number of iodine in I_2.
E

a. -1
b. 0
c. +1
d. +7

38. The oxidation numbers of P, S and Cl in $H_2PO_2^-$, H_2S and $KClO_4$ are, respectively
M

a. -1, -1, +3
b. +1, -2, +7
c. +1, +2, +7
d. -1, -2, +7
e. -1, -2, +3

39. Identify the oxidizing agent in the following redox reaction.
E

$$Hg^{2+}(aq) + Cu(s) \rightarrow Cu^{2+}(aq) + Hg(l)$$

a. $Hg^{2+}(aq)$
b. $Cu(s)$
c. $Cu^{2+}(aq)$
d. $Hg(l)$

40. Sodium thiosulfate, $Na_2S_2O_3$, is used as a "fixer" in black and white photography. Identify the reducing agent in the reaction of thiosulfate with iodine.
E

$$2S_2O_3^{2-}(aq) + I_2(aq) \rightarrow S_4O_6^{2-}(aq) + 2I^-(aq)$$

a. $I_2(aq)$
b. $I^-(aq)$
c. $S_2O_3^{2-}(aq)$
d. $S_4O_6^{2-}(aq)$

41. Which one of the following is not a redox reaction?
M

a. $2H_2(g) + O_2(g) \rightarrow 2H_2O(l)$
b. $Zn(s) + H_2SO_4(aq) \rightarrow ZnSO_4(aq) + H_2(g)$
c. $H_2O(l) + NH_3(g) \rightarrow NH_4^+(aq) + OH^-(aq)$
d. $6FeSO_4(aq) + K_2Cr_2O_7(aq) + 7H_2SO_4(aq) \rightarrow Cr_2(SO_4)_3(aq) + 3Fe_2(SO_4)_3(aq) + K_2SO_4(aq) + 7H_2O(l)$
e. $Cl_2(g) + 2KBr(aq) \rightarrow Br_2(l) + 2KCl(aq)$

42. Which one of the following is not a redox reaction?
M

a. $2H_2O_2(aq) \rightarrow 2H_2O(l) + O_2(g)$
b. $N_2(g) + 3H_2(g) \rightarrow 2NH_3(g)$
c. $BaCl_2(aq) + K_2CrO_4(aq) \rightarrow BaCrO_4(aq) + 2KCl(aq)$
d. $2Al(s) + Fe_2O_3(s) \rightarrow Al_2O_3(s) + 2Fe(s)$
e. $2H_2O(g) \rightarrow 2H_2(g) + O_2(g)$

43. Which one of the following is a redox reaction?
M

a. $2Na(g) + Cl_2(g) \rightarrow 2NaCl(s)$
b. $Ba^{2+}(aq) + SO_4^{2-}(aq) \rightarrow BaSO_4(s)$
c. $K_2Cr_2O_7(aq) + 2KOH(aq) \rightarrow 2K_2CrO_4(aq) + H_2O(l)$
d. $Na_2CO_3(s) + 2HCl(aq) \rightarrow 2NaCl(aq) + CO_2(g) + H_2O(l)$
e. $H_2O(l) \rightarrow H^+(aq) + OH^-(aq)$

44. Which one of the following is not a redox reaction?
M

a. $2Na(s) + 2H_2O(l) \rightarrow 2NaOH(aq) + H_2(g)$
b. $H_2(g) + Cl_2(g) \rightarrow 2HCl(g)$
c. $2H_2O_2(aq) \rightarrow 2H_2O(l) + O_2(g)$
d. $Fe_2O_3(s) + 3H_2SO_4(aq) \rightarrow Fe_2(SO_4)_3(aq) + 3H_2O(l)$
e. $2KMnO_4(aq) + 10FeSO_4(aq) + 8H_2SO_4(aq) \rightarrow$
$K_2SO_4(aq) + 2MnSO_4(aq) + 5Fe_2(SO_4)_3(aq) + 8H_2O(l)$

45. Identify all the spectator ions in the following reaction.
M

$$2KMnO_4(aq) + 10FeSO_4(aq) + 8H_2SO_4(aq) \rightarrow K_2SO_4(aq) + 2MnSO_4(aq) + 5Fe_2(SO_4)_3(aq) + 8H_2O(l)$$

a. only K^+
b. only SO_4^{2-}
c. only K^+ and SO_4^{2-}
d. only K^+, SO_4^{2-} and Fe^{2+}
e. only K^+, SO_4^{2-}, Fe^{2+} and Mn^{2+}

46. Balance the following redox equation using the smallest integers possible and select the correct coefficient for the iron(II) hydroxide, $Fe(OH)_2$.
M

$$Fe(OH)_2(s) + CrO_4^{2-}(aq) \rightarrow Fe_2O_3(s) + Cr(OH)_4^-(aq) + H_2O(l) + OH^-(aq)$$

a. 1
b. 2
c. 3
d. 6

47. Balance the following redox equation using the smallest integers possible and select the correct coefficient for the hydrogen sulfite ion, HSO_3^-.
M

$$MnO_4^-(aq) + HSO_3^-(aq) + H^+(aq) \rightarrow Mn^{2+}(aq) + SO_4^{2-}(aq) + H_2O(l)$$

a. 1
b. 2
c. 5
d. 10

48. Balance the following redox equation using the smallest integers possible and select the correct coefficient for the chromate anion, CrO_4^{2-}.
M

$$CrO_4^{2-}(aq) + CN^-(aq) + H_2O(l) \rightarrow CNO^-(aq) + Cr(OH)_4^-(aq) + OH^-(aq)$$

a. 1
b. 2
c. 3
d. 6

49. Balance the following redox equation using the smallest integers possible and select the correct coefficient for the bromide anion, Br^-.
M

$$Br_2(aq) + OH^-(aq) \rightarrow Br^-(aq) + BrO_3^-(aq) + H_2O(l)$$

a. 2
b. 5
c. 6
d. 10

50. The amount of calcium present in milk can be determined by adding oxalate to a sample and measuring the mass of calcium oxalate precipitated. What is the mass percent of calcium if 0.429 g of calcium oxalate forms in a 125-g sample of milk when excess aqueous sodium oxalate is added?
M

$$Na_2C_2O_4(aq) + Ca^{2+}(aq) \rightarrow CaC_2O_4(s) + 2Na^+(aq)$$

a. 0.107%
b. 0.343%
c. 1.10%
d. 1.37%

51. Select the classification for the following reaction.
E

$$2Na(s) + 2H_2O(l) \rightarrow 2NaOH(aq) + H_2(g)$$

a. precipitation
b. neutralization
c. redox
d. none of the above

52. Select the classification for the following reaction.
E

$$Fe^{2+}(aq) + 2OH^-(aq) \rightarrow Fe(OH)_2(s)$$

a. precipitation
b. neutralization
c. redox
d. none of the above

53. Select the classification for the following reaction.
M

$$NH_3(aq) + HNO_3(aq) \rightarrow NH_4NO_3(aq)$$

a. precipitation
b. neutralization
c. redox
d. none of the above

54. Select the classification for the following reaction.
E

$$KOH(aq) + HCl(aq) \rightarrow KCl(aq) + H_2O(l)$$

a. precipitation
b. neutralization
c. redox
d. none of the above

55. Select the classification for the following reaction.
M

$$Fe(s) + 2Fe^{3+}(aq) \rightarrow 3Fe^{2+}(aq)$$

a. precipitation
b. neutralization
c. redox
d. none of the above

56. Select the classification for the following reaction.
E

$$BaCl_2(aq) + K_2SO_4(aq) \rightarrow BaSO_4(s) + 2KCl(aq)$$

a. precipitation
b. neutralization
c. redox
d. none of the above

57. Select the classification for the following reaction
E

$$H_2(g) + Cl_2(g) \rightarrow 2HCl(g)$$

a. combination
b. decomposition
c. displacement
d. neutralization

58. Select the classification for the following reaction.
E

$$2Na(s) + Cl_2(g) \rightarrow 2NaCl(s)$$

a. combination
b. decomposition
c. displacement
d. neutralization

59. The compound P_4O_{10} is used in refining sugar. Select the classification for the reaction in which it is synthesized.
E

$$P_4(s) + 5O_2(g) \rightarrow P_4O_{10}(s)$$

a. combination
b. decomposition
c. displacement
d. neutralization

60. Calcium carbonate, $CaCO_3$, is used in the blast furnace during the production of steel. What is the classification for the reaction in which it forms from calcium oxide and carbon dioxide?
E

$$CaO(s) + CO_2(g) \rightarrow CaCO_3(s)$$

a. combination
b. decomposition
c. displacement
d. neutralization

61. Select the classification for the following reaction.
E

$$Na_2O(s) + H_2O(l) \rightarrow 2NaOH(aq)$$

a. combination
b. decomposition
c. displacement
d. neutralization

62. Select the classification for the following reaction.
E

$$CrO(s) + SO_2(g) \rightarrow CrSO_3(s)$$

a. combination
b. decomposition
c. displacement
d. neutralization

63. Select the classification for the following reaction.
E

$$2H_2O_2(aq) \rightarrow 2H_2O(l) + O_2(g)$$

a. combination
b. decomposition
c. displacement
d. neutralization

64. Select the classification for the following reaction.
E

$$CaCl_2 \cdot H_2O(s) + \text{heat} \rightarrow CaCl_2(s) + H_2O(g)$$

a. combination
b. decomposition
c. displacement
d. neutralization

65. Select the classification for the following reaction.
E

$$H_2CO_3(aq) \rightarrow H_2O(l) + CO_2(g)$$

a. combination
b. decomposition
c. displacement
d. neutralization

66. M Predict the product(s) for the following reaction.

$Cl_2O_7(g)$ + $H_2O(l)$

a. $HClO_4(aq)$
b. $H_2ClO_4(aq)$
c. $H(ClO_4)_2\ (aq)$
d. $HCl(aq) + O_2(g)$

67. M Predict the product(s) for the following reaction.

$BaO(s)$ + $CO_2(g)$ →

a. $Ba(s)$ + $CO_3(g)$
b. $BaCO_3(s)$
c. $BaO(s)$ + $C(s)$
d. No reaction occurs

68. E Predict the product(s) for the following reaction.

$H_2SO_4(aq)$ + $KOH(aq)$ →

a. $K_2SO_4(aq)$ + $H_2O(l)$
b. $K_2S(aq)$ + $H_2O(l)$
c. $K(s)$ + $H_2(g)$ + $SO_3(g)$
d. No reaction occurs

69. M Predict the product(s) for the following reaction.

$MgCO_3(s)$ + heat →

a. $MgO_2(s)$ + $CO(g)$
b. $MgO(s)$ + CO_2
c. $Mg(s)$ + $CO_2(g)$ + $O_2(g)$
d. No reaction occurs

70. M Pyrite, an impurity in some coals, reacts with oxygen to form the air pollutant sulfur dioxide. When 2.0×10^4 kg of coal containing 0.050 mass percent pyrite is burned, what mass of sulfur dioxide will form?

$2FeS_2(s)$ + $5O_2(g)$ → $4SO_2(g)$ + $2FeO(s)$

a. 11 g
b. 1.1×10^4 g
c. 1.1×10^6g
d. 2.1×10^7 g

Short Answer Questions

71. E An aqueous solution of lead nitrate, $Pb(NO_3)_2$, is mixed with one of sodium chromate, Na_2CrO_4, resulting in the formation of a precipitate of lead chromate. Write a balanced net ionic equation for this precipitation reaction, showing all phases.

72. M In both of the following reactions, a precipitate is formed. Complete and balance the equations, showing the phases of the products.

a. $AgNO_3(aq)$ + $CaCl_2(aq)$ →
b. $NaOH(aq)$ + $Fe(NO_3)_3(aq)$ →

73. In both of the following reactions, a precipitate is formed. Complete and balance the equations, showing the phases of the products.
M
a. $BaCl_2(aq)$ + $Na_2SO_4(aq)$ →
b. $Mg(NO_3)_2(aq)$ + $KOH(aq)$ →

74. Complete and balance the equation for the following acid-base reaction.
E

$Ca(OH)_2(aq)$ + $HCl(aq)$ →

75. a. You are provided with a 250 mL volumetric flask, deionized water and solid NaOH. How much NaOH should be weighed out in order to make 250. mL of 0.100 *M* solution?
M
b. 25.0 mL of the 0.100 *M* aqueous NaOH is titrated against sulfuric acid, H_2SO_4, according to the equation

$2NaOH(aq)$ + $H_2SO_4(aq)$ → $Na_2SO_4(aq)$ + $2H_2O(l)$

If the volume of sulfuric acid solution required to neutralize the NaOH is 18.62 mL, what is its concentration?

76. Sulfuric acid (H_2SO_4) reacts with potassium hydroxide (KOH) as follows.
M

$H_2SO_4(aq)$ + $2KOH(aq)$ → $K_2SO_4(aq)$ + $2H_2O(l)$

Calculate the volume of 0.100 *M* sulfuric acid required to neutralize 25.0 mL of 0.0821 *M* KOH. Show all your work.

77. A 0.1873 g sample of a pure, solid acid, H_2X was dissolved in water and titrated with 0.1052 *M* NaOH solution. The balanced equation for the neutralization reaction occurring is
M

$H_2X(aq)$ + $2NaOH(aq)$ → $Na_2X(aq)$ + $2H_2O(l)$

If the molar mass of H_2X is 85.00 g/mol, calculate the volume of NaOH solution needed in the titration.

78. In each of the following cases, write down the oxidation number of the indicated atom.
E

a. P in P_4
b. C in C_2H_6
c. S in H_2SO_4
d. Mn in MnO_4^-
e. S in $S_4O_6^{2-}$
f. P in Na_3PO_4

79. Write down the oxidation number of the indicated atom in each of the following formulas:
E

a. Si in SiO_2
b. Cl in ClO_2^-
c. Mn in $KMnO_4$
d. C in $C_6H_{12}O_6$

80. For each of the following species, write down, next to the formula, the oxidation number of the indicated atom.
E

a. P in $H_2PO_2^-$
b. S in $Na_2S_2O_3$
c. C in CH_2O

81. a. Explain or define what is meant by the term "oxidation."
E
b. Write down the oxidation numbers of the atoms in the following formulas.

$K_2Cr_2O_7$
K: Cr: O:

NaH
Na: H:

82. H Balance the redox equation

$$HNO_3(aq) + C_2H_6O(aq) + K_2Cr_2O_7(aq) \rightarrow KNO_3(aq) + C_2H_4O(aq) + H_2O(l) + Cr(NO_3)_3(aq)$$

83. H Balance the redox equation

$$K_2Cr_2O_7(aq) + HI(aq) \rightarrow KI(aq) + CrI_3(aq) + I_2(s) + H_2O(l)$$

84. H
a. Define "reduction"
b. Balance the following redox equation.

$$HNO_3(aq) + KI(aq) \rightarrow NO(g) + I_2(s) + H_2O(l) + KNO_3(aq)$$

85. E
a. Define "oxidation"
b. Identify the oxidizing and reducing agents in the following (unbalanced) equation.

$$HNO_3(aq) + C_2H_6O(aq) + K_2Cr_2O_7(aq) \rightarrow KNO_3(aq) + C_2H_4O(aq) + H_2O(l) + Cr(NO_3)_3(aq)$$

True/False Questions

86. M Some covalent compounds dissolve in water to produce conducting solutions.

87. E In a neutralization reaction the indicator will change color at the end point.

88. M In a neutralization reaction the equivalence point is the point where the indicator changes color.

89. M All acid base reactions produce a salt and water as the only products.

90. M A particular reaction may be both a precipitation and a neutralization.

91. E In a redox reaction, the reducing agent undergoes loss of electrons.

92. E In a redox reaction, the oxidizing agent undergoes loss of electrons.

93. M All combustion reactions are classified as combination reactions.

94. M The combustion of an element is always a combination reaction.

95. M A combination reaction may also be a displacement reaction.

96. M Chemical reactions generally reach equilibrium because one of the reactants is used up.

The Major Classes of Chemical Reactions
Chapter 4
Answer Key

1.	d	26.	a	51.	c
2.	a	27.	a	52.	a
3.	a	28.	b	53.	b
4.	d	29.	c	54.	b
5.	d	30.	c	55.	c
6.	c	31.	d	56.	a
7.	b	32.	b	57.	a
8.	b	33.	c	58.	a
9.	c	34.	d	59.	a
10.	a	35.	c	60.	a
11.	c	36.	b	61.	a
12.	b	37.	b	62.	a
13.	d	38.	b	63.	b
14.	e	39.	a	64.	b
15.	b	40.	c	65.	b
16.	b	41.	c	66.	a
17.	d	42.	c	67.	b
18.	a	43	a	68.	a
19.	c	44.	d	69.	b
20.	d	45.	c	70.	b
21.	a	46.	d		
22.	b	47.	c		
23.	c	48.	b		
24.	b	49.	b		
25.	b	50.	a		

71. $Pb^{2+}(aq) + CrO_4^{2-}(aq) \rightarrow PbCrO_4(s)$

72 a. $2AgNO_3(aq) + CaCl_2(aq) \rightarrow 2AgCl(s) + Ca(NO_3)_2(aq)$
b. $3NaOH(aq) + Fe(NO_3)_3(aq) \rightarrow Fe(OH)_3(s) + 3NaNO_3(aq)$

73. a. $BaCl_2(aq) + Na_2SO_4(aq) \rightarrow BaSO_4(s) + 2NaCl(aq)$
b. $Mg(NO_3)_2(aq) + 2KOH(aq) \rightarrow Mg(OH)_2(s) + 2KNO_3(aq)$

74. $Ca(OH)_2(aq) + 2HCl(aq) \rightarrow CaCl_2(aq) + 2H_2O(l)$

75. a. 1.00 g
b. 0.0671 *M*

76. 10.3 mL

77. 41.89 mL

78. a. 0
b. -3
c. +6
d. +7
e. +2.5
f. +5

79. a. +4
    b. +3
    c. +7
    d. 0

80. a. +1
    b. +2
    c. 0

81. a. Oxidation: loss of electrons or increase in oxidation number.
    b. K, +1 Cr, +6 O, -2 Na, +1 H, -1

82. $8HNO_3(aq) + 3C_2H_6O(aq) + K_2Cr_2O_7(aq) \rightarrow$
    $2KNO_3(aq) + 3C_2H_4O(aq) + 7H_2O(l) + 2Cr(NO_3)_3(aq)$

83. $K_2Cr_2O_7(aq) + 14HI(aq) \rightarrow 2KI(aq) + 2CrI_3(aq) + 3I_2(s) + 7H_2O(l)$

84. a. Reduction: gain of electrons or decrease in oxidation number
    b. $8HNO_3(aq) + 6KI(aq) \rightarrow 2NO(g) + 3I_2(s) + 4H_2O(l) + 6KNO_3(aq)$

85. a. Oxidation: loss of electrons or increase in oxidation number.
    b. Oxidizing agent is $K_2Cr_2O_7(aq)$; reducing agent is $C_2H_6O(aq)$.

86. T

87. T

88. F

89. F

90. T

91. F

92. T

93. F

94. T

95. F

96. F

Gases and the Kinetic-Molecular Theory
Chapter 5

Multiple Choice Questions

1. M

Mineral oil can be used in place of mercury in manometers when small pressure changes are to be measured. What is the pressure of an oxygen sample in mm of mineral oil if its pressure is 28.5 mm Hg? (*d* of mineral oil = 0.88 g/mL; *d* of Hg = 13.5 g/mL)

a. 1.9 mm mineral oil
b. 32 mm mineral oil
c. 380 mm mineral oil
d. 440 mm mineral oil

2. E

A flask containing helium gas is connected to an open-ended mercury manometer. The open end is exposed to the atmosphere, where the prevailing pressure is 752 torr. The mercury level in the open arm is 26 mm above that in the arm connected to the flask of helium. What is the helium pressure, in torr?

a. -26
b. 26
c. 726
d. 778
e. none of the above

3. E

A flask containing neon gas is connected to an open-ended mercury manometer. The open end is exposed to the atmosphere, where the prevailing pressure is 745 torr. The mercury level in the open arm is 50. mm below that in the arm connected to the flask of neon. What is the neon pressure, in torr?

a. -50.
b. 50.
c. 695
d. 795
e. none of the above

4. E

A flask containing argon gas is connected to an closed-ended mercury manometer. The closed end is under vacuum. If the mercury level in the closed arm is 230. mm above that in the arm connected to the flask, what is the argon pressure, in torr?

a. -230.
b. 230.
c. 530.
d. 790.
e. none of the above

5. E

Hydrogen gas exerts a pressure of 466 torr in a container. What is this pressure in atmospheres?

a. 0.217 atm
b. 0.613 atm
c. 1.63 atm
d. 4.60 atm

6. E

The pressure of hydrogen sulfide gas in a container is 35,650 Pa. What is this pressure in torr?

a. 267.4 torr
b. 351.8 torr
c. 3612 torr
d. 27090 torr

7. The pressure of sulfur dioxide in a container is 159 kPa. What is this pressure in atmospheres?
E

a. 0.209 atm
b. 0.637 atm
c. 1.57 atm
d. 21.2 atm

8. The air pressure in a volleyball is 75 psi. What is this pressure in torr?
E

a. 520 torr
b. 562 torr
c. 3900 torr
d. 7600 torr

9. "The volume of an ideal gas is directly proportional to the number of moles of the gas at constant
E temperature and pressure"is a statement of ____________ Law.

a. Charles's
b. Boyle's
c. Amonton's
d. Avogadro's

10. "The pressure of an ideal gas is inversely proportional to its volume at constant temperature and number of
E moles" is a statement of _________________ Law.

a. Charles's
b. Boyle's
c. Amonton's
d. Avogadro's

11. "The pressure of an ideal gas is directly proportional to its absolute temperature at constant volume and
E number of moles" is a statement of _______________ Law.

a. Charles's
b. Boyle's
c. Amonton's
d. Avogadro's

12. "The volume of an ideal gas is directly proportional to its absolute temperature at constant pressure and
E number of moles" is a statement of _______________ Law.

a. Charles's
b. Boyle's
c. Amonton's
d. Avogadro's

13. "The total pressure in a mixture of unreacting gases is equal to the sum of the partial pressures of the
E individual gases" is a statement of _________________ Law.

a. Charles's
b. Graham's
c. Dalton's
d. Avogadro's

14. “The rate of effusion of a gas is inversely proportional to the square root of its molar mass” is a statement of
E ______________________ Law.

a. Charles's
b. Graham's
c. Dalton's
d. Avogadro's

15. Which of the lines on the figure below is the best representation of the relationship between the
M volume of a gas and its pressure, other factors remaining constant?

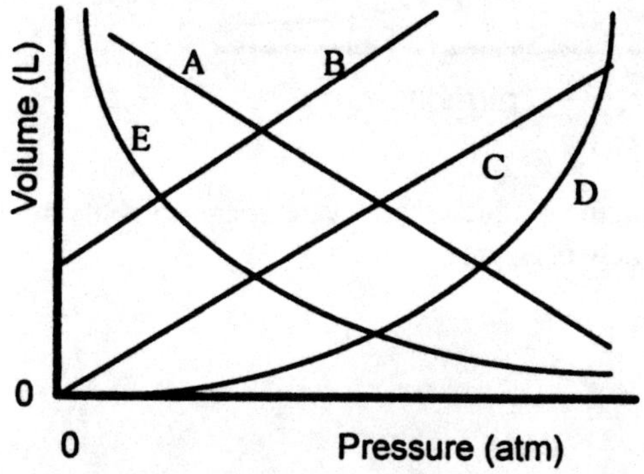

16. Which of the lines on the figure below is the best representation of the relationship between the
M volume of a gas and its absolute temperature, other factors remaining constant?

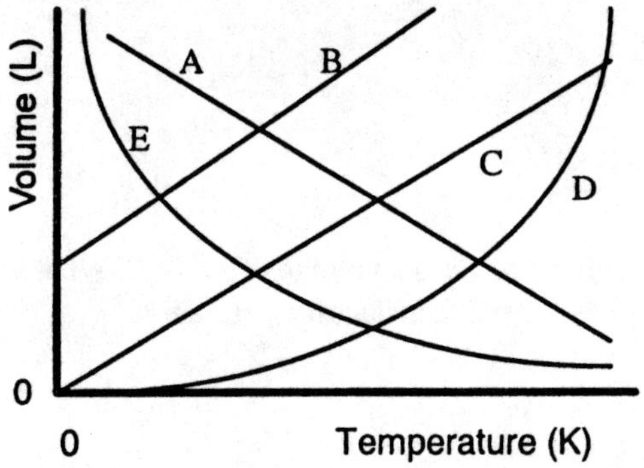

17. Which of the lines on the figure below is the best representation of the relationship between the
M volume of a gas and its Celsius temperature, other factors remaining constant?

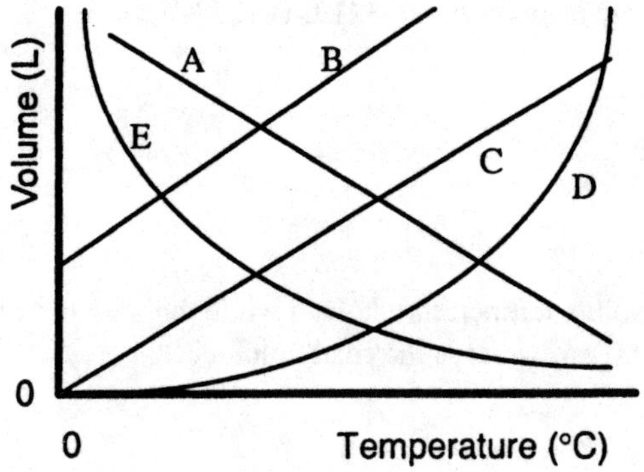

18. M Which of the lines on the figure below is the best representation of Avogadro's Law relating to the behavior of gases? (Pressure and temperature are assumed constant.)

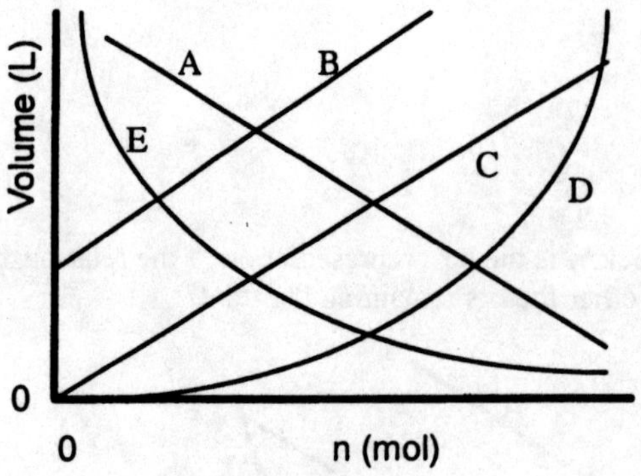

19. E A sample of an ideal gas has its volume doubled while its temperature remains constant. If the original pressure was 100 torr, what is the new pressure?

a. 10 torr
b. 50 torr
c. 100 torr
d. 200 torr

20. E A sample of the inert gas krypton has its pressure tripled while its temperature remained constant. If the original volume is 12 L, what is the final volume?

a. 4.0 L
b. 6.0 L
c. 9 L
d. 36 L

21. E A sample of nitrogen gas at 298 K and 745 torr has a volume of 37.42 L. What volume will it occupy if the pressure is increased to 894 torr at constant temperature?

a. 22.3 L
b. 31.2 L
c. 44.9 L
d. 380 L

22. E A sample of carbon dioxide gas at 125°C and 248 torr occupies a volume of 275 L. What will the gas pressure be if the volume is increased to 321 L at 125°C?

a. 212 torr
b. 356 torr
c. 441 torr
d. 359 torr

23. E A sample of oxygen gas has its absolute temperature halved while the pressure of the gas remained constant. If the initial volume is 400 mL, what is the final volume?

a. 20 mL
b. 200 mL
c. 400 mL
d. 800 mL

24. A sample container of carbon monoxide occupies a volume of 435 mL at a pressure of 785 torr and a
M temperature of 298 K. What would its temperature be if the volume were changed to 265 mL at a pressure of 785 torr?

a. 182 K
b. 387 K
c. 489 K
d. 538 K

25. A 0.850-mole sample of nitrous oxide, a gas used as an anesthetic by dentists, has a volume of 20.46 L at
M 123°C and 1.35 atm. What would be its volume at 468°C and 1.35 atm?

a. 5.38 L
b. 10.9 L
c. 38.3 L
d. 77.9 L

26. A sample of ammonia gas at 65.5°C and 524 torr has a volume of 15.31 L. What is its volume when the
M temperature is -15.8°C and its pressure is 524 torr?

a. 3.69 L
b. 11.6 L
c. 20.2 L
d. 63.5 L

27. A 500-mL sample of argon at 800 torr has its absolute temperature quadrupled. If the volume remains
E unchanged what is the new pressure?

a. 200 torr
b. 400 torr
c. 800 torr
d. 3200 torr

28. A 750-mL sample of hydrogen exerts a pressure of 822 torr at 325 K. What pressure does it exert if the
E temperature is raised to 475 K at constant volume?

a. 188 torr
b. 562 torr
c. 1.20×10^3 torr
d. 1.90×10^3 torr

29. A sample of methane gas, $CH_4(g)$, occupies a volume of 60.3 L at a pressure of 469 torr and a
E temperature of 29.3°C. What would be its temperature at a pressure of 243 torr and volume of 60.3 L?

a. -116.5°C
b. 15.5°C
c. 57.7°C
d. 310.6°C

30. Conditions at STP are
E

a. 0 K and 1 atm
b. 273.15 K and 760 torr
c. 0°C and 760 atm
d. 273.15°C and 760 torr

31. A sample of propane, a component of LP gas, has a volume of 35.3 L at 315 K and 922 torr. What is its
M volume at STP?

a. 25.2 L
b. 33.6 L
c. 37.1 L
d. 49.2 L

32. Nitrogen dioxide is a red-brown gas that is responsible for the color of photochemical smog. A sample of
M nitrogen dioxide has a volume of 28.6 L at 45.3°C and 89.9 kPa. What is its volume at STP?

a. 21.8 L
b. 27.6 L
c. 29.6 L
d. 37.6 L

33. Calculate the pressure of a helium sample at -207.3°C and 768 mL if it exerts a pressure of 175 kPa at
M 25.0°C and 925 mL.

a. 32.1 kPa
b. 46.6 kPa
c. 657 kPa
d. 953 kPa

34. Calculate the temperature of an argon sample at 55.4 kPa and 18.6 L if it occupies 25.8 L at 75.0°C and
M 41.1 kPa.

a. 65.2°C
b. 72.9°C
c. 77.2°C
d. 85.1°C

35. A carbon dioxide sample weighing 44.0g occupies 32.68 L at 65°C and 645 torr. What is its volume at STP?
M

a. 22.4 L
b. 31.1 L
c. 34.3 L
d. 47.7 L

36. A sample of nitrogen gas is confined to a 14.0 L container at 375 torr and 37.0°C. How many moles of
E nitrogen are in the container?

a. 0.271 mol
b. 2.27 mol
c. 3.69 mol
d. 206 mol

37. A compressed gas cylinder containing methane has a volume of 3.30 L. What pressure would 1.50 mol
E methane exert on the walls of the cylinder if its temperature were 25°C?

a. 9.00×10^{-2} atm
b. 0.933 atm
c. 1.70 atm
d. 11.1 atm

38. Assuming ideal behavior, what is the density of argon gas at STP, in g/L?
M

a. 0.0176
b. 0.0250
c. 0.0561
d. 1.78
e. 181.

39. What is the density of carbon dioxide gas at -25.2°C and 98.0 kPa?
M

a. 0.232 g/L
b. 0.279 g/L
c. 1.74 g/L
d. 2.09 g/L

40. Ima Chemist found the density of Freon-11 ($CFCl_3$) to be 5.58 g/L under her experimental conditions. Her measurements showed that the density of an unknown gas was 4.38 g/L under the same conditions. What is the molar mass of the unknown?
M

a. 96.7 g/L
b. 108 g/L
c. 165 g/L
d. 175 g/L

41. A flask with a volume of 3.16 L contains 9.33 grams of an unknown gas at 32.0°C and 1.00 atm. What is the molar mass of the gas?
M

a. 7.76 g/mol
b. 74.0 g/mol
c. 81.4 g/mol
d. 144 g/mol

42. Dr. I. M. A. Brightguy adds 0.1727 g of an unknown gas to a 125-mL flask. If Dr. B finds the pressure to be 736 torr at 20.0°C, is the gas likely to be methane, CH_4, oxygen, O_2, neon, Ne, or argon, Ar?
M

a. CH_4
b. O_2
c. Ne
d. Ar

43. A 250.0-mL sample of ammonia, $NH_3(g)$, exerts a pressure of 833 torr at 42.4°C. What mass of ammonia is in the container?
M

a. 0.0787 g
b. 0.180 g
c. 8.04 g
d. 59.8 g

44. What is the pressure in a 7.50-L flask if 0.15 mol of carbon dioxide is added to 0.33 mol of oxygen? The temperature of the mixture is 48.0°C.
M

a. 0.252 atm
b. 0.592 atm
c. 1.69 atm
d. 3.96 atm

45. If 0.750 L of argon at 1.50 atm and 177°C and 0.235 L of sulfur dioxide at 95.0 kPa and 63 .0°C are
H added to a 1 .00-L flask and the flask's temperature is adjusted to 25 .0°C, what is the resulting pressure in the flask?

a. 0.0851 atm
b. 0.244 atm
c. 0.946 atm
d. 1.86 atm

46. A gas mixture consists of equal masses of methane (molecular weight 16.0) and argon (atomic
M weight 40.0). If the partial pressure of argon is 200. torr, what is that of methane, in the same units?

a. 80.0
b. 200.
c. 256
d. 500.
e. 556

47. A gas mixture, with a total pressure of 300. torr, consists of equal masses of Ne (atomic weight 20.)
M and Ar (atomic weight 40.). The partial pressure of Ar, in torr, is

a. 75
b. 100.
c. 150.
d. 200.
e. none of the above

48. An unknown liquid is vaporized in a 273-mL flask by immersion in a water bath at 99°C. The barometric
H pressure is 753 torr. If the mass of the liquid retained in the flask is 1.362 g, what is its molar mass?

a. 20.4 g/mol
b. 40.9 g/mol
c. 154 g/mol
d. 184 g/mol

49. Magnesium metal (0.100 mol) and a volume of aqueous hydrochloric acid that contains 0.500 mol of
M HCl are combined and react to completion. How many liters of hydrogen gas, measured at STP, are produced?

$$Mg(s) + 2HCl(aq) \rightarrow MgCl_2(aq) + H_2(g)$$

a. 2.24
b. 4.48
c. 5.60
d. 11.2
e. 22.4

50. Hydrogen peroxide was catalytically decomposed and 75.3 mL of oxygen gas was collected over water at
M 25°C and 742 torr. What mass of oxygen was collected? (P_{water} = 24 torr @ 25°C)

a. 0.00291 g
b. 0.0931 g
c. 0.0962 g
d. 0.0993 g

51. M
Small quantities of hydrogen can be prepared by the addition of hydrochloric acid to zinc. A sample of 195 mL of hydrogen was collected over water at 25°C and 753 torr. What mass of hydrogen was collected? (P_{water} = 24 torr @ 25°C)

a. 0.00765 g
b. 0.0154 g
c. 0.0159 g
d. 0.0164 g

52. M
Lithium oxide is an effective absorber of carbon dioxide and can be used to purify air in confined areas such as space vehicles. What volume of carbon dioxide can be absorbed by 1.00 kg of lithium oxide at 25°C and 1.00 atm?

$$Li_2O(aq) + CO_2(g) \rightarrow Li_2CO_3(s)$$

a. 687 mL
b. 819 mL
c. 687 L
d. 819 L

53. H
Methane, $CH_4(g)$, reacts with steam to give *synthesis gas,* a mixture of carbon monoxide and hydrogen, which is used as starting material for the synthesis of a number of organic and inorganic compounds.

$$CH_4(g) + H_2O(g) \rightarrow CO(g) + H_2(g) \quad \text{[unbalanced]}$$

What mass of hydrogen is formed if 275 L of methane (measured at STP) is converted to synthesis gas?

a. 12.3 g
b. 24.7 g
c. 49.4 g
d. 74.2 g

54. M
Hydrochloric acid is prepared by bubbling hydrogen chloride gas through water. What is the concentration of a solution prepared by dissolving 225 L of HCl(*g*) at 37°C and 89.6 kPa in 5.25 L of water?

a. 1.49 *M*
b. 1.66 *M*
c. 7.82 *M*
d. 12.5 *M*

55. E
Which of the following gases effuses most rapidly?

a. nitrogen
b. oxygen
c. hydrogen chloride
d. ammonia

56. E
Which of the following gases will be the slowest to diffuse through a room?

a. methane, CH_4
b. hydrogen sulfide, H_2S
c. carbon dioxide, CO_2
d. water, H_2O

57. Arrange the following gases in order of increasing rate of effusion.
E

C_2H_6, Ar, HCl, PH_3

a. Ar < HCl < PH_3 < C_2H_6
b. C_2H_6 < PH_3 < HCl < Ar
c. Ar < PH_3 < C_2H_6 < HCl
d. C_2H_6 < HCl < PH_3 < Ar

58. A 3.0-L sample of helium was placed in container fitted with a porous membrane. Half of the helium effused through the membrane in 24 hours. A 3.0-L sample of oxygen was placed in an identical container. How long will it take for half of the oxygen to effuse through the membrane?
M

a. 8.5 h
b. 12 h
c. 48 h
d. 68 h

59. A compound composed of carbon, hydrogen, and chlorine effuses through a pinhole 0.411 times as fast as neon. Select the correct molecular formula for the compound.
M

a. $CHCl_3$
b. CH_2Cl_2
c. $C_2H_2Cl_2$
d. C_2H_3Cl

60. At what temperature in kelvin is the root mean square speed of helium atoms (atomic weight = 4.00) equal to that of oxygen molecules (molecular weight = 32.00) at 300. K?
H

a. 37.5
b. 75
c. 106
d. 292
e. 2400.

61. Select the gas with the highest average kinetic energy per mole at 298 K.
M

a. O_2
b. CO_2
c. H_2O
d. all have the same average kinetic energy

62. Select the substance with the largest root-mean-square molecular speed at 25°C.
E

a. NH_3
b. CO
c. H_2
d. N_2

63. Calculate the root-mean-square speed of methane, CH_4 (*g*), at 78°C.
M

a. 23 m/s
b. 350 m/s
c. 550 m/s
d. 740 m/s

64. M Freon-12, CF_2Cl_2, which has been widely used in air conditioning systems, is considered a threat to the ozone layer in the stratosphere. Calculate the root-mean-square velocity of Freon-12 molecules in the lower stratosphere where the temperature is -65°C.

a. 20 m/s
b. 120 m/s
c. 210 m/s
d. 260 m/s

65. M The temperature of the carbon dioxide atmosphere near the surface of Venus is 475°C. Calculate the average kinetic energy per mole of carbon dioxide molecules on Venus.

a. 2520 J/mol
b. 4150 J/mol
c. 5920 J/mol
d. 9330 J/mol

66. E Nitrogen will behave most like an ideal gas

a. at high temperature and high pressure
b. at high temperature and low pressure
c. at low temperature and high pressure
d. at low temperature and low pressure

67. H Use the van der Waals equation for real gases to calculate the pressure exerted by 1.00 mole of ammonia at 27°C in a 750-mL container. (a = 4.17 L^2 atm/mol^2, b =0.0371 L/mol)

a. 23.2 atm
b. 27.1 atm
c. 32.8 atm
d. 42.0 atm

68. M At moderate pressures (~ 200 atm), the measured pressure exerted by CO_2 gas is less than that predicted by the ideal gas equation. This is mainly because

a. such high pressures cannot be accurately measured
b. CO_2 will condense to a liquid at 200 atm pressure
c. gas phase collisions prevent CO_2 molecules from colliding with the walls of the container
d. of attractive intermolecular forces between CO_2 molecules
e. the volume occupied by the CO_2 molecules themselves becomes significant

69. M At very high pressures (~ 1000 atm), the measured pressure exerted by real gases is greater than that predicted by the ideal gas equation. This is mainly because

a. such high pressures cannot be accurately measured
b. real gases will condense to form liquids at 1000 atm pressure
c. gas phase collisions prevent molecules from colliding with the walls of the container
d. of attractive intermolecular forces between gas molecules
e. the volume occupied by the gas molecules themselves becomes significant

70. E The ozone layer is important because

a. ozone absorbs low energy radiation which warms the troposphere
b. ozone purifies the atmosphere by reacting with excess fluorocarbons
c. ozone absorbs ultraviolet radiation
d. ozone reflects high energy radiation such as X-rays and gamma rays.

Short Answer Questions

71. State Charles's Law and illustrate it with a graph, using standard x-y coordinate axes. Be sure to label the axes unambiguously with the correct gas variables.
M

72. State Boyle's Law and illustrate it with a graph, using standard x-y coordinate axes. Be sure to label the axes unambiguously with the correct gas variables.
M

73. State Avogadro's Law.
E

74. State Amontons's Law and illustrate it with a graph, using standard x-y coordinate axes. Be sure to label the axes unambiguously with the correct gas variables.
M

75. Briefly state the conditions corresponding to STP (standard temperature and pressure).
E

76. Starting from the Ideal Gas Equation, derive an equation corresponding to Charles's Law, stating all important assumptions or conditions.
M

77. An oxygen sample of 1.62 L is at 92.3 kPa and 30.0°C.
M

a. What is the pressure of oxygen, in torr?
b. What volume would the oxygen occupy if the pressure were 120.0 kPa and the temperature were 0.0°C?
c. How many moles of oxygen are in the sample?

78. Calculate the density in g/L of gaseous SF_6 at 50.0°C and 650. torr.
M

79. A 255-mL gas sample weighing 0.292 g is at 52810 Pa and 127°C.

a. How many moles of gas are present?
b. What is the molar mass of the gas?

80. A 1.00-L sample of a pure gas weighs 0.785 g and is at 0.965 atm and 29.2°C.
M

a. What is the molar mass of the gas?
b. If the volume and temperature are kept constant while 0.400 g of the same gas are added to that already in the container, what will the new pressure be?

81. A 1.30-L sample of argon gas is at 1.02 atm and 21.5°C.
M

a. What mass of argon is in the container?
b. If the temperature is raised to 500.0°C while the volume remains constant, calculate the new pressure, in atmospheres.

82. A 3.60-L gas sample occupies is at 95.5 kPa and 25.0°C.
M

a. Calculate the volume occupied by the gas at STP, assuming it behaves ideally.
b. If the gas sample weighs 6.10 g, calculate the molar mass of the gas.

83. Dry air is approximately 78% nitrogen, 21% oxygen and 1% argon, by number of molecules. What is the partial pressure of oxygen in a sample of dry air, if atmospheric pressure is 751 torr?
E

84. Aluminum metal shavings (10.0 g) are placed in 100. mL of 6.00 *M* hydrochloric acid. What is the maximum volume of hydrogen, measured at STP, which can be produced?
M

$$2Al(s) + 6HCl(aq) \rightarrow 2AlCl_3(aq) + 3H_2(g)$$

85. In the fermentation process, yeast converts glucose to ethanol and carbon dioxide. What volume of carbon dioxide, measured at 745 torr and 25.0°C, can be produced by the fermentation of 10.0 g of glucose?
M

$$C_6H_{12}O_6(aq) \rightarrow 2C_2H_5OH(aq) + 2CO_2(g)$$

86. Chlorofluorocarbons are important air pollutants implicated in global warming and ozone depletion. How fast does the chlorofluorocarbon CF_2Cl_2 diffuse relative to N_2, the main component of the atmosphere?
M

In other words, calculate the ratio of diffusion rates, $\dfrac{\text{Rate}_{CF_2Cl_2}}{\text{Rate}_{N_2}}$.

87. Use plots of PV/nRT versus P to show the behavior of an ideal gas and of typical real gases. Label the axes and show their scales.
M

88. A 20.0-L container holds 15.3 mol of Cl_2 gas at 227°C.
H

a. Calculate the pressure in atmospheres, assuming ideal behavior.
b. Calculate the pressure in atmospheres, assuming van der Waals behavior. The van der Waals constants for Cl_2 are $a = 6.49$ atm.L^2/mol^2 and $b = 0.0562$ L/mol.

True/False Questions

89. When a closed-ended manometer is used for pressure measurements, and the closed end is under vacuum, the level of manometer liquid in the closed arm can never be lower than that in the other arm.
E

90. For a gas obeying Boyle's Law, a plot of V versus $1/P$ will give a straight line passing through the origin.
M

91. For a gas obeying Charles's Law, a plot of V versus $1/T$ will give a straight line passing through the origin.
M

92. At a temperature of absolute zero, the volume of an ideal gas is zero.
E

93. According to the postulates of kinetic-molecular theory, the molecules of all gases at a given temperature have the same average speed.
E

94. According to the postulates of kinetic-molecular theory, the molecules of all gases at a given temperature have the same average kinetic energy.
E

95. According to the postulates of kinetic-molecular theory, the average kinetic energy of gas molecules is proportional to the absolute temperature.
E

96. For a pure gas sample, the average kinetic energy is also the most probable kinetic energy.
E

97. The rate of diffusion of a gas is inversely proportional to its molar mass.
M

98. For an ideal gas, a plot of PV/nRT versus P gives a straight line with a positive slope.
E

99. For real gases, $PV > nRT$, always.
E

100. For real gases, $PV < nRT$, always.
E

Gases and the Kinetic-Molecular Theory
Chapter 5
Answer Key

1.	d	26.	b	51.	b
2.	d	27.	d	52.	d
3.	c	28.	c	53.	d
4.	b	29.	a	54.	a
5.	b	30.	b	55.	d
6.	a	31.	c	56.	c
7.	c	32.	a	57.	a
8.	c	33.	b	58.	d
9.	d	34.	a	59.	a
10.	b	35.	a	60.	a
11.	c	36.	a	61.	d
12.	a	37.	d	62.	c
13.	c	38.	d	63.	d
14.	b	39.	d	64.	c
15.	e	40.	b	65.	d
16.	c	41.	b	66.	b
17.	b	42.	b	67.	b
18.	c	43.	b	68.	d
19.	b	44.	c	69.	e
20.	a	45.	c	70.	c
21.	b	46.	d		
22.	a	47.	b		
23.	b	48.	c		
24.	a	49.	a		
25.	c	50.	b		

71. At constant pressure, the volume of a fixed amount of gas is directly proportional to its absolute temperature. The plot should be *V* vs. *T*, showing a straight line of positive slope, passing through the origin.

72. At constant temperature, the volume of a fixed amount of gas is inversely proportional to its pressure. The plot should be *V* vs. 1/*P*, showing a straight line of positive slope, passing through the origin. Alternatively, a plot of *V* vs. *P*, showing appropriate curvature, is acceptable.

73. Equal volumes of gases at the same temperature and pressure contain the same number of particles.

74. At constant volume, the pressure of a fixed amount of gas is directly proportional to its absolute temperature. The plot should be *P* vs. *T*, showing a straight line of positive slope, passing through the origin.

75. At STP, the pressure is one atmosphere and the temperature is 0°C (273.15K)

76. Ideal Gas Equation is $PV = nRT$. Charles's Law refers to a fixed amount of gas (n is a constant) and constant pressure P. R is always constant. Rearrange the equation to $V = (nR/P)T$. The quantities in parentheses are all constant, so V = constant × T, which is Charles's Law.

77. a. 692 torr
b. 1.12 L
c. 5.93×10^{-2} mol

78. 4.71 g/L

79. a. 4.05×10^{-3} mol
b. 72.1 g/mol

80. a. 20.2 g/mol
b. 1.46 atm

81. a. 2.19 g
b. 2.68 atm

82. a. 3.11 L
b. 44.0 g/mol

83. 158 torr

84. 6.72 L

85. 2.77 L

86. 0.481

87. Ideal gas: the plot is a horizontal line with $PV/nRT = 1$. Real gases have curved plots, with $PV/nRT = 1$ at $P = 0$; at intermediate pressures PV/nRT may or may not be less than 1; at high pressures (e.g. 1000 atm) PV/nRT is greater than 1.

88. a. 31.4 atm
b. 29.0 atm

89. T

90. T

91. F

92. T

93. F

94. T

95. T

96. F

97. F

98. F

99. F

100. F

Thermochemistry: Energy Flow and Chemical Change
Chapter 6

Multiple Choice Questions

1. A system that does no work but which transfers heat to the surroundings has
E

a. $q < 0, \Delta E > 0$
b. $q < 0, \Delta E < 0$
c. $q > 0, \Delta E > 0$
d. $q > 0, \Delta E < 0$

2. A system that does no work but which receives heat from the surroundings has
E

a. $q < 0, \Delta E > 0$
b. $q > 0, \Delta E < 0$
c. $q = \Delta E$
d. $q = -\Delta E$

3. A system which undergoes an adiabatic change (i.e. $q = 0$) and does work on the surroundings has
E

a. $w < 0, \Delta E < 0$
b. $w < 0, \Delta E > 0$
c. $w > 0, \Delta E < 0$
d. $w > 0, \Delta E > 0$

4. A system which undergoes an adiabatic change (i.e. $q = 0$) and has work on it by the surroundings has
E

a. $w = \Delta E$
b. $w = -\Delta E$
c. $w > 0, \Delta E < 0$
d. $w < 0, \Delta E > 0$

5. A system receives 575 J of heat and delivers 425 J of work. Calculate the change in the internal energy, ΔE,
E of the system.

a. -150 J
b. 150 J
c. -1000 J
d. 1000 J

6. A system delivers 225 J of heat to the surroundings while delivering 645 J of work. Calculate the change in
E the internal energy, ΔE, of the system.

a. -420 J
b. 420 J
c. -870 J
d. 870 J

7. A system delivers 1275 J of heat while the surroundings perform 855 J of work on it. Calculate ΔE in J.
E

a. -2130
b. -420
c. 420
d. 2130

8. A system absorbs 21.6 kJ of heat while performing 6.9 kJ of work on the surroundings.
M If the initial internal energy, E, is 61.2 kJ, what is the final value of E?

a. 32.7 J
b. 46.5 J
c. 75.9 J
d. 89.7 J

9. A system initially has an internal energy E of 501 J. It undergoes a process during which it releases
M 111 J of heat energy to the surroundings, and does work of 222 J. What is the final energy of the system, in J?

a. 168
b. 390
c. 612
d. 834
e. cannot be calculated without more information

10. A system expands from a volume of 1.00 L to 2.00 L against a constant external pressure of 1.00
M atm. The work (w) done by the system, in J, is

a. 1.00
b. 2.00
c. 1.01×10^2
d. 1.01×10^5
e. none of the above

11. A system contracts from an initial volume of 15.0 L to a final volume of 10.0 L under a constant
M external pressure of 0.800 atm. The value of w, in J, is
a. -4.0
b. 4.0
c. -405
d. 405
e. 4.05×10^3

12. An ideal gas (the system) is contained in a flexible balloon at a pressure of 1 atm and is initially at a
H temperature of 20.°C. The surrounding air is at the same pressure, but its temperature is 25°C. When the system has equilibrated with its surroundings, both systems and surroundings are at 25°C and 1 atm. In changing from the initial to the final state, which one of the following relationships regarding the system is correct?

a. $\Delta E < 0$
b. $\Delta E = 0$
c. $\Delta H = 0$
d. $w > 0$
e. $q > 0$

13. Which one of the following relationships is always correct?
E

a. potential energy + kinetic energy = constant
b. $E = q + w$
c. $\Delta E = \Delta H - P\Delta V$
d. $H = E + PV$
e. $\Delta H = q_v$

14. In which of the following processes is $\Delta H = \Delta E$?
M

a. Two moles of ammonia gas are cooled from 325°C to 300°C at 1.2 atm.
b. One gram of water is vaporized at 100°C and 1 atm.
c. Two moles of hydrogen iodide gas react to form hydrogen gas and iodine gas in a 40-L container.
d. Calcium carbonate is heated to form calcium oxide and carbon dioxide in a container with variable volume.

15. For which one of the following reactions will ΔH be approximately (or exactly) equal to ΔE?
M

a. $H_2(g) + Br_2(g) \rightarrow 2HBr(g)$
b. $H_2O(l) \rightarrow H_2O(g)$
c. $CaCO_3(s) \rightarrow CaO(s) + CO_2(g)$
d. $2H(g) + O(g) \rightarrow H_2O(l)$
e. $CH_4(g) + 2O_2(g) \rightarrow CO_2(g) + 2H_2O(l)$

16. In which one of the following reactions would you expect ΔH to be substantially greater than ΔE
M (i.e. $\Delta H > \Delta E$)?

a. $H_2(g) + Br_2(g) \rightarrow 2HBr(g)$
b. $CO_2(s) \rightarrow CO_2(g)$
c. $C_2H_2(g) + H_2(g) \rightarrow C_2H_4(g)$
d. $H_2O(s) \rightarrow H_2O(l)$
e. $HCl(aq) + NaOH(aq) \rightarrow NaCl(aq) + H_2O(l)$

17. Cold packs, whose temperatures are lowered when ammonium nitrate dissolves in water, are carried
E by athletic trainers when transporting ice is not possible. Which of the following is true of this reaction?

a. $\Delta H < 0$, process is exothermic
b. $\Delta H > 0$, process is exothermic
c. $\Delta H < 0$, process is endothermic
d. $\Delta H > 0$, process is endothermic

18. The dissolution of barium hydroxide in water is an exothermic process. Which of the following statements
M is correct?

a. The energy of solid barium hydroxide plus pure water is less than that of the solution, at the same temperature.
b. The energy of solid barium hydroxide plus pure water is greater than that of the solution, at the same temperature.
c. The energy of solid barium hydroxide plus pure water is the same as that of the solution, at the same temperature.
d. The temperature of the solution is lower than of the barium hydroxide and water before mixing.

19. H Two solutions (the system), each of 25.0 mL volume and at 25.0°C, are mixed in a beaker. A reaction occurs between them, and the temperature rises to 35.0°C. After the products have equilibrated with the surroundings, the temperature is again 25.0°C and the total volume is 50.0 mL. No gases are involved in the reaction. Which one of the following relationships concerning the change from initial to final states (both at 25.0°C) is correct?

a. $\Delta E = 0$
b. $\Delta H = 0$
c. $w = 0$
d. $q = 0$
e. $\Delta E > 0$

20. M When they react with oxygen fats release more energy when they react with oxygen than carbohydrates do because

a. fats contain more bonds to oxygen than carbohydrates do
b. fats contain fewer bonds to oxygen than carbohydrates do
c. the total energy of the carbon-carbon and carbon-hydrogen bonds in fats is greater than the energy content of the carbon-oxygen and oxygen-hydrogen bonds in the reaction products (carbon dioxide and water)
d. the total energy of the carbon-carbon and carbon-hydrogen bonds in fats is greater than the energy content of the bonds in carbohydrates

21. M When one mole of each of the following liquids is burned, which will produce the most heat energy?

a. C_6H_{14}
b. $C_6H_{14}O$
c. $C_6H_{12}O$
d. $C_6H_{10}O_3$

22. E A Snickers® candy bar contains 280 Calories, of which the fat content accounts for 120 Calories. What is the energy of the fat content, in kJ?

a. 5.0×10^{-1} kJ
b. 29 kJ
c. 5.0×10^{2} kJ
d. 5.0×10^{5} kJ

23. M Your favorite candy bar, Gummy Beakers, contains 1.2×10^{6} J of energy while your favorite soft drink, Bolt, contains 6.7×10^{5} J. If you eat two packs of Gummy Beakers a day and drink 3 cans of Bolt, what percent of your 2000 Calorie daily food intake is left for broccoli, beans, beef, etc?

a. 27%
b. 47%
c. 53%
d. 73%

24. E Natural gas, or methane, is an important fuel. Combustion of one mole of methane releases 802.3 kilojoules of energy. How much energy does that represent in kilocalories?

a. 1.92×10^2 kcal
b. 3.36×10^3 kcal
c. 1.92×10^5 kcal
d. 3.36×10^6 kcal

25. E Which of the following is not a state function?

a. internal energy
b. volume
c. work
d. pressure

26. E Calculate *q* when 28.6 g of water is heated from 22.0°C to 78.3°C.

a. 0.385 kJ
b. 1.61 kJ
c. 6.74 kJ
d. 9.37 kJ

27. E If, as a pioneer, you wished to warm your room by taking an object heated on top of a pot-bellied stove to it, which of the following 15-pound objects would be the best choice? The specific heat capacity (in J/(g·K)) for each substance is given in parentheses. Iron (0.450), copper (0.387), granite (0.79), gold (0.129)

a. iron
b. copper
c. granite
d. gold

28. E Ethylene glycol, used as a coolant in automotive engines, has a specific heat capacity of 2.42 J/(g·K). Calculate *q* when 3.65 kg of ethylene glycol is cooled from 132°C to 85°C.

a. -1900 kJ
b. -420 kJ
c. -99 kJ
d. -0.42 kJ

29. M A 275-g sample of nickel at 100.0°C is placed in 100.0 mL of water at 22.0°C. What is the final temperature of the water? Assume that no heat is lost to or gained from the surroundings. Specific heat capacity of nickel = 0.444 J/(g·K)

a. 39.6°C
b. 40.8°C
c. 79.2°C
d. 82.4°C

30. E

Benzene is a starting material in the synthesis of nylon fibers and polystyrene (styrofoam). Its specific heat capacity is 1.74 J/(g·K). If 16.7 kJ of energy is absorbed by a 225-g sample of benzene at 20.0°C, what is its final temperature?

a. -22.7°C
b. 36.7°C
c. 42.7°C
d. 62.7°C

31. E

When Karl Kaveman adds chilled grog to his new granite mug, he removes 10.9 kJ of energy from the mug. If it has a mass of 625 g and was at 25°C, what is its new temperature? Specific heat capacity of granite = 0.79 J/(g·K)

a. 3°C
b. 14°C
c. 22°C
d. 47°C

32. M

The Starship Enterprise is caught in a time warp and Spock is forced to use the primitive techniques of the 20th century to determine the specific heat capacity of an unknown mineral. The 307-g sample was heated to 98.7°C and placed into a calorimeter containing 72.4 g of water at 23.6°C. The heat capacity of the calorimeter was 15.7 J/K. The final temperature in the calorimeter was 32.4°C. What is the specific heat capacity of the mineral?

a. 0.124 J/(g·K)
b. 0.131 J/(g·K)
c. 0.138 J/(g·K)
d. 0.145 J/(g·K)

33. M

A piece of copper metal is initially at 100.0°C. It is dropped into a coffee cup calorimeter containing 50.0 g of water at a temperature of 20.0°C. After stirring, the final temperature of both copper and water is 25.0°C. Assuming no heat losses, and that the specific heat (capacity) of water is 4.18 J/(g·K), what is the heat capacity of the copper in J/K?

a. 2.79
b. 3.33
c. 13.9
d. 209
e. None of the above

34. M

A common laboratory reaction is the neutralization of an acid with a base. When 50.0 mL of 0.500 *M* HCl at 25.0°C is added to 50.0 mL of 0.500 *M* NaOH at 25.0°C in a coffee cup calorimeter, the temperature of the mixture rises to 28.2°C. What is the heat of reaction per mole of acid? Assume the mixture has a specific heat capacity of 4.18 J/(g·K) and that the densities of the reactant solutions are both 1.00 g/mL.

a. 670 J
b. 1300 J
c. 27 kJ
d. 54 kJ

35. Sand is converted to pure silicon in a three step process. The third step is
E

$$SiCl_4(g) + 2Mg(s) \rightarrow 2MgCl_2(s) + Si(s) \qquad \Delta H = -625.6\text{ kJ}$$

What is the enthalpy change when 25.0 mol of silicon tetrachloride is converted to elemental silicon?

a. -25.0 kJ
b. -7820 kJ
c. -1.56×10^4 kJ
d. -3.13×10^4 kJ

36. Calcium hydroxide, which reacts with carbon dioxide to form calcium carbonate, was used by the ancient
E Romans as mortar in stone structures. The reaction for this process is

$$Ca(OH)_2(s) + CO_2(g) \rightarrow CaCO_3(s) + H_2O(g) \qquad \Delta H = -69.1\text{ kJ}$$

What is the enthalpy change if 3.8 mol of calcium carbonate is formed?

a. -18 kJ
b. -69 kJ
c. -73 kJ
d. -260 kJ

37. Galena is the ore from which elemental lead is extracted. In the first step of the extraction process, galena is
M heated in air to form lead(II) oxide.

$$2PbS(s) + 3O_2(g) \rightarrow 2PbO(s) + 2SO_2(g) \qquad \Delta H = -827.4\text{ kJ}$$

What mass of galena is converted to lead oxide if 975 kJ of heat are liberated?

a. 203 g
b. 282 g
c. 406 g
d. 564 g

38. The highly exothermic thermite reaction, in which aluminum reduces iron(III) oxide to elemental iron, has
M been used by railroad repair crews to weld rails together.

$$2Al(s) + Fe_2O_3(s) \rightarrow 2Fe(s) + Al_2O_3(s) \qquad \Delta H = -850\text{ kJ}$$

What mass of iron is formed when 725 kJ of heat are released?

a. 47 g
b. 65 g
c. 95 g
d. 130 g

39. Use Hess's Law to calculate the enthalpy change for the reaction
M

$WO_3(s) + 3H_2(g) \rightarrow W(s) + 3H_2O(g)$

from the following equations:

$2W(s) + 3O_2(g) \rightarrow 2WO_3(s)$ $\Delta H = -1685.4$ kJ
$2H_2(g) + O_2(g) \rightarrow 2H_2O(g)$ $\Delta H = -477.84$ kJ

a. 125.9 kJ
b. 252.9 kJ
c. 364.9 kJ
d. 1207.6 kJ

40. Calculate the enthalpy change for the reactions
M

$NO(g) + O(g) \rightarrow NO_2(g)$

from the following reactions:

$NO(g) + O_3(g) \rightarrow NO_2(g) + O_2(g)$ $\Delta H = -198.9$ kJ
$O_3(g) \rightarrow 1.5O_2(g)$ $\Delta H = -142.3$ kJ
$O_2(g) \rightarrow 2O(g)$ $\Delta H = 495.0$ kJ

a. -551.6 kJ
b. -304.1 kJ
c. 153.8 kJ
d. 438.4 kJ

41. Use the following equations to calculate the molar enthalpy of formation for manganese(IV)
M oxide, $MnO_2(s)$.

$2MnO_2(s) \rightarrow 2MnO(s) + O_2(g)$ $\Delta H = 264$ kJ
$MnO_2(s) + Mn(s) \rightarrow 2MnO(s)$ $\Delta H = -240$ kJ

a. -504 kJ
b. -24 kJ
c. 24 kJ
d. 504 kJ

42. Which one of the following statements about standard states is incorrect?
E

a. The standard state of a solid compound is the pure solid
b. The standard state of a liquid compound is the pure liquid
c. The standard state of a gaseous compound is the gas at a pressure of 1 atmosphere
d. The standard state of an aqueous solute is a saturated solution in water
e. The standard state of an element is the form in which it is stable at 1 atm and a specified temperature, usually 25°C

43. Which one of the following equations represents the formation reaction of $CH_3OH(l)$?
E

a. $C(g) + 2H_2(g) + \frac{1}{2}O_2(g) \rightarrow CH_3OH(l)$
b. $C(g) + 4H(g) + O(g) \rightarrow CH_3OH(l)$
c. $C(\text{graphite}) + 4H(g) + O(g) \rightarrow CH_3OH(l)$
d. $C(\text{diamond}) + 4H(g) + O(g) \rightarrow CH_3OH(l)$
e. $C(\text{graphite}) + 2H_2(g) + \frac{1}{2}O_2(g) \rightarrow CH_3OH(l)$

44. Which one of the following is not a correct formation reaction? (products are correct)
E

a. $H_2(g) + O(g) \rightarrow H_2O(l)$
b. $\frac{1}{2}H_2(g) + \frac{1}{2}Cl_2(g) \rightarrow HCl(g)$
c. $6C(\text{graphite}) + 3H_2(g) \rightarrow C_6H_6(l)$
d. $C(\text{graphite}) \rightarrow C(\text{diamond})$
e. $6C(\text{graphite}) + 6H_2(g) + 3O_2(g) \rightarrow C_6H_{12}O_6(s)$

45. Which one of the following is a correct formation reaction?
E

a. $C(\text{diamond}) \rightarrow C(\text{graphite})$
b. $H_2(g) + O(g) \rightarrow H_2O(l)$
c. $C(\text{graphite}) + 4H(g) \rightarrow CH_4(g)$
d. $6C(\text{graphite}) + 6H_2O(s) \rightarrow C_6H_{12}O_6(s)$
e. $2C(\text{graphite}) + 3H_2(g) + \frac{1}{2}O_2(g) \rightarrow C_2H_5OH(l)$

46. Calculate the $\Delta H°_{rxn}$ for the decomposition of calcium carbonate to calcium oxide and carbon dioxide.
E $\Delta H^\circ_f\,[CaCO_3(s)] = -1206.9$ kJ/mol; $\Delta H^\circ_f\,[CaO(s)] = -635.1$ kJ/mol; $\Delta H^\circ_f\,[CO_2(g)] = -393.5$ kJ/mol

$$CaCO_3(s) \rightarrow CaO(s) + CO_2(g)$$

a. -2235.5 kJ
b. -178.3 kJ
c. 178.3 kJ
d. 2235.5 kJ

47. Nitric acid, which is among the top 15 chemicals produced in the United States, was first prepared over
E 1200 years ago by heating naturally occurring sodium nitrate (called saltpeter) with sulfuric acid and collecting the vapors produced. Calculate $\Delta H°_{rxn}$ for this reaction. $\Delta H^\circ_f\,[NaNO_3(s)] = 467.8$ kJ/mol; $\Delta H^\circ_f\,[NaHSO_4(s)] = -1125.5$ kJ/mol; $\Delta H^\circ_f\,[H_2SO_4(l)] = -814.0$ kJ/mol; $\Delta H^\circ_f\,[HNO_3(g)] = -135.1$ kJ/mol

$$NaNO_3(s) + H_2SO_4(l) \rightarrow NaHSO_4(s) + HNO_3(g)$$

a. -644.2 kJ
b. -21.2 kJ
c. 21.2 kJ
d. 644.2 kJ

48. E An important step in the synthesis of nitric acid is the conversion of ammonia to nitric oxide. ΔH°_f [$NH_3(g)$] = -45.9 kJ/mol; ΔH°_f [$NO(g)$] = 90.3 kJ/mol; ΔH°_f [$H_2O(g)$] = -241.8 kJ/mol

$$4NH_3(g) + 5O_2(g) \rightarrow 4NO(g) + 6H_2O(g)$$

Calculate ΔH°_{rxn} for this reaction.

a. -906.0 kJ
b. -197.4 kJ
c. 197.4 kJ
d. 906.0 kJ

49. E Calculate the ΔH°_{rxn} for the following reaction. ΔH°_f [$SiO_2(s)$] = -910.9 kJ/mol; ΔH°_f [$SiCl_4(g)$] = -657.0 kJ/mol; ΔH°_f [$HCl(g)$] = -92.3 kJ/mol; ΔH°_f [$H_2O\ (g)$] = -241.8 kJ/mol

$$SiO_2(s) + 4HCl(g) \rightarrow SiCl_4(g) + 2H_2O(g)$$

a. -139.5 kJ
b. -104.4 kJ
c. 104.4 kJ
d. 139.5 kJ

50. E Calculate the ΔH°_{rxn} for the following reaction. ΔH°_f [$AsH_3(g)$] = 66.4 kJ/mol; ΔH°_f [$H_3AsO_4(aq)$] = -904.6 kJ/mol; ΔH°_f [$H_2O(l)$] = -285.8 kJ/mol;

$$H_3AsO_4(aq) + 4H_2(g) \rightarrow AsH_3(g) + 4H_2O(l)$$

a. -685.2 kJ
b. -172.2 kJ
c. 172.2 kJ
d. 685.2 kJ

51. E Ethanol, C_2H_5OH, is being promoted as a clean fuel and is used as an additive in many gasoline mixtures. Calculate the ΔH°_{rxn} for the combustion of ethanol. ΔH°_f [$C_2H_5OH(l)$] = -277.7 kJ/mol; ΔH°_f [$CO_2(g)$] = -393.5 kJ/mol; ΔH°_f [$H_2O(g)$] = -241.8 kJ/mol

a. -1234.7 kJ
b. -357.6 kJ
c. 357.6 kJ
d. 1234.7 kJ

Short Answer Questions

52. E Consider the equation $\Delta E = q + w$
Explain fully the meaning of all three terms in the equation, and also the implied sign convention for q and w.

53. M Starting from equations relating pressure to force and force to work, derive the relationship $w = -P\Delta V$, explaining the steps in your argument.

54. M
a. A gas sample absorbs 53 kJ of heat and does 18 kJ of work. Calculate the change in its internal energy.
b. A system expands against a constant pressure of 1.50 atm, from an initial volume of 1.00 L to a final volume of 10.0 L. Calculate the work (w) involved in this process, in kJ.

55.
H

Calculate, in J, the work done by 10.0 g of CO_2 when it sublimes against a pressure of 1.00 atm to form gaseous CO_2 at 0.0°C. The volume of $CO_2(s)$ can be neglected; $CO_2(g)$ can be assumed to behave ideally. The process occurring is

$$CO_2(s) \rightarrow CO_2(g)$$

56.
H

1.00 mol of an ideal gas (the system) is heated from 20.0°C to 100.0°C at a constant pressure of 1.00 atm.

a. Given that the internal energy of an ideal gas is $E = \frac{3}{2}RT$, calculate ΔE for this change, in J.
b. Calculate w for this change, in J.
c. Hence, calculate q for this change, in J.

57.
M

a. Starting from the equation $H = E + PV$, show how the relationship $\Delta H = q_p$ is derived. Clearly indicate any necessary assumptions or conditions.
b. In one sentence, state in full what is meant by the equation: $\Delta H = q_p$.

58.
M

Although internal energy (E) is more fundamental and conceptually easier than enthalpy (H), in most chemical applications ΔH is more relevant and useful than ΔE. Why?

59.
E

a. Explain fully what is meant by the term "state function."
b. (i) Give two examples of thermodynamic quantities which state functions.
(ii) Give two examples of thermodynamic quantities which are not state functions.

60.
M

When 1.00 g of solid NH_4Cl is dissolved in 25.00 g of water contained in a coffee cup calorimeter, both reagents initially being at 25.0°C, the temperature falls to 22.4°C. Assuming that the heat capacity of the ammonium chloride solution is 4.18 J/(g·K), calculate the heat (enthalpy) of solution of NH_4Cl, (a) in J/g and (b) in kJ/mol

61.
M

2.53 g of solid NaOH is dissolved in 100.00 g of water in a coffee cup calorimeter, all the reagents initially being at 20.0°C. Calculate the final temperature of the solution obtained, given the following information:

$$NaOH(s) \rightarrow NaOH(aq) \qquad \Delta H^\circ = -43.0 \text{ kJ}$$

Heat capacity of NaOH solution = 4.18 J/(g·K)

62.
H

A mass of 1.250 g of benzoic acid ($C_7H_6O_2$) was completely combusted in a bomb calorimeter. If the heat capacity of the calorimeter was 10.134 kJ/K and the heat of combustion of benzoic acid is -3226 kJ/mol, calculate (to three decimal places) the temperature increase that should have occurred in the apparatus.

63.
H

The reaction

$$2NaOH(aq) + H_2SO_4(aq) \rightarrow Na_2SO_4(aq) + 2H_2O(l)$$

was studied in a coffee cup calorimeter. 100. mL portions of 1.00 M aqueous NaOH and H_2SO_4, each at 24.0°C, were mixed. The maximum temperature achieved was 30.6°C. Neglect the heat capacity of the cup and the thermometer, and assume that the solution of products has a density of exactly 1 g/mL and a specific heat capacity of 4.18 J/(g·K).

a. Calculate the heat of reaction, q, in J.
b. Calculate ΔH, the heat (enthalpy) of reaction, in kJ/mol of Na_2SO_4 produced.

64. M

a. State Hess's Law.

b. Use the $\Delta H°$ data given below to calculate $\Delta H°$ for the reaction:

$$C_2H_4(g) + H_2(g) \rightarrow C_2H_6(g)$$

Data:

	$\Delta H°$ (kJ)
$C_2H_6(g) + \frac{7}{2}O_2(g) \rightarrow 2CO_2(g) + 3H_2O(l)$	-1560
$C_2H_4(g) + 3O_2(g) \rightarrow 2CO_2(g) + 2H_2O(l)$	-1411
$2H_2(g) + O_2(g) \rightarrow 2H_2O(l)$	-572

65. M

Given the following data:

	$\Delta H°$ (kJ)
$2O_3(g) \rightarrow 3O_2(g)$	- 427
$O_2(g) \rightarrow 2O(g)$	495
$NO(g) + O_3(g) \rightarrow NO_2(g) + O_2(g)$	- 199

calculate $\Delta H°$ for the reaction

$$NO(g) + O(g) \rightarrow NO_2(g)$$

66. M

Diborane (B_2H_6) has been considered as a possible rocket fuel. Calculate $\Delta H°$ for the reaction

$$B_2H_6(g) \rightarrow 2B(s) + 3H_2(g)$$

using the following data:

	$\Delta H°$ (kJ)
$2B(s) + \frac{3}{2}O_2(g) \rightarrow B_2O_3(s)$	-1273
$B_2H_6(g) + 3O_2(g) \rightarrow B_2O_3(s) + 3H_2O(g)$	-2035
$H_2(g) + \frac{1}{2}O_2(g) \rightarrow H_2O(g)$	-242

67. E

Clearly state the thermodynamic standard state of

a. an element or compound

b. a solute

68. M

a. Define, or explain fully what is meant by the standard enthalpy of formation of a substance, $\Delta H_f°$.

b. What is the standard state of the element oxygen?

c. Write down in full the formation reaction for liquid ethanol, $C_2H_5OH(l)$. The equation should be balanced and should indicate the physical state of each substance.

69. M

a. Write a balanced equation for the combustion of benzene, $C_6H_6(l)$ in oxygen.

b. The standard heat of combustion of benzene is -3271 kJ/mol. Calculate its standard heat of formation, $\Delta H_f°$, given the data:

$$\Delta H_f°[CO_2(g)] = -394 \text{ kJ}; \quad \Delta H_f°[H_2O(l)] = -286 \text{ kJ}$$

True/False Questions

70. H

The only way in which a system can do work on the surroundings is by expansion against the external pressure.

71. E

Different chemical bonds have different potential energies.

72. M

For a reaction in a sealed, rigid container, ΔH is always greater than ΔE.

73. ΔH does not depend on the path of a reaction, but ΔE does.
E

74. The enthalpy (*H*) of liquid water is greater than that of the same quantity of ice at the same temperature.
E

75. In an endothermic reaction, in going from the reactants to the products at the same temperature, the value of *q* is negative.
M

76. The more C-O and O-H bonds there are in a substance, the greater will be the amount of heat released when a fixed mass of the substance is burned.
E

77. ΔE values obtained by bomb calorimetry can be converted to give accurate ΔH values.
E

78. The standard state of a substance in aqueous solution is a 1 *M* solution.
E

79. The standard heat (enthalpy) of formation of graphite is zero.
E

80. Standard heats (enthalpies) of formation of compounds, ΔH°_f, may be positive or negative.
E

Thermochemistry: Energy Flow and Chemical Change
Chapter 6
Answer Key

1. b
2. c
3. a
4. a
5. b
6. c
7. b
8. c
9. a
10. c
11. d
12. e
13. d
14. c
15. a
16. b
17. d
18. b
19. c
20. b
21. a
22. c
23. b
24. a
25. c
26. c
27. c
28. b
29. a
30. d
31. a
32. c
33. c
34. d
35. c
36. d
37. d
38. c
39. a
40. b
41. a
42. d
43. e
44. a
45. e
46. c
47. c
48. a
49. d
50. b
51. a

52. E is the internal energy of a system, i.e. the total energy of all forms of energy in the system. So, ΔE is the change internal energy of the system. q is heat transferred to or from the system, w is work done on or by the system. The sign convention is that q and w are positive when the system gains energy; i.e. heat transferred to the system, or work done on the system, are both positive.

53. Consider the system being contained in a cylinder by a piston of area A. The piston exerts a pressure P on the system. If the system expands by pushing back the piston a distance Δh against the pressure P, the system does work on the surroundings equal to the force times the distance moved. i.e. $w = F \times \Delta h = P \times A \times \Delta h$, since pressure = force/area. $A \times \Delta h$ is the increase in volume of the system, ΔV. So, in magnitude, $w = P\Delta V$. With the sign convention that w is negative if the system does work, $w = -P\Delta V$.

54. a. $\Delta E = 35$ kJ
b. $w = -1.37$ kJ

55. $w = -516$ J

56. a. $\Delta E = 998$ J
b. $w = -665$ J
c. $q = 1.663 \times 10^3$ J

57. a. Since $H = E + PV$ (defining equation for H), $\Delta H = \Delta E + \Delta(PV)$. For a process at constant pressure, this becomes $\Delta H = \Delta E + P\Delta V$. Since $\Delta E = q + w$ and $w = -P\Delta V$, this becomes $\Delta H = q + w - w$. For a constant pressure process, this reduces to $\Delta H = q_p$.
b. For a reaction occurring at constant pressure, the enthalpy change of the system is equal to the heat of reaction.

58. Most chemical processes of interest occur under conditions of constant (or very nearly constant) pressure. Constant volume processes are relatively rare. So, since ΔH is the heat of reaction at constant pressure, it is more useful than ΔE, the heat of reaction at constant volume.

59. a. A state function depends only on the present state of a system, not on its history or the path followed in a process.
b. (i) internal energy, enthalpy (and possibly T, P, etc.)
(ii) heat and work

60. a. 283 J/g
b. 15.1 kJ/mol

61. 26.3°C

62. Temperature increase is 3.259 K (or °C)

63. a. 5.5×10^3 J
b. 110 kJ/mol Na_2SO_4

64. a. The enthalpy change for an overall process is equal to the sum of the enthalpy changes of its individual steps.
b. $\Delta H° = -137$ kJ

65. $\Delta H° = -233$ kJ

66. $\Delta H° = -36$ kJ

67. a. Standard state is the stable form of the substance at 1 atm and a specified temperature, usually 298 K.
b. Standard state is a solution of 1 M concentration.

68. a. ΔH_f° is the enthalpy change accompanying the formation of one mole of a substance from its elements, all substances being in their standard states.
b. Pure O_2 gas at a pressure of 1 atm and a specified temperature.
c. $2C(graphite) + 3H_2(g) + \frac{1}{2}O_2(g) \rightarrow C_2H_5OH(l)$

69. a. $C_6H_6(l) + \frac{15}{2}O_2(g) \rightarrow 6CO_2(g) + 3H_2O(l)$ (or with coefficients doubled)
b. $\Delta H_f^\circ[C_6H_6(l)] = 49$ kJ

70. F

71. T

72. F

73. F

74. T

75. F

76. F

77. T

78. T

79. T

80. T

Quantum Theory and Atomic Structure
Chapter 7

Multiple Choice Questions

1. The first scientist to propose that the atom had a dense nucleus which occupied only a small fraction of the volume of the atom was
E

a. Planck.
b. Bohr.
c. Rydberg.
d. Rutherford.

2. The first scientist to propose that an object could emit only certain amounts of energy was
E

a. Planck.
b. Einstein.
c. Bohr.
d. Rydberg.

3. A model that successfully explained the photoelectric effect was proposed by
E

a. Planck.
b. Einstein.
c. Compton.
d. Rydberg.

4. The scientist who developed an equation which relates the line spectra of atoms with transitions between specific energy levels was
E

a. Planck.
b. de Broglie.
c. Bohr.
d. Rydberg.

5. The scientist who was first to propose that electrons in an atom could have only certain energies was
E

a. Planck.
b. Einstein.
c. Bohr.
d. Rydberg.

6. The concept that particles of matter could have wave properties was proposed by
M

a. Einstein.
b. Planck.
c. de Broglie.
d. Compton.

7. The scientist who demonstrated that photons transferred momentum during collisions with matter was
M

a. Bohr.
b. de Broglie.
c. Planck.
d. Compton.

8. The principle which states that one cannot know the exact position and velocity of a particle
E simultaneously was proposed by

a. Einstein.
b. Planck.
c. Heisenberg.
d. Compton.

9. Which word best describes the phenomenon which gives rise to a rainbow?
M

a. reflection
b. dispersion
c. diffraction
d. interference

10. Contact lenses can focus light due to the _______________ of the waves.
E

a. diffraction
b. reflection
c. refraction
d. dispersion

11. The interference pattern seen when light passes through narrow, closely spaced slits, is due to
M

a. diffraction.
b. reflection.
c. refraction.
d. dispersion.

12. Interference of light waves
E

a. separates light into its component colors.
b. creates a pattern of light and dark regions.
c. focuses a broad beam of light into a point.
d. bends light as it passes the edge of an object.

13. Select the arrangement of electromagnetic radiation which starts with the lowest energy and increases to
M greatest energy.

a. radio, visible, infrared, ultraviolet
b. microwave, infrared, visible, ultraviolet
c. visible, ultraviolet, infrared, gamma rays
d. X-radiation, visible, infrared, microwave

14. Select the arrangement of electromagnetic radiation which starts with the lowest energy and increases to greatest energy.
M

a. radio, infrared, ultraviolet, gamma rays
b. radio, ultraviolet, infrared, gamma rays
c. gamma rays, infrared, radio, ultraviolet
d. gamma rays, ultraviolet, infrared, radio

15. Select the arrangement of electromagnetic radiation which starts with the lowest wavelength and increases to greatest wavelength.
M

a. radio, infrared, ultraviolet, gamma rays
b. radio, ultraviolet, infrared, gamma rays
c. gamma rays, ultraviolet, infrared, radio
d. gamma rays, infrared, radio, ultraviolet

16. Select the arrangement of electromagnetic radiation which starts with the lowest wavelength and increases to greatest wavelength.
M

a. microwave, visible, infrared, gamma rays
b. radio, infrared, visible, ultraviolet
c. visible, X-rays, infrared, microwave
d. ultraviolet, visible, infrared, microwave

17. The FM station KDUL broadcasts music at 99.1 MHz. Find the wavelength of these waves.
M

a. 1.88×10^{-2} m
b. 0.330 m
c. 3.03 m
d. 5.33×10^{2} m

18. The AM station KBOR plays your favorite music from the 20's and 30's at 1290 kHz. Find the wavelength of these waves.
M

a. 4.30×10^{-2} m
b. 0.144 m
c. 6.94 m
d. 232 m

19. An infrared wave has a wavelength of 6.5×10^{-4} cm. What is this distance in angstroms, Å?
E

a. 6.5×10^{-4} Å
b. 2.2×10^{-4} Å
c. 4.6×10^{3} Å
d. 6.5×10^{4} Å

20. A radio wave has a frequency of 8.6×10^{8} Hz. What is the energy of one photon of this radiation?
M

a. 7.7×10^{-43} J
b. 2.3×10^{-34} J
c. 5.7×10^{-25} J
d. 1.7×10^{-16} J

21. Infrared radiation from the sun has a wavelength of 6200 nm. Calculate the energy of one photon of
M that radiation.

a. 4.1×10^{-39} J
b. 4.1×10^{-30} J
c. 3.2×10^{-29} J
d. 3.2×10^{-20} J

22. Green light has a wavelength of 5200Å. Calculate the energy of one photon of green light.
M

a. 3.4×10^{-40} J
b. 3.4×10^{-30} J
c. 3.8×10^{-29} J
d. 3.8×10^{-19} J

23. Calculate the frequency of a photon absorbed when the hydrogen atom undergoes a transition from $n = 2$
H to $n = 4$. ($R = 1.096776 \times 10^7$ m^{-1})

a. 2.056×10^6 s^{-1}
b. 2.742×10^6 s^{-1}
c. 6.165×10^{14} s^{-1}
d. 8.226×10^{14} s^{-1}

24. If the energy of a photon is 1.32×10^{-18} J, what is its wavelength in nm?
E

a. 1.50×10^{-7}
b. 150.
c. 1.99×10^{15}
d. 1.99×10^{24}
e. None of the above

25. A photon has an energy of 5.53×10^{-17} J. What is its frequency in s^{-1}?
E

a. 3.66×10^{-50}
b. 3.59×10^{-9}
c. 8.35×10^{16}
d. 1.20×10^{-17}
e. 2.78×10^{8}

26. For potassium metal, the work function ϕ (the minimum energy needed to eject an electron from the
M metal surface) is 3.68×10^{-19} J. Which is the longest wavelength of the following which could excite photoelectrons?

a. 550. nm
b. 500. nm
c. 450. nm
d. 400. nm
e. 350. nm

27. Platinum, which is widely used as a catalyst, has a work function ϕ (the minimum energy needed to eject an
M electron from the metal surface) of 9.05×10^{-19} J. What is the longest wavelength of light which will cause electrons to be emitted?

a. 2.196×10^{-7} m
b. 4.553×10^{-6} m
c. 5.654×10^{2} m
d. 1.370×10^{15} m

28. Consider the following adjectives used to describe types of spectrum:
M continuous line atomic emission absorption
How many of them are appropriate to describe the spectrum of radiation given off by a black body?

a. none
b. one
c. two
d. three
e. four

29. Consider the following adjectives used to describe types of spectrum:
M continuous line atomic emission absorption
How many of them are appropriate to describe the spectrum of radiation absorbed by a sample of mercury vapor?

a. one
b. two
c. three
d. four
e. five

30. Line spectra from all regions of the electromagnetic spectrum, including the Paschen series of infrared
M lines for hydrogen, are used by astronomers to identify elements present in the atmospheres of stars. Calculate the wavelength of the photon emitted when the hydrogen atom undergoes a transition from $n = 5$ to $n = 3$. ($R = 1.096776 \times 10^{7}$ m^{-1})

a. 205.1 nm
b. 384.6 nm
c. 683.8 nm
d. 1282 nm

31. According to the Rydberg equation, the line with the shortest wavelength in the emission spectrum
M of atomic hydrogen is predicted to lie at a wavelength (in nm) of

a. 91.2
b. 1.10×10^{-2}
c. 1.10×10^{2}
d. 1.10×10^{16}
e. none of the above

32. According to the Rydberg equation, the longest wavelength (in nm) in the series of H-atom lines
M with $n_l = 3$ is

a. 1875
b. 1458
c. 820.
d. 656
e. 365

33. An electron in the $n = 6$ level emits a photon with a wavelength of 410.2 nm. To what energy level does
M the electron move?

a. $n = 1$
b. $n = 2$
c. $n = 3$
d. $n = 4$

34. The Bohr theory of the hydrogen atom predicts the energy difference (in J) between the $n = 3$ and the
E $n = 5$ state to be

a. 8.72×10^{-20}
b. 1.36×10^{-19}
c. 2.42×10^{-19}
d. 1.55×10^{-19}
e. 1.09×10^{-18}

35. According to the Bohr theory of the hydrogen atom, the minimum energy (in J) needed to ionize a
M hydrogen atom from the $n = 2$ state

a. 2.18×10^{-18}
b. 1.64×10^{-18}
c. 5.45×10^{-19}
d. 3.03×10^{-19}
e. none of the above

36. A sprinter must average 24.0 mi/h to win a 100-m dash in 9.30 s. What is his wavelength at this speed
M if his mass is 84.5 kg?

a. 7.29×10^{-37} m
b. 3.26×10^{-37} m
c. 5.08×10^{-30} m
d. 1.34×10^{-30} m

37. The de Broglie equation predicts that the wavelength (in m) of a proton moving at 1000. m/s is
E

a. 3.96×10^{-10}
b. 3.96×10^{-7}
c. 2.52×10^{9}
d. 2.52×10^{6}
e. none of the above

38. According to the Heisenberg uncertainty principle, if the uncertainty in the speed of an electron is
E 3.5×10^3 m/s, the uncertainty in its position (in m) is at least

a. 1.7×10^{-8}
b. 6.6×10^{-8}
c. 17
d. 66
e. none of the above

39. The size of an atomic orbital is associated with
E

a. the principal quantum number (n)
b. the angular momentum quantum number (l)
c. the magnetic quantum number (m_l)
d. the spin quantum number (m_s)

40. The shape of an atomic orbital is associated with
E

a. the principal quantum number (n)
b. the angular momentum quantum number (l)
c. the magnetic quantum number (m_l)
d. the spin quantum number (m_s)

41. The orientation in space of an atomic orbital is associated with
E

a. the principal quantum number (n)
b. the angular momentum quantum number (l)
c. the magnetic quantum number (m_l)
d. the spin quantum number (m_s)

42. Atomic orbitals developed using quantum mechanics
M

a. describe regions of space in which one is most likely to find an electron
b. describe exact paths for electron motion
c. give a description of the atomic structure which is essentially the same as the Bohr model
d. allow scientists to calculate an exact volume for the hydrogen atom

43. The energy of an electron in the hydrogen atom is determined by
M

a. the principal quantum number (n) only
b. the angular momentum quantum number (l) only
c. the principal and angular momentum quantum numbers (n & l)
d. the principal and magnetic quantum numbers (n & m_l)

44. Which of the following is a correct set of quantum numbers for an electron in a 3d orbital?
H

a. $n = 3, l = 0, m_l = -1$
b. $n = 3, l = 1, m_l = +3$
c. $n = 3, l = 2, m_l = 0$
d. $n = 3, l = 3, m_l = +2$

45. H Which of the following is a correct set of quantum numbers for an electron in a 5*f* orbital?

a. $n = 5, l = 3, m_l = +1$
b. $n = 5, l = 2, m_l = +3$
c. $n = 4, l = 3, m_l = 0$
d. $n = 4, l = 2, m_l = +1$

46. E In the quantum mechanical treatment of the hydrogen atom, which one of the following combinations of quantum numbers is not allowed?

	n	l	m_l
a.	3	0	0
b.	3	1	-1
c.	3	2	2
d.	3	2	-1
e.	3	3	2

47. E Which one of the following sets of quantum numbers can correctly represent a 3*p* orbital?

a. $n = 1$, $l = 3$, $m_l = 0$
b. $n = 1$, $l = 3$, $m_l = 3$
c. $n = 3$, $l = 2$, $m_l = 1$
d. $n = 3$, $l = 1$, $m_l = -1$
e. $n = 3$, $l = 0$, $m_l = 1$

Short Answer Questions

48. M For the following equations
a. name the scientist to whom the equation is attributed
b. in not more than three lines, explain clearly what the equation means or represents.

1. $E = nh\nu$
2. $l = h/mu$
3. $H\psi = E\psi$
4. $\Delta x \cdot m\Delta u \geq h/4\pi$
5. $E_{photon} = h\nu$

49. M In not more than three lines for each answer, briefly outline one important scientific contribution of each of the following
a. Planck
b. de Broglie
c. Heisenberg

50. E
a. What is the frequency of microwave radiation which has a wavelength of 10.7 cm?
b. What is the energy of one photon of this radiation?

51. M Use the Rydberg equation to calculate the wavelength, in nm, of the least energetic (longest wavelength) line in the visible series ($n_l = 2$) of the spectrum of atomic hydrogen.

52. H
a. Calculate the wavelength in nm of a photon whose energy is 6.00×10^{-19} J.
b. Would the photon in (a) have enough energy to ionize a hydrogen atom in its ground state (i.e. to separate the proton and electron completely)? Use the Bohr equation to explain your answer.

53. M
a. Use Bohr's equation to calculate how much energy (in J) is needed to promote an electron from the H-atom ground state to the $n = 4$ level.
b. If a photon provides the energy in (a), what is its wavelength in nm?

54. M
a. Use the Bohr equation to calculate the energy needed to ionize a hydrogen atom from its ground state.
b. What is the minimum wavelength of a photon needed for it to have the energy needed in (a)?

55. Use the Bohr equation to calculate the energy of

H a. the largest energy absorption or emission process involving the $n = 2$ state of the hydrogen atom
b. the smallest energy absorption or emission process involving the $n = 2$ state of the hydrogen atom

56. What is the speed of an electron in m/s if its wavelength is 0.155 nm?
E

57. a. Calculate the momentum of a photon of green light, wavelength 515 nm.
M b. If this photon is traveling in a vacuum, what is its "mass"?

58. What is the minimum uncertainty in the position of a neutron if the uncertainty in its speed is 0.0250 m/s?
E

59. In the quantum mechanical treatment of the hydrogen atom, the functions ψ and ψ^2 both feature
M prominently. Briefly explain (in principle) how they are obtained and what, if anything, their physical meanings are.

60. Explain the context and meanings of the terms "orbit" and "orbital", making a clear distinction
M between them.

61. The following combinations of quantum numbers are not allowed. Correct each set by changing only
E one quantum number, and write in an appropriate corrected value.

a. $n = 2$	$l = 2$	$m_l = 2$	Corrected:	______ = ______
b. $n = 4$	$l = -2$	$m_l = 0$	Corrected:	______ = ______

62. What are the possible values for the following quantum numbers in an atom?
E
a. n
b. l
c. m_l

63. For the following orbitals, state the values or n, l and m_l which apply, and draw a sketch showing the
E shape and orientation of the orbital.
a. $3s$
b. $2p_x$

True/False Questions

64. The energy of a photon is directly proportional to the wavelength of the radiation.
E

65. Line spectra are characteristic of atoms in the gas phase.
E

66. Continuous spectra are characteristic of heated solids.
E

67. Continuous spectra are characteristic of molecules in the gas phase.
E

68. In the Rydberg equation, for a fixed value of n_1, the longest wavelength line has $n_2 = \infty$.
M

69. The Rydberg equation is an example of an empirical equation.
E

70. The Rydberg equation, giving the wavelengths of lines in the spectrum of the hydrogen atom, was
E obtained by assuming that energy is quantized.

71. M In the Bohr model of the hydrogen atom, the electron moves in a circular path which Bohr referred to as an orbital.

72. E Other factors being constant, a heavy object will have a longer de Broglie wavelength than a light object.

73. M In the quantum mechanical treatment of the hydrogen atom, the probability of finding an electron at any point is proportional to the wave function ψ.

74. E In the quantum mechanical treatment of the hydrogen atom, the energy depends on the principal quantum number n but not on the values of l or m_l.

Quantum Theory and Atomic Structure
Chapter 7
Answer Key

1.	d	17.	c	33.	b
2.	a	18.	d	34.	d
3.	b	19.	d	35.	c
4.	d	20.	c	36.	a
5.	c	21.	d	37.	a
6.	c	22.	d	38.	a
7.	d	23.	c	39.	a
8.	c	24.	b	40.	b
9.	b	25.	c	41.	c
10.	c	26.	b	42.	a
11.	a	27.	a	43.	a
12.	b	28.	c	44.	c
13.	b	29.	c	45.	a
14.	a	30.	d	46.	e
15.	c	31.	a	47.	d
16.	d	32.	a		

48. 1. a. Planck
b. A black body can only emit or absorb certain amounts of energy, i.e. whole number multiples of $h\nu$. The energy emitted/absorbed is quantized.

2. a. de Broglie
b. The equation is a quantitative representation of wave-particle duality. A wavelength (wave property) *l* corresponds to a momentum (particle property) of *mu*.

3. a. Schrodinger
b. When a wave function ψ which is a solution to the Schrodinger equation is operated on by *H*, the result is an energy times the wave function. Functions which do not satisfy this requirement are not solutions, and not acceptable wave functions.

4. a. Heisenberg
b. We cannot simultaneously know both the position (x) and the momentum (mu) of a particle exactly. There is a minimum uncertainty, $h/4\pi$, in their product, which is a fundamental property of matter, not an instrumental limitation.

5. a. Einstein
b. The energy of electromagnetic radiation is quantized, the energy being $h\nu$. In this sense, electromagnetic radiation behaves like a stream of particles rather than waves.

49. a. Planck is responsible for first proposing that energy is quantized. He proposed that a black body can emit or absorb electromagnetic energy E only in exact multiples of $h\nu$, giving rise to the equation $E = nh\nu$ where n is a positive integer.
b. de Broglie proposed that particles can have some wavelike properties, and developed a relationship which allows the wavelength (a wave property) to be calculated from the momentum of a particle.
c. Heisenberg is responsible for the uncertainty principle which states that we cannot simultaneously know both the position (x) and the momentum (mu) of a particle exactly. There is a minimum uncertainty, $h/4\pi$, in their product, which is a fundamental property of matter, not an instrumental limitation.

50. a. 2.80×10^9 s^{-1}
b. 1.86×10^{-24} J

51. 656 nm

52. a. 331 nm
b. No. 2.18×10^{-18} J is needed, the difference between the $n = 1$ and $n = \infty$ levels

53. a. 2.04×10^{-18} J
b. 97.2 nm

54. a. 2.18×10^{-18} J
b. 91.2 nm

55. a. 1.64×10^{-18} J ($n = 1$ to $n = 2$)
b. 3.03×10^{-19} J ($n = 3$ to $n = 2$)

56. 4.69×10^6 m/s

57. a. 1.29×10^{-27} kg·m/s
b. 4.29×10^{-36} kg

58. 1.26×10^{-6} m

59. The wave function, ψ, is a solution to the Schrodinger equation. It describes a wave, but has no physical meaning of its own. The square of the wave function, ψ^2, represents the probability of finding the electron at any point in the atom.

60. Orbit is a term from the Bohr theory, and refers to an exact, circular path followed by an electron. It has been superseded by the concept of an orbital, which arises from the Schrodinger equation. An orbital is a wave function, ψ or, more loosely, a region of space close to the nucleus where an electron is likely to be found.

61. a. $n = 3$
b. $l = 0$ (or 1, 2 or 3)

62. a. $n = 1, 2, 3, ...$
b. $l = 0, 1, ..., n\text{-}1$
c. $m_l = -l, -l+1, ..., +l$

63. a. $3s$ $n = 3, l = 0, m_l = 0$
b. $2p_x$ $n = 2, 1 = 1, m_l = 1$ (or 0 or -1)

64. F

65. T

66. T

67. F

68. F

69. T

70. F

71. F

72. F

73. F

74. T

Electron Configuration and Chemical Periodicity
Chapter 8

Multiple Choice Questions

1. "Each electron in an atom must have its own unique set of quantum numbers" is a statement of
E

a. the aufbau principle.
b. the Pauli exclusion principle.
c. Hund's rule.
d. the periodic law.

2. Energy states of atoms containing more than one electron arise from nucleus-electron and electron-electron
E interactions. Which of the following statements describes these effects?

a. larger nuclear charge lowers energy, more electrons in an orbital lowers energy
b. larger nuclear charge lowers energy, more electrons in an orbital increases energy
c. smaller nuclear charge lowers energy, more electrons in an orbital lowers energy
d. smaller nuclear charge lowers energy, more electrons in an orbital increases energy

3. Which one of the following statements about orbital energies is incorrect?
E

a. In the hydrogen atom, the energy of an orbital depends only on the value of the quantum number *n*.
b. In many-electron atoms the energy of an orbital depends on both *n* and *l*.
c. Inner electrons shield outer electrons more effectively than do electrons in the same orbital.
d. The splitting of sublevels in many-electron atoms is explained in terms of the penetration effect
e. The energy of a given orbital increases as the nuclear charge *Z* increases.

4. The effective nuclear charge for an atom is less than the actual nuclear charge due to
M

a. shielding.
b. penetration.
c. paramagnetism.
d. electron-pair repulsion.

5. The ____________________ quantum numbers are associated with the energy of an electron in a many-
E electron atom.

a. n and l
b. n and m_l
c. l and m_l
d. n and m_s

6. "Electrons added to atomic orbitals of the same energy will remain unpaired with parallel spins until the
E subshell is more than half-filled" is a statement of

a. the aufbau principle.
b. the Pauli exclusion principle.
c. Hund's rule.
d. the periodic law.

7. Which one of the following statements about atomic structure and quantum numbers is incorrect?
M

a. In a given atom, the maximum number of electrons having principal quantum number $n = 3$, is 18.
b. The number of orbitals in a given *f* subshell is 7.
c. For $n = 4$, the largest possible value of l is 3.
d. For $n = 4$, the largest possible value of m_l is 2.
e. The following set of quantum numbers for a single orbital is not allowed: $n = 3, l = 1, m_l = -2$.

8. In a single atom, what is the maximum number of electrons which can have quantum number $n = 4$?
M

a. 16
b. 18
c. 32
d. 36
e. none of the above

9. Select the correct set of quantum numbers (n, l, m_l, m_s) for the highest energy electron in the ground state of potassium, K.
M

a. $4, 1, -1, \frac{1}{2}$
b. $4, 1, 0, \frac{1}{2}$
c. $4, 0, 1, \frac{1}{2}$
d. $4, 0, 0, \frac{1}{2}$

10. Select the correct set of quantum numbers (n, l, m_l, m_s) for the highest energy electron in the ground state of tin, Sn.
M

a. $5, 2, -1, \frac{1}{2}$
b. $5 ,2, 0, \frac{1}{2}$
c. $5, 1, 2, \frac{1}{2}$
d. $5, 1, 0, \frac{1}{2}$

11. Select the correct set of quantum numbers (n, l, m_l, m_s) for the first electron removed in the formation of a cation for strontium, Sr.
M

a. $5, 1 , 0, -\frac{1}{2}$
b. $5, 1, 0, \frac{1}{2}$
c. $5, 0, 0, -\frac{1}{2}$
d. $5, 0, 1, \frac{1}{2}$

12. Select the correct electron configuration for sulfur ($Z = 16$).
E

a. $1s^2 1p^6 2s^2 2p^6$
b. $1s^2 2s^2 2p^6 2d^6$
c. $1s^2 2s^2 2p^6 3s^2 3p^4$
d. $1s^2 2s^2 2p^6 3s^2 3d^4$

13. Select the correct electron configuration for Cu ($Z = 29$).
H

a. $[Ar]4s^2 3d^9$
b. $[Ar]4s^1 3d^{10}$
c. $[Ar]4s^2 4p^6 3d^3$
d. $[Ar]4s^2 4d^9$

14. M Select the correct electron configuration for Te ($Z = 52$).

a. $[Kr]5s^2 5p^6 4d^8$
b. $[Kr]5s^2 5d^{10} 5p^4$
c. $[Kr]5s^2 4d^{10} 5p^4$
d. $[Kr]5s^2 4f^{14}$

15. M The electronic structure $1s^2 2s^2 2p^6 3s^2 3p^6 4s^2 3d^8$ refers to the ground state of

a. Kr
b. Fe
c. Ni
d. Pd
e. none of the above

16. M According to the Schrodinger model of the atom, in a ground state atom of silver (Ag), how many electrons will there be with the quantum number $l = 1$? (The n, m_l and m_s quantum numbers may have any appropriate values.)

a. 9
b. 12
c. 18
d. 24
e. 36

17. M An atom of element number 33 (As) is in its ground electronic state. Which one of the following sets of quantum numbers could not apply to any of its electrons?

a.	b.	c.	d.	e.
$n = 2$	$n = 3$	$n = 3$	$n = 4$	$n = 4$
$l = 1$	$l = 0$	$l = 2$	$l = 0$	$l = 2$
$m_l = -1$	$m_l = 0$	$m_l = 2$	$m_l = 0$	$m_l = -2$
$m_s = +\frac{1}{2}$	$m_s = -\frac{1}{2}$	$m_s = +\frac{1}{2}$	$m_s = -\frac{1}{2}$	$m_s = +\frac{1}{2}$

18. M An element with the electron configuration [noble gas] $ns^2(n-1)d^{10}np^2$ has ___________ valence electrons.

a. 2
b. 4
c. 10
d. 16

19. M An element with the electron configuration [noble gas]$ns^2(n-1)d^8$ has ___________ valence electrons.

a. 2
b. 6
c. 8
d. 10

20. M Select the valence shell electron configuration for selenium ($Z = 34$).

a. $[Ar]4s^2 3d^{10} 4p^4$
b. $[Ar]4s^2 4p^4$
c. $[Ar]4s^2 4d^{10} 4p^4$
d. $[Ar]4s^2 4p^6 4d^8$

21. Which of the following electron configurations represents the ground state for an element?
E

a. [Ne]$3s^1 3p^1$
b. [He]$2s^1 2p^3$
c. [Ne]$3s^2 3p^3$
d. [Ne]$3s^2 3p^3 3d^1$

22. Which of the following electron configuration is correct for the excited state of an element?
E

a. [He]$2s^2 2p^5$
b. [Ne]$3s^2 3p^1$
c. [ArJ$4s^1 4p^1$
d. [Kr]$5s^2 4d^7$

23. Select the element with the largest atomic size.
M

a. S
b. Ca
c. Ba
d. Po

24. Select the element with the smallest atomic size.
M

a. Na
b. Al
c. Ga
d. K

25. Arrange potassium, rubidium, calcium and barium in order of increasing atomic size.
M

a. K < Ca < Rb < Ba
b. Ca < K < Rb < Ba
c. Ca < K < Ba < Rb
d. K < Ca < Ba < Rb

26. Arrange silicon, antimony, selenium and bismuth in order of decreasing atomic size.
M

a. Si > Sb > Bi > Se
b. Se > Si > Sb > Bi
c. Bi > Sb > Si > Se
d. Bi > Sb > Se > Si

27. Arrange strontium, cesium, iodine, and lead in order of increasing atomic size.
M

a. Cs < Sr < Pb < I
b. Sr < I < Cs <Pb
c. I < Sr < Pb < Cs
d. Sr < Cs < I < Pb

28. Which of the following has the smallest atomic radius?
M

a. Li
b. Ne
c. Rb
d. Sr
e. Xe

29. M Which of the following has the greatest atomic radius?

a. Li
b. Ne
c. Rb
d. Sr
e. Xe

30. H Which one of the following equations correctly represents the process relating to the ionization energy of X?

a. $X(s) \rightarrow X^+(g) + e^-$
b. $X_2(g) \rightarrow X^+(g) + X^-(g)$
c. $X(g) + e^- \rightarrow X^-(g)$
d. $X^-(g) \rightarrow X(g) + e^-$
e. $X(g) \rightarrow X^+(g) + e^-$

31. M Select the element with the largest first ionization energy.

a. Na
b. Ca
c. Cl
d. Te

32. M Select the element with the smallest first ionization energy.

a. Rb
b. Mg
c. I
d. As

33. M Arrange sodium, oxygen, fluorine, and strontium in order of increasing first ionization energy.

a. Na < Sr < O < F
b. Sr < Na < O < F
c. Sr < Na < F < O
d. Na < Sr < F < O

34. M Arrange calcium, cesium, selenium, and sulfur in order of decreasing first ionization energy.

a. S > Se > Ca > Cs
b. Se > S > Ca > Cs
c. S > Se > Cs > Ca
d. Se > S > Cs > Ca

35. M Select the element with the largest second ionization energy (IE_2).

a. K
b. Ca
c. Sr
d. Al

36. Elements with the highest first ionization energies are found in the ___________ corner of the periodic table.
E

a. upper left
b. lower left
c. upper right
d. lower right

37. Which one of the following equations correctly represents the process involved in the electron affinity of X?
H

a. $X(g) \rightarrow X^+(g) + e^-$
b. $X^+(g) \rightarrow X^+(aq)$
c. $X^+(g) + e^- \rightarrow X(g)$
d. $X(g) + e^- \rightarrow X^-(g)$
e. $X^+(g) + Y^-(g) \rightarrow XY(s)$

38. Elements with _______________ first ionization energies and ___________ electron affinities generally form cations.
H

a. low, very negative
b. high, positive or slightly negative
c. low, positive or slightly negative
d. high, very negative

39. Elements with _______________ first ionization energies and ___________ electron affinities generally form anions.
H

a. low, very negative
b. high, positive or slightly negative
c. low, positive or slightly negative
d. high, very negative

40. Select the element with the greatest metallic character.
M

a. Rb
b. Ca
c. Cs
d. Pb

41. Select the element with the least metallic character.
M

a. L
b. Sr
c. Tl
d. Ge

42. The most acidic oxides are formed from elements found in the _______________ corner of the periodic table.
E

a. upper right
b. upper left
c. lower right
d. lower left

43. The most basic oxides are formed from elements found in the ________________ corner of the
E periodic table.

a. upper right
b. upper left
c. lower right
d. lower left

44. Select the most acidic compound from the following.
M

a. SO_2
b. Al_2O_3
c. CaO
d. PbO

45. Select the most basic compound from the following.
M

a. Bi_2O_3
b. SiO_2
c. Cs_2O
d. Na_2O

46. Which of the following elements will from a cation with a +2 charge?
E

a. Si
b. Sr
c. Ga
d. Cs

47. Which of the following ions will be most likely to form when selenium ionizes?
E

a. Se^{6+}
b. Se^{4+}
c. Se^{2-}
d. Se^{4-}

48. Which of the following atoms will be diamagnetic?
M

a. Cr
b. Ru
c. Cd
d. Pt

49. Which of the following elements is paramagnetic?
M

a. Kr
b. Zn
c. Sr
d. Fr

50. Select the paramagnetic ion.
M

a. Cu^+
b. Ag^+
c. Fe^{3+}
d. Cd^{2+}

51. Select the diamagnetic ion.
M

a. Cu^{2+}
b. Ni^{2+}
c. Cr^{3+}
d. Sc^{3+}

52. Consider the set of isolectronic atoms and ions A^{2-}, B^-, C, D^+, and E^{2+}. Which arrangement of relative radii is correct?
M

a. $A^{2-} > B^- > C > D^+ > E^{2+}$
b. $E^{2+} > D^+ > C > B^- > A^{2-}$
c. $A^{2-} > B^- > C < D^+ < E^{2+}$
d. $A^{2-} < W < C > D^+ > E^{2+}$

53. Arrange these ions in order of decreasing size.
M

a. $P^{3-} > Cl^- > K^+ > Ca^{2+}$
b. $Ca^{2+} > K^+ > Cl^- > P^3$-
c. $K^+ > Cl^- > Ca^{2+} > P^3$-
d. $K^+ > Cl^- > P^3 > Ca^{2+}$

Short Answer Questions

54. What is the maximum number of electrons in an atom which can have
H
a. quantum number $n = 4$
b. orbital designation $3d$
c. orbital designation $2p_z$

55. In a single atom, write down the maximum possible number of electrons:
H
a. with $n = 3$
b. with the designation $4d_{xy}$
c. with $n = 5, l = 3, m_s = +\frac{1}{2}$

56. State Hund's rule, and show how it applies to the ground state of phosphorus atoms.
M

57. Write down the full electron configuration for ground state atoms of:
M
a. S (element 16)
b. Pd (element 46)

58. Write down the full electron configuration for ground state atoms of:
M
a. element number 12
b. element number 23
c. element number 32

59. With the aid of a diagram, describe how the atomic radii of main group elements vary with position in the periodic table.
E

60. Define what is meant by ionization energy, and write a balanced chemical equation to represent the
H relevant process for element X.

61. Describe how the ionization energies of main group elements vary across a period in the periodic
M table, using a diagram if necessary.

62. Briefly describe how the atomic radii and ionization energies of group 1A(1) elements compare with
H those of group 8A(18). Also, explain why the values of these properties are so different between these two groups.

63. Define what is meant by electron affinity, and write a balanced chemical equation to represent the
H relevant process for element Y.

True/False Questions

64. In Mendeleev's version of the periodic table, the elements were arranged in order of increasing
M atomic number.

65. Moseley's measurements of nuclear charges of the elements provided the basis for arranging the
M elements of the periodic table in order of increasing atomic number.

66. According to the exclusion principle, two is the maximum number of electrons in an atom which can
E share the same four quantum numbers.

67. The difference in energies between the 1*s* and 2*s* orbitals is due to the penetration effect.
M

68. In neutral atoms, the 3*d* orbitals have higher energy than the 4*s* orbitals.
E

69. Elements in which the outermost electron has the same principal quantum number *n*, show similar
M chemical properties.

70. The maximum number of electrons in an atom with the same value of *n* is $2n^2$.
M

71. Atomic size decreases across a period due to an increase in the effective nuclear charge, Z_{eff}.
M

72. First ionization energies of neutral atoms may be positive or negative.
E

73. Electron affinities of neutral atoms may be positive or negative.
M

74. In moving down a group in the periodic table, the oxides of the elements become more acidic innature.
M

75. In forming ions of the first series of transition metals, the 4*s* electrons are lost before the 3*d* electrons.
M

Electron Configuration and Chemical Periodicity
Chapter 8
Answer Key

1.	b	19.	a	37.	d
2.	b	20.	a	38.	c
3.	e	21.	c	39.	d
4.	a	22.	c	40.	c
5.	a	23.	c	41.	d
6.	c	24.	b	42.	a
7.	d	25.	b	43.	d
8.	c	26.	d	44.	a
9.	d	27.	c	45.	c
10.	c	28.	b	46.	b
11.	c	29.	c	47.	c
12.	c	30.	e	48.	c
13.	b	31.	c	49.	d
14.	c	32.	a	50.	c
15.	c	33.	a	51.	d
16.	c	34.	a	52.	a
17.	e	35.	a	53.	a
18.	b	36.	c		

54.
a. 32
b. 10
c. 2

55.
a. 18
b. 2
c. 7

56. **Electrons occupy orbitals within the same subshell singly, with parallel spins, before pairing.**
P: $Z = 15$ $1s^2 2s^2 2p^6 3s^2 3p^3$ The last three electrons occupy the $3p_x$, $3p_y$ and $3p_z$ orbitals singly, with parallel spins.

57.
a. $1s^2 2s^2 2p^6 3s^2 3p^4$
b. $1s^2 2s^2 2p^6 3s^2 3p^6 4s^2 3d^{10} 4p^6 5s^2 4d^8$

58.
a. $1s^2 2s^2 2p^6 3s^2$
b. $1s^2 2s^2 2p^6 3s^2 3p^6 4s^2 3d^3$
c. $1s^2 2s^2 2p^6 3s^2 3p^6 4s^2 3d^{10} 4p^2$

59. **Among the main group elements, atomic radii decrease from left to right across periods, and increase from the top of the table to the bottom, i.e. down groups.**

60. **Ionization energy is the minimum energy needed to remove an electron from a gaseous atom, usually expressed on a per mole basis.**

$$X(g) \rightarrow X^+(g) + e^-$$

61. **Ionization energies generally increase from left to right across a period, but there are some small exceptions to this trend. The exceptions occur with the group 3A (3) and 6A (6) elements.**

62. Group 1A (1) atoms are much larger than those of group 8A (18) atoms, when elements within the same period are compared (e.g., Na and Ar). This is because electrons are being added in the same shell as one moves across the period, and Z_{eff} is increasing due to poor shielding. Thus, the outermost electron in Ar is more tightly held than that in Na. For the same reason, the ionization energy of group 1A (1) atoms is much less than that of group 8A (18) atoms.

63. Electron affinity is the energy change when a gaseous atom gains an electron, usually expressed on a per mole basis.

$$X(g) + e^- \rightarrow X^-(g)$$

64. F

65. T

66. F

67. F

68. T

69. F

70. T

71. T

72. F

73. T

74. F

75. T

Models of Chemical Bonding
Chapter 9

Multiple Choice Questions

1. Which of the following contains ionic bonding?
E

a. CO
b. SrF_2
c. Al
d. OCl_2

2. Which of the following is an ionic compound?
E

a. H_2S
b. NH_3
c. I_2
d. KI

3. Which of the following is an ionic compound?
E

a. SO_2
b. $BaBr_2$
c. NI_3
d. PH_3

4. Which of the following is a covalent compound?
E

a. Na_2O
b. $CaCl_2$
c. Cl_2O
d. CsCl

5. Which of the following contains covalent bonds?
E

a. BaO
b. IBr
c. Mg
d. LiBr

6. In which of these compounds are the atoms held together by polar covalent bonding?
E

a. $SrCl_2$
b. CsCl
c. ClF
d. TiF_2

7. In which of these substances are the atoms held together by metallic bonding?
E

a. CO_2
b. Cr
c. Br_2
d. S_8

8. Select the element whose Lewis symbol is correct.
E

a. ·Fr·
b. ·Ra·
c. ·Pb·
d. :Te:

9. Select the element whose Lewis symbol is correct.
E

a. ·Ga·
b. ·Al·
c. $:\overset{\cdot\cdot}{\underset{\cdot\cdot}{\mathrm{Br}}}\cdot$
d. $\cdot\overset{\cdot}{\underset{\cdot}{\mathrm{Tl}}}\cdot$

10. Select the correct formula for a compound formed from calcium and chlorine.
M

a. CaCl
b. $CaCl_2$
c. Ca_2Cl
d. Ca_2Cl_2

11. Select the correct formula for a compound formed from barium and nitrogen.
M

a. BaN
b. BaN_2
c. Ba_2N_3
d. Ba_3N_2

12. The lattice energy for ionic crystals increases as the charge on the ions ____________ and the size of the
M ions _________________ .

a. increases, increases
b. increases, decreases
c. decreases, increases
d. decreases, decreases

13. Select the compound with the highest (i.e. most negative) lattice energy.
M

a. CaS(*s*)
b. MgO(*s*)
c. NaI(*s*)
d. LiBr(*s*)

14. Select the compound with the lowest (i.e. least negative) lattice energy.
M

a. CsBr(*s*)
b. NaCl(*s*)
c. SrO(*s*)
d. CaO(*s*)

15. M

The lattice energy of CaF_2 is the energy change for which one, if any, of the following processes?

a. $Ca^{2+}(s) + 2F^{-}(g) \rightarrow CaF_2(g)$
b. $CaF_2(g) \rightarrow CaF_2(s)$
c. $Ca(g) + 2F(g) \rightarrow CaF_2(s)$
d. $CaF_2(aq) \rightarrow CaF_2(s)$
e. None of the above

16. M

The lattice energy of $MgCl_2$ is the energy change for which one of the following processes?

a. $Mg(s) + Cl_2(g) \rightarrow MgCl_2(s)$
b. $Mg(g) + 2Cl(g) \rightarrow MgCl_2(s)$
c. $Mg^{2+}(s) + 2Cl^{-}(g) \rightarrow MgCl_2(g)$
d. $Mg^{2+}(g) + 2Cl^{-}(g) \rightarrow MgCl_2(s)$
e. $MgCl_2(aq) \rightarrow MgCl_2(s)$

17. H

Calculate the lattice energy of magnesium sulfide.

$Mg(s) \rightarrow Mg(g)$	$\Delta H° = 148$ kJ/mol
$Mg(g) \rightarrow Mg^{2+}(g) + 2e^{-}$	$\Delta H° = 2186$ kJ/mol
$\frac{1}{8}S_8(s) \rightarrow S(g)$	$\Delta H° = 279$kJ/mol
$S(g) + 2e^{-} \rightarrow S^{2-}(g)$	$\Delta H° = 450$ kJ/mol
$Mg(s) + \frac{1}{8}S_8(s) \rightarrow MgS(s)$	$\Delta H°_f = -343$ kJ/mol
$Mg^{2+}(g) + S^{2-}(g) \rightarrow MgS(s)$	$\Delta H°_{MgS} = ?$

a. -3406 kJ/mol
b. -2720. kJ/mol
c. 2720. kJ/mol
d. 3406 kJ/mol

18. M

Analysis of an unknown substance showed that it has a high boiling point and is brittle. It is an insulator as a solid but conducts electricity when melted. Which of the following substances would have those characteristics?

a. HCl
b. Al
c. KBr
d. SiF_4

19. M

Which one of the following properties is least characteristic of typical ionic compounds?

a. high melting point
b. high boiling point
c. brittleness
d. poor electrical conductor when solid
e. poor electrical conductor when molten

20. M

Arrange the following bonds in order of increasing bond strength.

a. C–F < C–Cl < C–Br < C–I
b. C–I < C–Br < C–Cl < C–F
c. C–Br < C–I < C–Cl < C–F
d. C–I < C–Br < C–F< C–Cl

21. Select the strongest bond in the following group.
E

a. C–S
b. C–O
c. C=C
d. C≡N

22. Which one of the following properties is least characteristic of substances composed of small, covalently-bonded molecules?
M

a. low melting point
b. low boiling point
c. weak bonds
d. poor electrical conductor when solid
e. poor electrical conductor when molten

23. Electronegativity is a measure of
M

a. the energy needed to remove an electron from an atom
b. the energy released when an electron is added to an atom
c. the attraction by an atom for electrons in a chemical bond
d. the magnitude of the negative charge on an electron

24. Select the element with the highest electronegativity.
M

a. S
b. Ru
c. Si
d. Te

25. Select the element with the lowest electronegativity.
M

a. Si
b. Se
c. Sr
d. Sc

26. Which of the following elements is the most electronegative?
E

a. Ne
b. Rb
c. P
d. I
e. Cl

27. Arrange aluminum, nitrogen, phosphorus and indium in order of increasing electronegativity.
M

a. Al < In < N < P
b. Al < In < P < N
c. In < Al < P < N
d. In < P < Al < N

28. Arrange calcium, rubidium, sulfur, and arsenic in order of decreasing electronegativity.
M

a. S > As > Rb > Ca
b. S > As > Ca > Rb
c. As > S > Rb > Ca
d. As > S > Ca > Rb

29. Which of the following lists is in the correct sequence of decreasing electronegativity of the atoms
E concerned?

a. O > S > Ca > Rb > K
b. O > S > Ca > K > Rb
c. O > S > Rb > K > Ca
d. O > S > Rb > Ca > K
e. None of the above

30. Based on electronegativity trends in the periodic table, predict which of the following compounds
M will have the greatest % ionic character in its bonds.

a. H_2O
b. LiI
c. RbF
d. CaO
e. HCl

31. Select the most polar bond.
M

a. C–O
b. Si–F
c. Cl–F
d. C–F

32. Which of the following compounds displays the greatest ionic character in its bonds?
M

a. NO_2
b. CO_2
c. H_2O
d. HF

33. Analysis of an unknown substance showed that it has a moderate melting point and is a good conductor of
M heat and electricity in the solid phase. Which of the following substances would have those characteristics?

a. NaCl
b. Ga
c. CCl_4
d. I_2

34. Which one of the following properties is least characteristic of typical metals?
M

a. moderately high melting point
b. high boiling point
c. brittleness
d. good electrical conductor when solid
e. good electrical conductor when molten

Short Answer Questions

35. Ionic bonding typically occurs when a _____________ bonds with a _____________ .
E

36. Covalent bonding typically occurs when a _____________ bonds with a _____________ .
E

37. When an atom is represented in a Lewis electron dot symbol, the element symbol represents _____________ and the dots represent _____________.
H

38. In not more than three sentences, describe the electron arrangement responsible for bonding in solid $SrCl_2$.
M

39. The lattice energy of rubidium chloride is the energy change accompanying the process
M

$$Rb^+(g) + Cl^-(g) \rightarrow RbCl(s)$$

Calculate the lattice energy of RbCl using the following data:

		$\Delta H°$ (kJ/mol)
$Rb(s)$	$\rightarrow Rb(g)$	86
$Rb(g)$	$\rightarrow Rb^+(g) + e^-$	409
$Cl_2(g)$	$\rightarrow 2Cl(g)$	242
$Cl(g) + e^-$	$\rightarrow Cl^-(g)$	-355
$Rb(s) + \frac{1}{2}Cl_2(g)$	$\rightarrow RbCl(s)$	-435

40. Describe, with appropriate explanations, the key factors which affect the magnitude of the lattice energy of an ionic substance.
M

41. A hypothetical ionic substance will not form merely because it has a high lattice energy. Explain why, using energy-based arguments.
H

42. In not more than three sentences, describe the electron arrangement responsible for bonding in Cl_2 molecules.
M

43. Using appropriate, real examples to illustrate your answer, describe the correlation between bond energy and bond length for a series of single bonds.
M

44. Using appropriate, real examples to illustrate your answer, describe the correlation between bond energy and bond length for a series of varying bond order.
M

45. Give a clear and concise definition of the term "electronegativity"; i.e., what does it measure?
E

46. Describe in brief how electronegativity values can be used to predict the percent ionic character of a bond between two atoms.
E

47. Most of the copper sold in major metal markets is highly purified, typically to 99.99%. Why is this?
M

48. In not more than three sentences, describe the key features of bonding in solid aluminum.
M

True/False Questions

49. The majority of elements are good conductors when in solid form.
M

50. A single covalent bond consists of a single delocalized electron pair.
M

51. The lattice energy is the energy released when separated ions in the gas phase combine to form ionic molecules in the gas phase.
E

52. The lattice energy of large ions is greater in magnitude than that of small ions of the same charge.
E

53. The electrostatic energy of two charged particles is inversely proportional to the square of the distance between them.
M

54. The electrostatic energy of two charged particles is inversely proportional to the distance between them.
M

55. Bond energy increases as bond order increases, for bonding between a given pair of atoms.
E

56. Covalently bonded substances do not necessarily exist as separate molecules.
E

57. Electronegativities on Pauling's scale are calculated from ionization energies and electron affinities.
M

58. No real bonds are 100% ionic in character.
E

59. As a measure of the strength of metallic bonding, the boiling point of a metal is a better indicator than its melting point.
M

Models of Chemical Bonding
Chapter 9
Answer Key

1.	b	13.	b	25.	c
2.	d	14.	a	26.	e
3.	b	15.	e	27.	c
4.	c	16.	d	28.	b
5.	b	17.	a	29.	b
6.	c	18.	c	30.	c
7.	b	19.	e	31.	b
8.	b	20.	b	32.	d
9.	c	21.	d	33.	b
10.	b	22.	c	34.	c
11.	d	23.	c		
12.	b	24.	a		

35. metal — non-metal

36. non-metal — non-metal

37. nucleus and inner electrons — valence electrons

38. This is an example of ionic bonding in which Sr, from group 2A (2) will be present as Sr^{2+} ions, while Cl, from group 7A (7) will be present as Cl^- ions. The cations and anions will be arranged in a crystalline lattice so that nearest neighbors will be ions of opposite charge, thus achieving a net coulombic attraction. There are no molecules present; each ion is equally attracted to all its nearest neighbors.

39. Lattice energy = - 696 kJ/mol

40. By Coulomb's law, the energy of two electrical charges is proportional to the product of their charges and inversely proportional to the distance separating them. If one approximates the lattice energy by considering only nearest neighbor interactions, the energy will be proportional to the product of the charges and inversely proportional to their separation. Thus a lattice consisting of small ions with multiple charges will have the greatest (negative) lattice energy.

41. In order for an ionic substance to form, the overall energy change accompanying its formation, from, say, its elements, should be favorable. A Born-Haber cycle can be used to separate and identify the various steps and energy changes involved. These include atomization and the ionization of the metal, atomization (e.g., bond dissociation) and electron affinity of the non-metal, and the lattice energy. Only if the sum of all these energy changes is favorable, is formation of the ionic substance likely to occur.

42. This is covalent bonding. The chlorine atoms each have 7 valence electrons, and can achieve stable, octet valence structures by sharing one electron each. This shared electron pair constitutes a single, pure covalent bond.

43. The bonds between carbon, C, and the halogens, X (X = F, Cl, Br, I) can be used as an examples. The size of X increases as one moves down the group. This means that the C–X bond length will become correspondingly longer. As this occurs the bond becomes weaker. Thus, the C–I bond is the longest and weakest, C–F the shortest and strongest of the series.

44. Carbon and oxygen form single, double and triple bonds. The C≡O bond in carbon monoxide is roughly three times as strong as the C–O single bond, while the C=O bond is about twice as strong as the single bond. The bond energy is approximately proportional to the bond order.

45. Electronegativity measures an atom's ability to attract shared (bonding) electron pairs to itself. High electronegativity implies strong attraction of such shared pairs.

46. **Take the electronegativity difference (ΔEN) of the bonded atoms. The larger the difference between their electronegativities, the more ionic will be the nature of the bond. If $\Delta EN = 0$, the bond will be pure covalent.**

47. **The main use of copper is as an electrical conductor. Copper's electrical conductivity is significantly decreased by the presence of impurities, hence the need for high purity.**

48. **This is metallic bonding. The aluminum atoms pool their valence electrons, and the metal can be viewed as consisting of Al^{3+} ions in a sea of the pooled, highly mobile electrons. There are no specific, directional bonds, but the arrangement of cations and electrons is such that there is a very strong net attraction.**

49. **T**

50. **F**

51. **F**

52. **F**

53. **F**

54. **T**

55. **T**

56. **T**

57. **F**

58. **T**

59. **T**

The Shapes of Molecules
Chapter 10

Multiple Choice Questions

1. Which one of the following Lewis structures is definitely incorrect?
E

a. BF_3 b. XeO_3 c. N_2 d. $AlCl_4^-$ e. NH_4^+

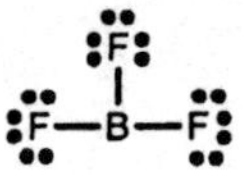
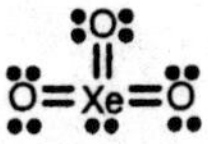

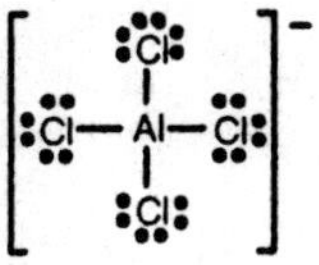
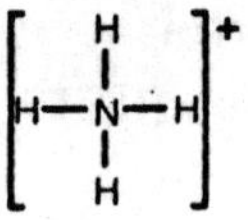

2. Which one of the following Lewis structures is definitely incorrect?
M

a. NO b. HCN c. NO_2^- d. SO_3^{2-} e. PCl_5

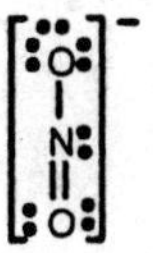
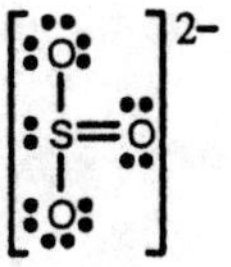
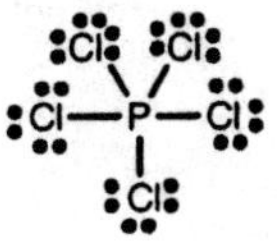

3. Which one of the following Lewis structures is definitely incorrect?
M

a. NO_2 b. $BeCl_2$ c. CO_3^{2-} d. CH_4 e. SO_2

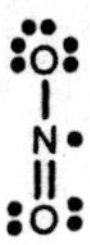
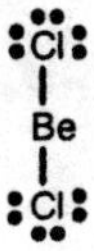
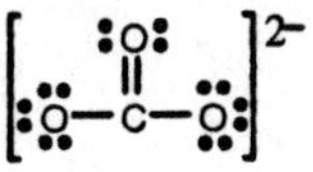
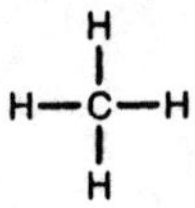

4. Select the best Lewis structure for ClCN.
M

a. 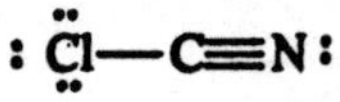

b. 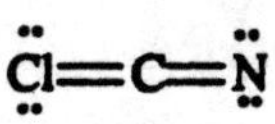

c. 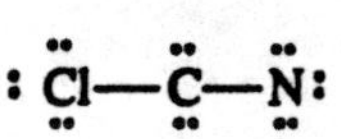

d. 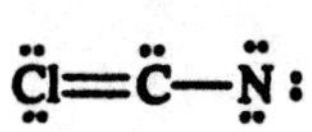

5. M Hydrazine, N_2H_4, is a good reducing agent that has been used as a component in rocket fuels. Select its Lewis structure.

a. 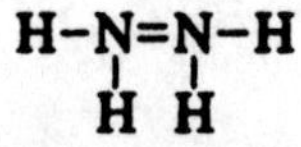

b. 

c. 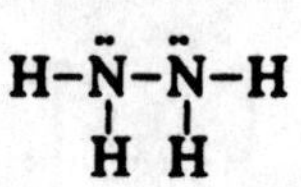

d. 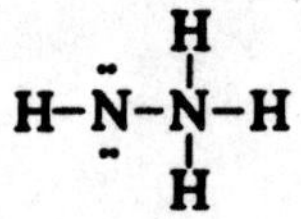

6. E Select the correct Lewis structure for nitrogen trifluoride, NF_3.

a. 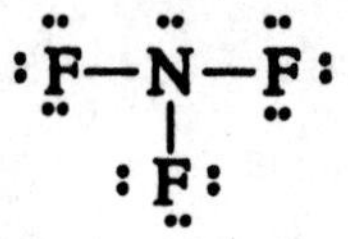

b. 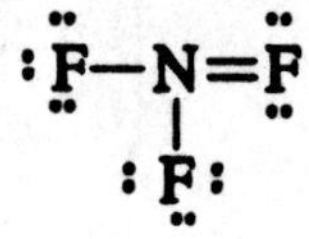

c. 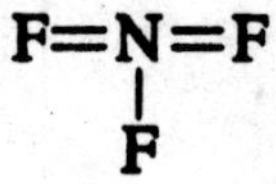

d. 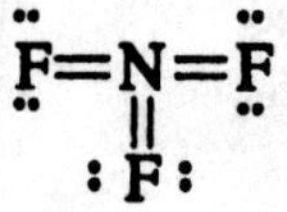

7. M Select the correct Lewis structure for NOCl, a reactive material used as an ionizing solvent.

a.` 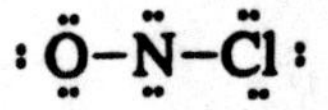

b. 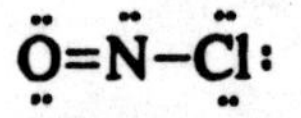

c. 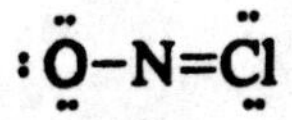

d. 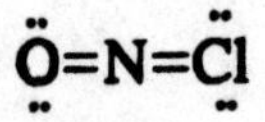

8. E Oxygen difluoride is a powerful oxidizing and fluorinating agent. Select its Lewis structure.

a. 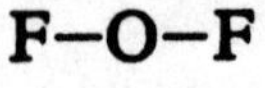

b. 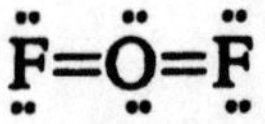

c. 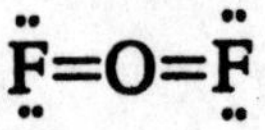

d. 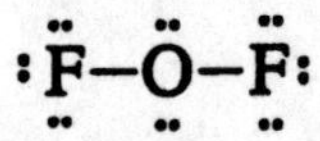

9. Select the best Lewis structure for P_2I_4.
M

a. 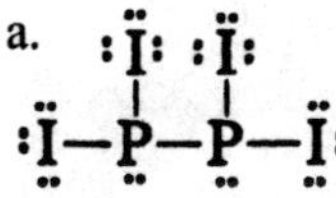
b. 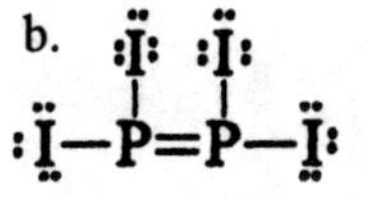
c. 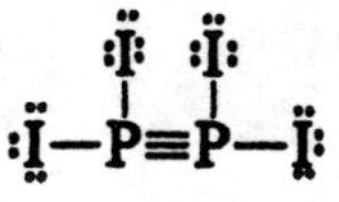
d. 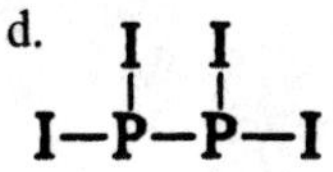

10. Thionyl chloride is used as an oxidizing and chlorinating agent in organic chemistry. Select the best Lewis structure for $SOCl_2$.
H

a. 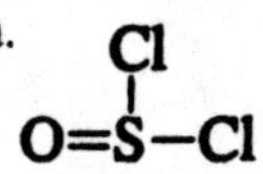
b. 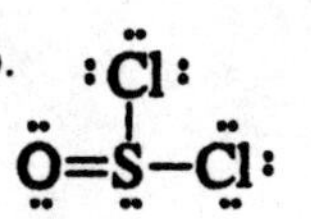
c. 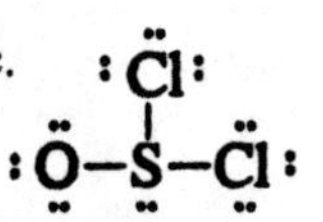
d. 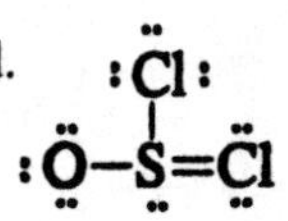

11. Select the correct Lewis structure for $TeBr_2$.
M

a. 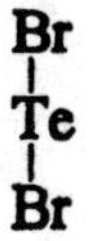
b. 
c. 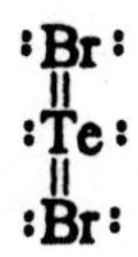
d. 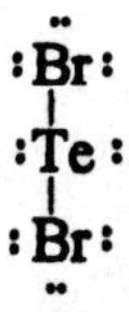

12. Which one of the following exhibits resonance?
H

a. CO_2
b. ClO_3^-
c. $COCl_2$
d. NO_2^+

13. Which one of the following exhibits resonance?
H

a. SO_3
b. SO_3^{2-}
c. I_3^-
d. SCO (C = central atom)

14. Select the Lewis structure in which formal charges are minimized for the periodate anion, IO_4^-.
M

a. 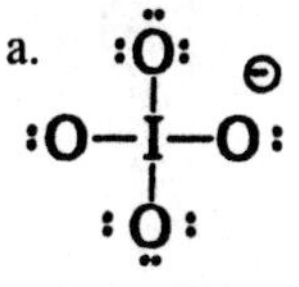
b. 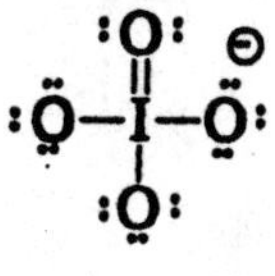
c. 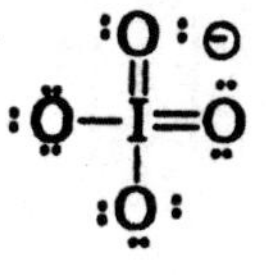
d. 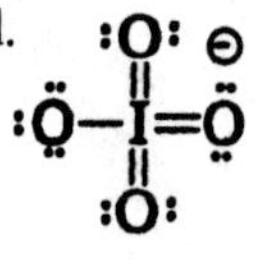

15. Phosphoryl iodide is used in the preparation of organophosphorus derivatives and phosphate esters. Select the Lewis structure for POI_3 which minimizes formal charges.
M

a. 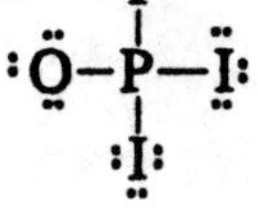
b. 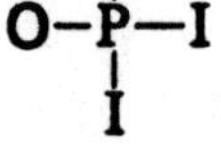
c. 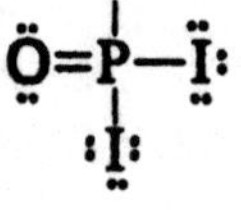
d. 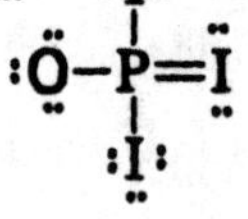

16. Select the Lewis structure for XeO_2F_2 which minimizes formal charges.
M

a. 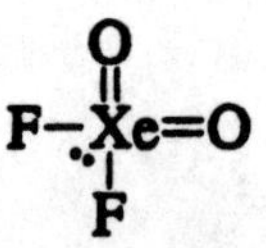
b. 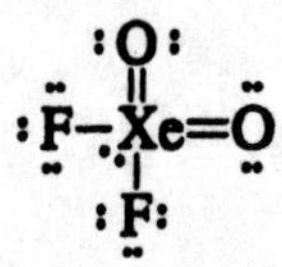
c. 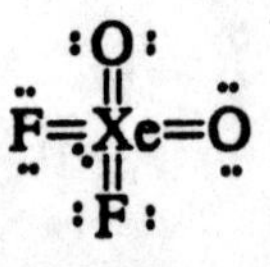
d. 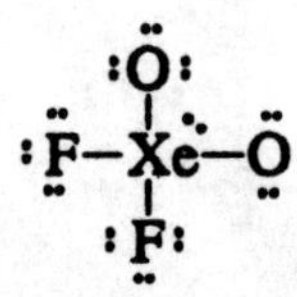

17. In the following Lewis structure for ClO_3F, chlorine has a formal charge of ____ and an oxidation number of ____.
M

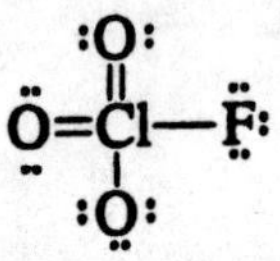

a. 1, 7
b. 1, -1
c. 1, 1
d. 7, -1

18. In the following Lewis structure for phosphate, phosphorus has a formal charge of ____ and an oxidation number of ____.
M

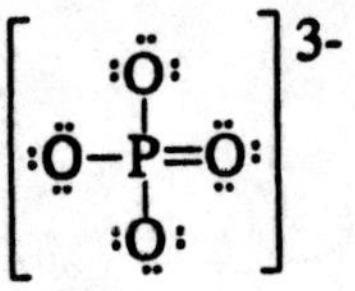

a. 0, -3
b. 0, 5
c. 5, -3
d. 5, 5

19. In which of the following does the nitrogen atom have a formal charge of -1?
M

a. 
b. 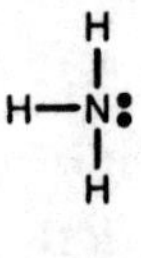
c. 
d. 
e. 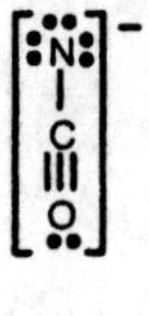

20. The formal charges on Cl and O in the structure shown for the ClO^- ion are, respectively
M

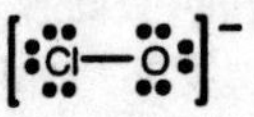

a. 0 and -1
b. -1 and 0
c. 1 and -2
d. -2 and 1
e. none of the above

21. M In which one of the following structures does the central atom have a formal charge of +2?

a. SF_6

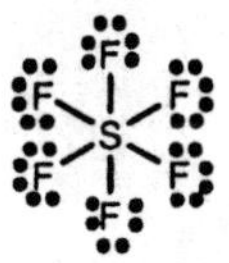

b. SO_4^{2-}

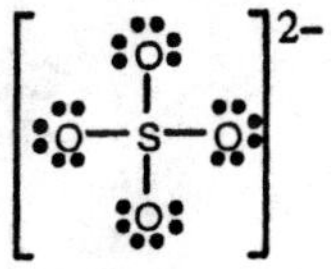

c. O_3

d. $BeCl_2$

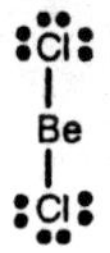

e. $AlCl_4^-$

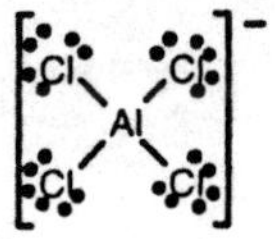

22. M The formal charge on Cl in the structure shown for the perchlorate ion is

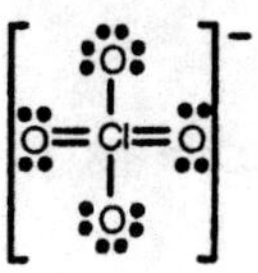

a. -2
b. -1
c. 0
d. +1
e. +2

23. E In which one of the following species is the central atom (the first atom in the formula) an exception to the octet rule?

a. NH_3
b. NH_4^+
c. I_2
d. BH_4^-
e. SF_6

24. M In which one of the following species is the central atom (the first atom in the formula) likely to violate the octet rule?

a. BF_4^-
b. XeO_3
c. $SiCl_4$
d. NH_3
e. CH_2Cl_2

25. M Which of the following atoms can expand its valence shell when bonding?

a. N
b. C
c. O
d. P

26. H Nitrogen and hydrogen combine to form ammonia in the Haber process. Calculate (in kJ) the standard enthalpy change $\Delta H°$ for the reaction written below, using the bond energies given.

$$N_2(g) + 3H_2(g) \rightarrow 2NH_3(g)$$

Bond:	N≡N	H–H	N–H
Bond energy (kJ/mol):	945	432	391

a. -969
b. -204
c. -105
d. 204
e. 595

27.
H

Hydrogenation of double and triple bonds is an important industrial process. Calculate (in kJ) the standard enthalpy change $\Delta H°$ for the hydrogenation of ethyne (acetylene) to ethane.

$$H–C≡C–H(g) + 2H_2(g) \rightarrow H_3C–CH_3(g)$$

Bond:	C–C	C≡C	C–H	H–H
Bond energy (kJ/mol):	347	839	413	432

a. -296
b. -51
c. 51
d. 296
e. 381

28.
H

Acetone can be easily converted to isopropyl alcohol by addition of hydrogen to the carbon–oxygen double bond. Calculate the enthalpy of reaction using the bond energies given.

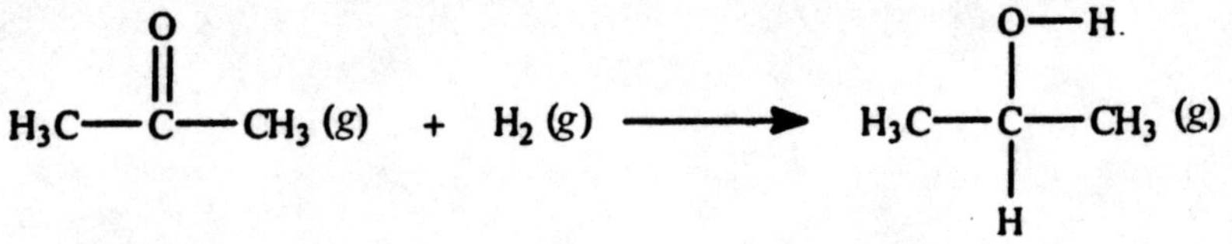

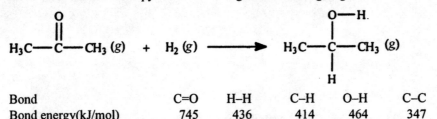

Bond	C=O	H–H	C–H	O–H	C–C	C–O
Bond energy(kJ/mol)	745	436	414	464	347	351

a. -366 kJ/mol
b. -48 kJ/mol
c. +48 kJ/mol
d. +366 kJ/mol

29.
E

According to VSEPR theory, a molecule with the general formula AX_2 will have a ___ molecular shape.

a. linear
b. bent
c. trigonal planar
d. tetrahedral

30.
E

According to VSEPR theory, a molecule with the general formula AX_3 will have a ___ molecular shape.

a. linear
b. bent
c. trigonal planar
d. tetrahedral

31.
E

According to VSEPR theory, a molecule with the general formula AX_4 will have a ___ molecular shape.

a. bent
b. trigonal planar
c. trigonal pyramidal
d. tetrahedral

32.
E

According to VSEPR theory, a molecule with the general formula AX_5 will have a ___ molecular shape.

a. tetrahedral
b. trigonal planar
c. trigonal pyramidal
d. trigonal bipyramidal

33. According to VSEPR theory, a molecule with the general formula AX_6 will have a _____ molecular shape.
E

a. tetrahedral
b. trigonal planar
c. trigonal bipyramidal
d. octahedral

34. According to VSEPR theory, a molecule with the general formula AX_2E will have a _____ molecular shape.
M

a. bent
b. see-saw
c. trigonal planar
d. T-shaped

35. According to VSEPR theory, a molecule with the general formula AX_2E_2 will have a _____ molecular shape.
M

a. linear
b. bent
c. trigonal planar
d. tetrahedral

36. According to VSEPR theory, a molecule with the general formula AX_2E_3 will have a _____ molecular shape.
M

a. bent
b. linear
c. trigonal planar
d. T-shaped

37. According to VSEPR theory, a molecule with the general formula AX_3E will have a __ molecular shape.
M

a. bent
b. trigonal planar
c. trigonal pyramidal
d. tetrahedral

38. According to VSEPR theory, a molecule with the general formula AX_3E_2 will have a _____ molecular shape.
H

a. trigonal pyramidal
b. trigonal bipyramidal
c. trigonal planar
d. T-shaped

39. According to VSEPR theory, a molecule with the general formula AX_4E will have a __ molecular shape.
H

a. bent
b. see-saw
c. trigonal planar
d. T shaped

40. According to VSEPR theory, a molecule with the general formula AX_4E_2 will have a _____ molecular shape.
M

a. tetrahedral
b. square pyramidal
c. square planar
d. octahedral

41. According to VSEPR theory, a molecule with the general formula AX_5E will have a _____ molecular shape.
M

a. tetrahedral
b. trigonal bipyramidal
c. square pyramidal
d. octahedral

42. What is the molecular shape of N_2O as predicted by the VSEPR theory?
E

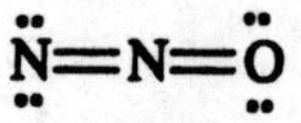

a. linear
b. bent
c. angular
d. trigonal

43. What is the molecular shape of the thiocyanate anion, SCN^-, as predicted by the VSEPR theory? (Carbon is the central atom.)
M

a. linear
b. bent
c. angular
d. trigonal

44. What is the molecular shape of ClCN as predicted by the VSEPR theory? (Carbon is the central atom.)
M

a. linear
b. bent
c. angular
d. trigonal

45. What is the molecular shape of BeH_2 as predicted by the VSEPR theory?
E

a. linear
b. bent
c. angular
d. trigonal

46. What is the molecular shape of NOCl as predicted by the VSEPR theory?
E

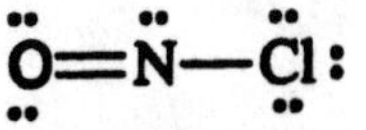

a. linear
b. trigonal planar
c. bent
d. tetrahedral

47. What is the molecular shape of BCl_3 as predicted by the VSEPR theory?
M

a. linear
b. trigonal planar
c. bent
d. tetrahedral

48. What is the molecular shape of NO_2^- as predicted by the VSEPR theory?
M

a. linear
b. trigonal planar
c. bent
d. tetrahedral

49. What is the molecular symmetry around the carbons in CCl_2CH_2 as predicted by the VSEPR theory?
M

a. linear
b. trigonal planar
c. V-shaped
d. tetrahedral

50. What is the molecular shape of ClO_3F as predicted by the VSEPR theory?
E

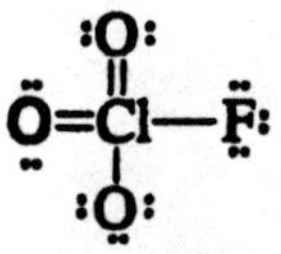

a. trigonal pyramidal
b. square planar
c. square pyramidal
d. tetrahedral

51. What is the molecular shape of HOF as predicted by the VSEPR theory?
M

a. trigonal pyramidal
b. trigonal
c. tetrahedral
d. bent

52. What is the molecular shape of NH_2Cl as predicted by the VSEPR theory?
M

a. trigonal pyramidal
b. tetrahedral
c. T-shaped
d. see-saw

53. What is the molecular shape of XeO_2F_2 as predicted by the VSEPR theory?
M

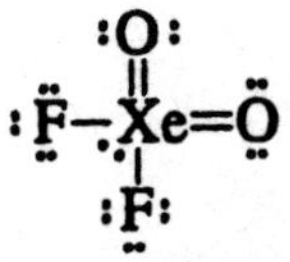

a. square planar
b. tetrahedral
c. square pyramidal
d. see-saw

54. What is the molecular shape of ClF_2^- as predicted by the VSEPR theory?
M

a. linear
b. bent
c. see-saw
d. T-shaped

55. What is the molecular shape of SCl_3F as predicted by the VSEPR theory?
M

a. linear
b. bent
c. see-saw
d. T-shaped

56. What is the molecular shape of SiF_6^{2-} as predicted by the VSEPR theory?
E

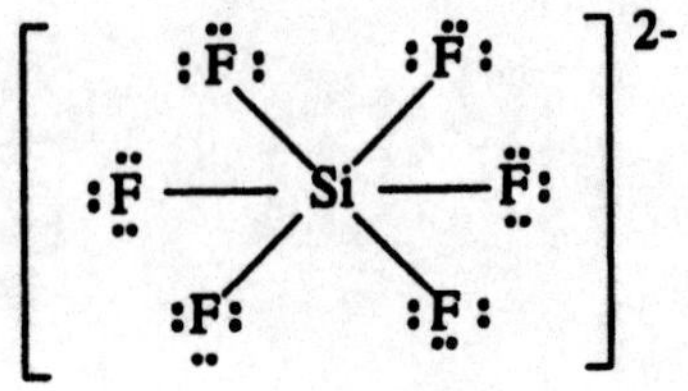

a. trigonal bipyramidal
b. hexagonal
c. tetrahedral
d. octahedral

57. What is the molecular shape of ClF_4^- as predicted by the VSEPR theory?
M

a. square pyramidal
b. square planar
c. see-saw
d. octahedral

58. Which one of the following molecules and ions will have a planar geometry?
H

a. PCl_3
b. BF_4^-
c. XeF_4
d. BrF_5
e. H_3O^+

59. Use VSEPR theory to decide which one of the following ions and molecules is likely to be planar.
H (The central atom is always first in the formula.)

a. BrF_3
b. H_3O^+
c. PCl_3
d. SO_4^{2-}
e. SF_4

60. Use VSEPR theory to decide which one of the following molecules and ions will have a trigonal
M pyramidal geometry. (The central atom is always first in the formula.)

a. PCl_3
b. BF_3
c. SO_3
d. BrF_3
e. CO_3^{2-}

61. Use VSEPR theory to predict the electron group arrangement around iodine, the central atom in the
M ion IF_2^-.

a. octahedral
b. trigonal bipyramidal
c. tetrahedral
d. trigonal planar
e. bent

62. Use VSEPR theory to decide which one of the following molecules and ions will definitely have at
M least one 90° bond angle in it. (In each case except water, the central atom is the first one in the formula.)

a. $AlCl_4^-$
b. NH_3
c. PCl_5
d. CO_2
e. H_2O

63. Predict the ideal bond angles in $GeCl_4$ using the molecular shape given by the VSEPR theory.
M

a. 90°
b. 109°
c. 120°
d. 180°

64. Predict the ideal bond angles in $AsCl_3$ using the molecular shape given by the VSEPR theory.
M

a. 90°
b. 109°
c. 120°
d. 180°

65. Predict the ideal bond angles in FNO using the molecular shape given by the VSEPR theory.
M

a. 90°
b. 109°
c. 120°
d. 180°

66. Predict the ideal bond angles around nitrogen in N_2F_2 using the molecular shape given by the VSEPR theory.
M

a. 90°
b. 109°
c. 120°
d. 180°

67. Predict the ideal bond angles around carbon in C_2I_2 using the molecular shape given by the VSEPR theory.
M

a. 90°
b. 109°
c. 120°
d. 180°

68. Predict the ideal bond angles in IF_2^- using the molecular shape given by the VSEPR theory.
M

a. 90°
b. 109°
c. 120°
d. 180°

69. Predict the actual bond angle in $SeCl_2$ using the VSEPR theory.
H

a. more than 120°
b. between 109° and 120°
c. between 90° and 109°
d. less than 90°

70. Predict the actual bond angles in BrF_3 using the VSEPR theory.
H

a. more than 120°
b. between 109° and 120°
c. between 90° and 109°
d. less than 90°

71. Predict the actual bond angles in SF_3^+ using the VSEPR theory.
H

a. more than 120°
b. between 109° and 120°
c. between 90° and 109°
d. less than 90°

72. Which of the following molecules has a net dipole moment?
M

a. $BeCl_2$
b. SF_2
c. KrF_2
d. CO_2

73. Which of the following has no net dipole moment?
M

a. N_2O
b. NF_3
c. H_2Se
d. TeO_3

74. Which one of the following molecules does not have a dipole moment?
M

a. CS_2
b. H_2S
c. CH_2Cl_2
d. PH_3
e. CH_2O

Short Answer Questions

75. Draw Lewis structures, showing all valence electrons, for:
E

a. N
b. Br^-
c. O_2
d. SO_4^{2-}

76. Draw Lewis structures, showing all valence electrons, for the following species:
M

a. S^{2-}
b. CO
c. SO_2
d. CH_3OH

77. Draw Lewis structures which obey the octet rule, for the following atoms, molecules and ions,
M showing all valence electrons. Central atoms are shown in bold.

a. NH_3
b. O_3 (Hint: O_3 is not cyclic)
c. HCN
d. SO_3

78. For the chlorate ion, ClO_3^-, draw two different valid Lewis structures, as follows:
M

a. a structure in which the octet rule is obeyed
b. a structure in which formal charges are minimized

79. Name and outline the concept which is introduced when more than one valid Lewis structure can
H be drawn for a given molecule or ion. Use appropriate diagrams of the formate ion (HCO_2^-; carbon is the central atom) to illustrate.

80. Draw all important resonance structures of the nitrate ion, NO_3^-
M

81. Using SO_2 as an example, describe the sort of experimental data which might suggest that no single Lewis
M structure is an accurate representation of its bonding.

82. List the three important ways in which molecules can violate the octet rule, and in each case draw
M one Lewis structure of your choice as an example.

83. Oxygen difluoride is an unstable molecule that reacts readily with water. Calculate the bond
M energy of the O-F bond using the standard enthalpy of reaction and the bond energy data provided.

$$OF_2(g) + H_2O(g) \rightarrow O{=}O(g) + 2HF(g) \qquad \Delta H^\circ = -318 \text{ kJ}$$

Bond:	O-H	O=O	H-F
Bond energy (kJ/mol):	467	498	565

84. Ethanol is sometimes used as an additive in oxygenated gasoline. Calculate its enthalpy of
M combustion using the bond energies given.

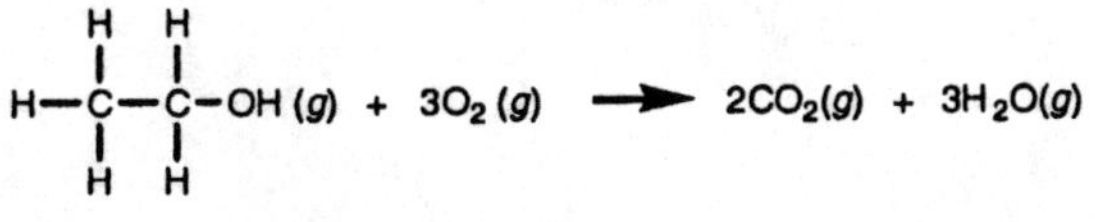

Bond:	C-C	C-H	C-O	C=O	O-H	O=O
Bond energy (kJ/mol):	347	413	358	799	467	498

85. The Lewis structure of formaldehyde, CH_2O, is shown. Use VSEPR theory to predict the molecular
M geometry and the H-C-H bond angle. Outline your reasoning.

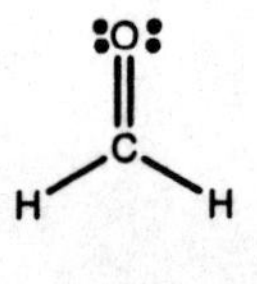

86. What is the shape of the PF_3 molecule? Explain your answer, using VSEPR theory.
M

87. Draw the Lewis structure of XeF_4. Use this structure, in conjunction with VSEPR theory, to
M predict the shape of this molecule. Outline your reasoning.

88. a. Draw and name three molecular shapes for molecules having the VSEPR formulas AX_3, AX_3E
M and AX_3E_2, respectively.
b. If the three X groups in the above formulas are identical, which of the three shapes would result in a molecule with a dipole moment?

89. Explain what is meant by "dipole moment", and give an example of a molecule which has polar
E bonds but which does not itself have a dipole moment.

True/False Questions

90. All possible resonance structures contribute equally to the resonance hybrid.
M

91. When resonance occurs, the bond lengths in a molecule fluctuate rapidly.
E

92. In a Lewis structure for a molecule or ion, the sum of the formal charges on the atoms is equal to the
E charge on the molecule or ion.

93. In formaldehyde, CH_2O, both the formal charge and the oxidation number of carbon are zero.
M

94. Boron never achieves an octet in any of its compounds.
M

95. The Lewis structure of NO_2 violates the octet rule.
M

96. In order for a non-cyclic triatomic molecule to be bent, VSEPR theory requires that there must be
M two lone pairs on the central atom.

97. According to VSEPR theory, a molecule with the general formula AX_3E_2 (where E represents a lone
M pair on A) will be trigonal planar.

98. The molecule AX_2, where A and X are different elements, will have a dipole moment if the molecule
E is bent.

99. A molecule which contains polar bonds will always have a dipole moment.
E

The Shapes of Molecules
Chapter 10
Answer Key

1.	c	26.	c	51.	d
2.	b	27.	a	52.	a
3.	e	28.	b	53.	d
4.	a	29.	a	54.	a
5.	c	30.	c	55.	c
6.	a	31.	d	56.	d
7.	b	32.	d	57.	b
8.	d	33.	d	58.	c
9.	a	34.	a	59.	a
10.	b	35.	b	60.	a
11.	d	36.	b	61.	b
12.	b	37.	c	62.	c
13.	b	38.	d	63.	b
14.	d	39.	b	64.	b
15.	c	40.	c	65.	c
16.	b	41.	c	66.	c
17.	a	42.	a	67.	d
18.	b	43.	a	68.	d
19.	c	44.	a	69.	c
20.	a	45.	a	70.	d
21.	b	46.	c	71.	c
22.	d	47.	b	72.	b
23.	e	48.	c	73.	d
24.	b	49.	b	74.	a
25.	d	50.	d		

75. a. 

b. 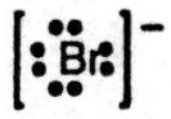

c. 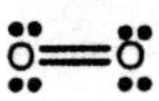

d. 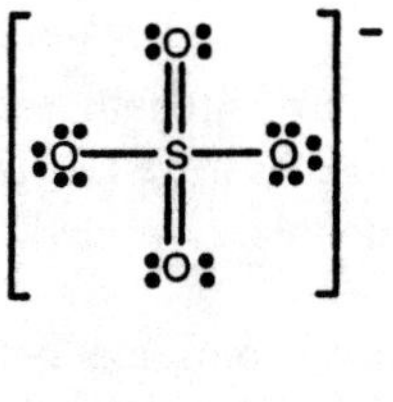

76. a. 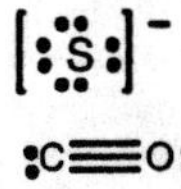

b. 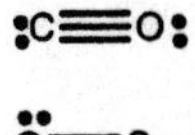

c. 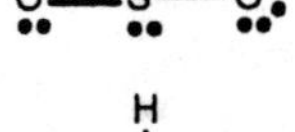

d. 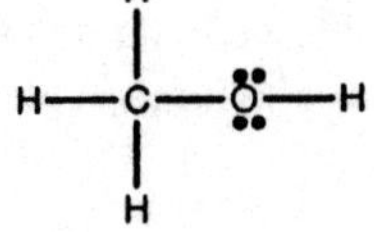

77. a. 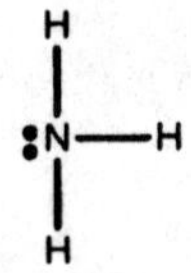

b. 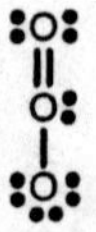

c. 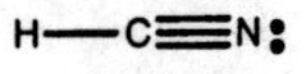

d. 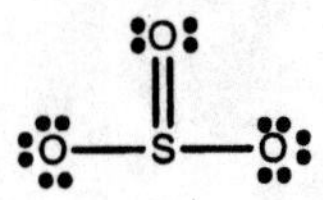

78. a. 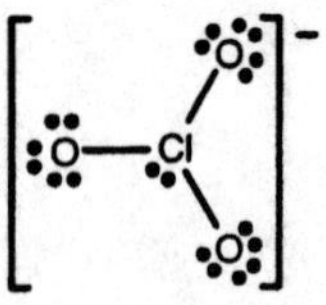 b. 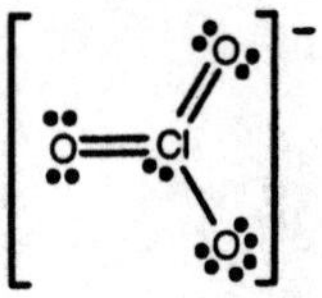

79. The concept is resonance. In this situation no single Lewis structure can adequately represent the bonding in a molecule. An average of the different Lewis structures is a better representation of the bonding than any single structure. The two important resonance structures are shown below.

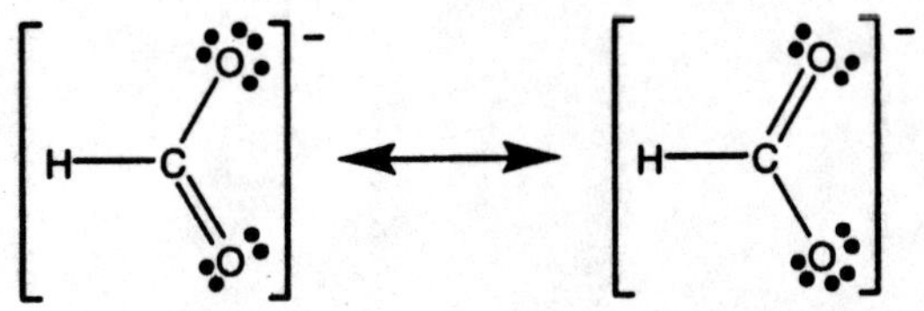

80.

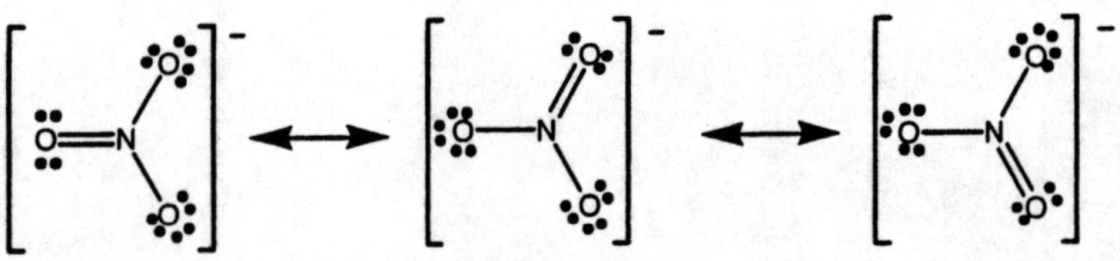

81. In the SO_2 molecule, the two sulfur-oxygen bonds would be identical in length and strength, and these values would be intermediate between those of sulfur-oxygen single and double bonds. A single Lewis structure would show two different types of bond in the molecule, one single and one double.

82. Electron-deficient molecules have fewer than 8 electrons in the valence shells of atoms, e.g. boron in BF_3. Odd-electron molecules cannot obey the octet rule. Examples are NO, NO_2 and ClO_2. Atoms from period 3 and beyond can expand their valence shells to exceed the octet count. Example: SF_6.

83. 188 kJ

84. -1267 kJ

85. There are three electron groups around the central atom, carbon. These are a double bond and two single bonds. The molecule is thus of the AX_3 type, and its geometry will be trigonal planar. The bond angles will be 120°.

86. The Lewis structure has a lone pair on the phosphorus atom, and the VSEPR formula is thus AX_3E. There are four electron groups, giving a tetrahedral electron group arrangement. The molecular shape is trigonal pyramidal.

87. The Lewis structure is shown alongside. The VSEPR formula is AX_4E_2, and the electron group arrangement is therefore octahedral. The lone pairs will lie at opposite vertices, resulting in a square planar molecular geometry

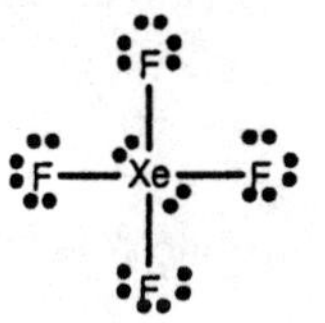

88. (a) The three structures and their molecular shapes are shown below.

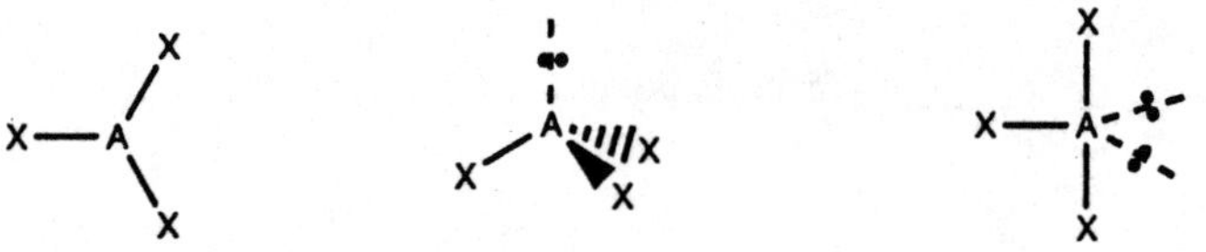

trigonal planar   trigonal pyramidal   T-shaped

(b) The trigonal pyramidal and T-shaped molecules will have dipole moments.

89. A dipole moment arises in a molecule when the "centers of gravity" of the positive and negative charges do not coincide. There is thus a separation of charge. The dipole moment is the product of this charge and the distance of separation. Carbon dioxide has two polar carbon-oxygen bonds. However, because the molecule is linear, the two bond dipoles are exactly opposite in direction, and they cancel each other out. The CO_2 molecule has no dipole moment.

90. F

91. F

92. T

93. T

94. F

95. T

96. F

97. F

98. T

99. F

Theories of Covalent Bonding
Chapter 11

Multiple Choice Questions

1. A molecule with the formula AX_2 uses ________ to form its bonds.
E

a. *sp* hybrid orbitals
b. sp^2 hybrid orbitals
c. sp^3 hybrid orbitals
d. sp^3d hybrid orbitals

2. A molecule with the formula AX_3 uses ________ to form its bonds.
E

a. *sp* hybrid orbitals
b. sp^2 hybrid orbitals
c. sp^3 hybrid orbitals
d. sp^3d hybrid orbitals

3. A molecule with the formula AX_4 uses ________ to form its bonds.
E

a. *sp* hybrid orbitals
b. sp^2 hybrid orbitals
c. sp^3 hybrid orbitals
d. sp^3d hybrid orbitals

4. A molecule with the formula AX_3E uses ________ to form its bonds.
M

a. *s* and *p* atomic orbitals
b. sp^3 hybrid orbitals
c. sp^2 hybrid orbitals
d. *sp* hybrid orbitals

5. A molecule with the formula AX_4E uses ________ to form its bonds.
M

a. sp^2 hybrid orbitals
b. sp^3 hybrid orbitals
c. sp^3d hybrid orbitals
d. sp^3d^2 hybrid orbitals

6. A molecule with the formula AX_4E_2 uses ________ to form its bonds.
M

a. sp^2 hybrid orbitals
b. sp^3 hybrid orbitals
c. sp^3d hybrid orbitals
d. sp^3d^2 hybrid orbitals

7. Carbon uses ______ hybrid orbitals in ClCN.
M

a. *sp*
b. sp^2
c. sp^3
d. sp^3d

8. Valence bond theory predicts that carbon will use _____ hybrid orbitals in the carbonate anion, CO_3^{2-}.
M

a. *sp*
b. sp^2
c. sp^3
d. sp^3d

9. Valence bond theory predicts that sulfur will use _____ hybrid orbitals in sulfur dioxide, SO_2.
M

a. *sp*
b. sp^2
c. sp^3
d. sp^3d

10. When PCl_5 solidifies it forms PCl_4^+ cations and PCl_6^- anions. According to valence bond theory, what hybrid orbitals are used by phosphorus in the PCl_4^+ cations?
M

a. *sp*
b. sp^2
c. sp^3
d. sp^3d

11. Valence bond theory predicts that tin will use _____ hybrid orbitals in $SnCl_3^-$.
M

a. *sp*
b. sp^2
c. sp^3
d. sp^3d

12. Valence bond theory predicts that tin will use _____ hybrid orbitals in SnF_5^-.
M

a. sp^2
b. sp^3
c. sp^3d
d. sp^3d^2

13. Valence bond theory predicts that iodine will use _____ hybrid orbitals in ICl_2^-.
M

a. sp^2
b. sp^3
c. sp^3d
d. sp^3d^2

14. Valence bond theory predicts that bromine will use _____ hybrid orbitals in BrF_5.
M

a. sp^2
b. sp^3
c. sp^3d
d. sp^3d^2

15. M Valence bond theory predicts that xenon will use _____ hybrid orbitals in $XeOF_4$.

a. sp^2
b. sp^3
c. sp^3d
d. sp^3d^2

16. M Which one of the following statements about orbital hybridization is incorrect?

a. The carbon atom in CH_4 is sp^3 hybridized.
b. The carbon atom in CO_2 is sp hybridized.
c. The nitrogen atom in NH_3 is sp^2 hybridized.
d. sp^2 hybrid orbitals are coplanar, and at 120° to each other.
e. sp hybrid orbitals lie at 180° to each other.

17. M For which one of the following molecules is the indicated type of hybridization not appropriate for the central atom?

a. $BeCl_2$ sp^2
b. SiH_4 sp^3
c. BF_3 sp^2
d. C_2H_2 sp
e. H_2O sp^3

18. M According to valence bond theory, which of the following molecules involves sp^2 hybridization of orbitals on the central atom (underlined and bold)?

a. $\underline{\mathbf{C}}_2H_2$
b. $\underline{\mathbf{C}}_2H_4$
c. $\underline{\mathbf{C}}_2H_6$
d. $\underline{\mathbf{C}}O_2$
e. $H_2\underline{\mathbf{O}}$

19. M Determine the shape (geometry) of PCl_3 and then decide on the appropriate hybridization of phosphorus in this molecule. (Phosphorus is the central atom.)

a. sp^3
b. sp^2
c. sp
d. sp^3d
e. sp^3d^2

20. E According to valence bond theory, the triple bond in ethyne (acetylene, C_2H_2) consists of

a. three σ bonds and no π bonds
b. two σ bonds and one π bond
c. one σ bond and two π bonds
d. no σ bonds and three π bonds
e. none of the above

21. According to molecular orbital (MO) theory, the twelve outermost electrons in the O_2 molecule are distributed as follows:
H

a. 12 in bonding MOs, 0 in antibonding MOs.
b. 10 in bonding MOs, 2 in antibonding MOs.
c. 9 in bonding MOs, 3 in antibonding MOs.
d. 8 in bonding MOs, 4 in antibonding MOs.
e. 7 in bonding MOs, 5 in antibonding MOs.

22. According to molecular orbital theory, what is the bond order in the O_2^+ ion?
H

a. 5.5
b. 5
c. 4
d. 2.5
e. 1.5

23. Which of the following statements relating to molecular orbital (MO) theory is incorrect?
M

a. Combination of two atomic orbitals produces one bonding and one antibonding MO.
b. A bonding MO is lower in energy than the two atomic orbitals from which it is formed.
c. Combination of two 2*p* orbitals may result in either σ or π MOs.
d. A species with a bond order of zero will not be stable.
e. In a stable molecule having an even number of electrons, all electrons must be paired.

24. One can safely assume that the 3*s*- and 3*p*-orbitals will form molecular orbitals similar to those formed when 2*s*- and 2*p*-orbitals interact. According to molecular orbital theory, what will be the bond order for the Cl_2^+ ion?
H

a. 0.5
b. 1
c. 1.5
d. 2

25. The nitrosonium ion, NO^+, forms a number of interesting complexes with nickel, cobalt, and iron. According to molecular orbital theory, which of the following statements about NO^+ is correct?
H

a. NO^+ has a bond order of 2 and is paramagnetic.
b. NO^+ has a bond order of 2 and is diamagnetic.
c. NO^+ has a bond order of 3 and is paramagnetic.
d. NO^+ has a bond order of 3 and is diamagnetic.

Short Answer Questions

26. In one sentence state the basic principle of valence bond theory.
M

27. In not more than two sentences, explain when and why chemists make use of the concept of hybridization.
M

28. What type of hybridization is needed to explain why ethyne, C_2H_2, is linear?
E

29. Describe the bonding in ethylene (ethene, C_2H_4) according to valence bond theory. Be sure to
H indicate the orbital hybridization on the carbon atoms. Draw a diagram clearly showing at least one bond of each type which occurs in the molecule.

30. a. In the context of valence bond theory, explain the difference in geometry between a σ and a π
H bond. Use a real molecule to illustrate your answer.
b. What two important differences are there in the properties of σ and π bonds, in terms of how they affect the structure and reactivity of molecules?

31. In one sentence state how molecular orbitals are usually obtained.
M

32. Sketch the shapes of the σ_{1s} and σ_{1s}* molecular orbitals formed by the overlap of two hydrogen 1*s*
M atomic orbitals.

33. Explain what is meant by a node (or nodal plane) in a molecular orbital and draw sketches of the
M following orbitals, indicating at least one nodal plane in each one.
a. a σ* orbital
b. a π* orbital

34. In the context of molecular orbital (MO) theory, explain how atomic *p* orbitals can give rise to:
M a. a σ MO
b. a π MO

35. Explain what is meant by the term "bond order" and describe how it can be calculated using the
M information in a molecular orbital energy level diagram.

36. a. What simple experiment could you perform to show that a substance is paramagnetic?
M b. What microscopic (atomic/molecular) feature must a substance possess in order to be paramagnetic?
c. Can it be predicted whether or not all homonuclear diatomic ions, X_2^+, will be paramagnetic? Explain.

True/False Questions

37. According to valence bond theory, overlap of bonding orbitals of atoms will weaken a bond, due
H to electron-electron repulsion.

38. Valence bond theory explains the bonding in diatomic molecules such as HCl without resorting to
E the use of hybrid orbitals.

39. Hybrid orbitals of the sp^3d type occur in sets of four.
E

40. The angles between sp^2 hybrid orbitals are 109.5°.
E

41. Atoms of period 3 and beyond can undergo sp^3d^2 hybridization, but atoms of period 2 cannot.
M

42. In the valence bond treatment, overlap of an *s* orbital on one atom with an sp^3 orbital on another
E atom can give rise to a σ bond.

43. Overlap of two sp^2 hybrid orbitals produces a π bond.
M

44. In the valence bond treatment, a π bond is formed when two $2p$ orbitals overlap side to side.
M

45. A carbon-carbon double bond in a molecule may give rise to the existence of *cis* and *trans* isomers.
M

46. A carbon-carbon triple bond in a molecule may give rise to the existence of *cis* and *trans* isomers.
M

47. In molecular orbital theory, combination of two atomic orbitals produces two molecular orbitals.
E

48. In applying molecular orbital theory, the bond order is calculated in the same way as with Lewis structures.
M

49. In molecular orbital theory, molecules with an even number of electrons will have bond orders which are whole numbers.
M

50. In molecular orbital theory, combination of two $2p$ atomic orbitals may give rise to either σ or π type molecular orbitals.
M

51. According to molecular orbital theory, all diatomic molecules with an even number of electrons will be diamagnetic.
M

Theories of Covalent Bonding
Chapter 11
Answer Key

1.	a	10.	c	19.	a
2.	b	11.	c	20.	c
3.	c	12.	c	21.	d
4.	b	13.	c	22.	d
5.	c	14.	d	23.	e
6.	d	15.	d	24.	c
7.	a	16.	c	25.	d
8.	b	17.	a		
9.	b	18.	b		

26. A covalent bond forms when the orbitals of two atoms overlap and a pair of electrons occupies the region between the nuclei.

27. Chemists postulate hybridization when the observed geometry of a molecule cannot be rationalized in terms of overlap of the *s*, *p* and/or *d* orbitals of the atoms concerned.

28. *sp*

29. The carbon atoms are sp^2 hybridized, leaving the $2p_z$ orbital on each carbon unhybridized. Overlap of one sp^2 hybrid on each carbon and occupation of this region by an electron pair produces a C-C σ bond. Overlap of the remaining four hybrid orbitals with 1*s* orbitals of four hydrogen atoms, produces four C-H σ bonds. The unhybridized $2p_z$ orbitals overlap laterally to produce a π bond which has two regions of electron density, above and below the C-C σ bond. (Diagram needed.)

30. a. A σ bond has its region of highest electron density between the bonded atoms, along the axis joining them. A π bond has two regions of overlap, between the atoms but above and below the axis joining them.
b. A σ bond is stronger than a π bond. There is free rotation about a σ bond but not about a π bond, giving rise to the possibility of geometrical isomers in the latter case.

31. By the linear combination (addition or subtraction) of atomic orbitals.

32.

σ_{1s} σ_{1s}^*

33. A nodal plane is a plane of zero electron density. Nodal planes are the dashed lines shown below.

a. b.

34. a. A σ MO arises when atomic *p* orbitals which are directed towards each other, are combined.
b. A π MO arises from the combination of atomic *p* orbitals which lie parallel to each other, perpendicular to the internuclear axis.

35. The bond order in MO theory is the net number of bonds, where a pair of electrons in a bonding MO constitutes a bond and a pair in an antibonding MO constitutes an antibond. It is equal to the total number of electrons in bonding MOs minus the total number in antibonding MOs, divided by two.

36. a. Place a sample of the substance near a magnet; if it is attracted to the magnet, it is paramagnetic.
b. The substance must have unpaired electrons in order to be paramagnetic.
c. Yes, they will necessarily be paramagnetic since each will have an odd number of electrons.

37. F

38. T

39. F

40. F

41. T

42. T

43. F

44. T

45. T

46. F

47. T

48. F

49. T

50. T

51. F

Intermolecular Forces: Liquids, Solids, and Phase Changes
Chapter 12

Multiple Choice Questions

1.
M
Pentane, C_5H_{12}, boils at 35°C. Which of the following is true about kinetic energy, E_k, and potential energy, E_p, when liquid pentane at 35°C is compared with pentane vapor at 35°C?

a. $E_k(g) < E_k(l)$; $E_p(g) \approx E_p(l)$
b. $E_k(g) > E_k(l)$; $E_p(g) \approx E_p(l)$
c. $E_p(g) < E_p(l)$; $E_k(g) \approx E_k(l)$
d. $E_p(g) > E_p(l)$; $E_k(g) \approx E_k(l)$

2.
M
Which of the following is true about kinetic energy, E_k, and potential energy, E_p, when ethyl alcohol at 40°C is compared with ethyl alcohol at 20°C?

a. $E_k(40°C) < E_k(20°C)$; $E_p(40°C) \approx E_p(20°C)$
b. $E_k(40°C) > E_k(20°C)$; $E_p(40°C) \approx E_p(20°C)$
c. $E_p(40°C) < E_p(20°C)$; $E_k(40°C) \approx E_k(20°C)$
d. $E_p(40°C) > E_p(20°C)$; $E_k(40°C) \approx E_k(20°C)$

3.
M
A sample of octane in equilibrium with its vapor in a closed 1 .0-L container has a vapor pressure of 50.0 torr at 45°C. The container's volume is increased to 2.0 L at constant temperature and the liquid/vapor equilibrium is reestablished. What is the vapor pressure?

a. > 50.0 torr
b. 50.0 torr
c. < 50.0 torr
d. the mass of the octane vapor is needed to calculate the vapor pressure

4.
E
Which one of the following quantities is generally not obtainable from a single heating or cooling curve of a substance, measured at atmospheric pressure?

a. melting point
b. boiling point
c. triple point
d. heat of fusion
e. heat of vaporization

5.
M
The phase diagram for xenon has a solid-liquid curve with at positive slope. Which of the following is true?

a. solid xenon has a higher density than liquid xenon
b. solid xenon has a lower density than liquid xenon
c. solid xenon has the same density as liquid xenon
d. freezing xenon is an endothermic process

6. M Liquid ammonia (boiling point = -33.4°C) can be used as a refrigerant and heat transfer fluid. How much energy is needed to heat 25.0 g of $NH_3(l)$ from -65.0°C to -12.0°C?

Specific heat capacity, $NH_3(l)$	4.7 J/(g·K)
Specific heat capacity, $NH_3(g)$	2.2 J/(g·K)
Heat of vaporization	23.5 kJ/mol
Molar mass, $\mathcal{M}$	17.0 g/mol

a. 5.5 kJ
b. 39 kJ
c. 340 kJ
d. 590 kJ

7. M Diethyl ether, used as a solvent for extraction of organic compounds from aqueous solutions, has a high vapor pressure which makes it a potential fire hazard in laboratories in which it is used. How much energy is released when 100.0 g is cooled from 53.0°C to 10.0°C?

boiling point	34.5°C
heat of vaporization	351 J/g
specific heat capacity, $(CH_3)_2O(l)$	3.74 J/(g·K)
specific heat capacity, $(CH_3)_2O(g)$	2.35 J/(g·K)

a. 10.1 kJ
b. 13.1 kJ
c. 16.1 kJ
d. 48.6 kJ

8. H Octane has a vapor pressure of 40. torr at 45.1°C and 400. torr at 104.0°C. What is its heat of vaporization?

a. 39.0 kJ/mol
b. 46.0 kJ/mol
c. 590 kJ/mol
d. 710 kJ/mol

9. H Liquid sodium can be used as a heat transfer fluid. Its vapor pressure is 40.0 torr at 633°C and 400.0 torr at 823°C. Calculate its heat of vaporization.

a. 43.4 kJ/mol
b. 52.5 kJ/mol
c. 70.6 kJ/mol
d. 1.00×10^2 kJ/mol

10. Examine the phase diagram for the substance Bogusium (Bo) and select the correct statement.
E

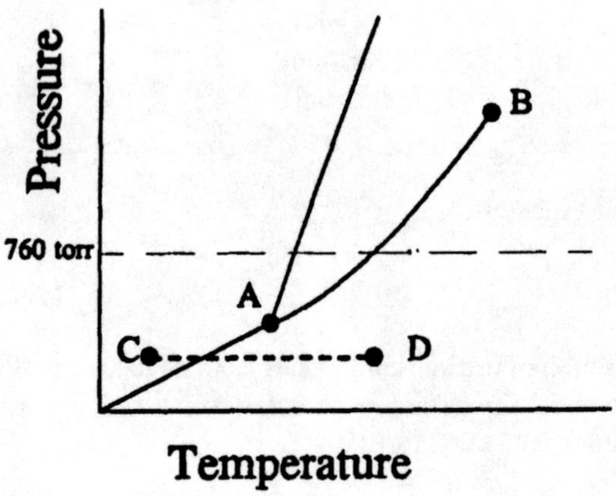

a. Bo(s) has a lower density than Bo(l)
b. The triple point for Bo is at a higher temperature than the melting point for Bo
c. Bo changes from a solid to a liquid as one follows the line from C to D
d. Point B represents the critical temperature and pressure for Bo

11. Examine the following phase diagram and identify the feature represented by point A.
E

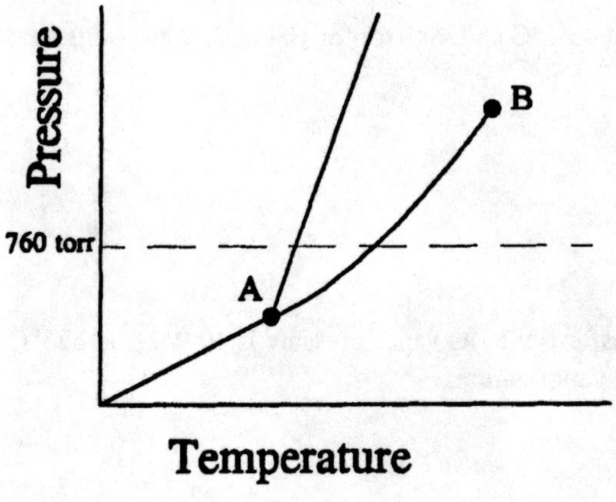

a. melting point
b. critical point
c. triple point
d. sublimation point

12. Examine the following phase diagram and identify the feature represented by point B.
E

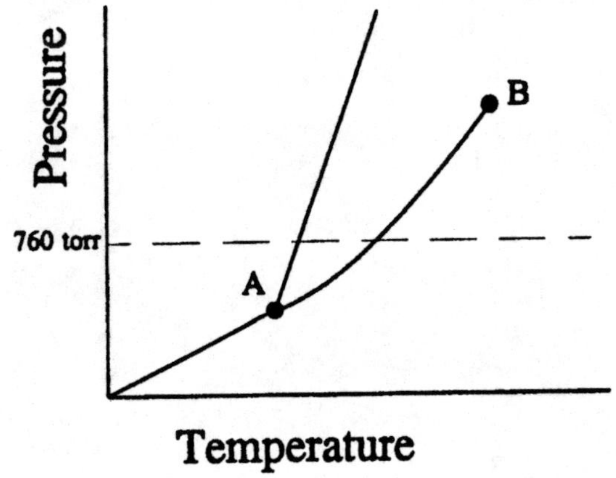

a. melting point
b. triple point
c. critical point
d. sublimation point

13. Consider the following phase diagram and identify the process occurring as one goes from point C to point D.
M

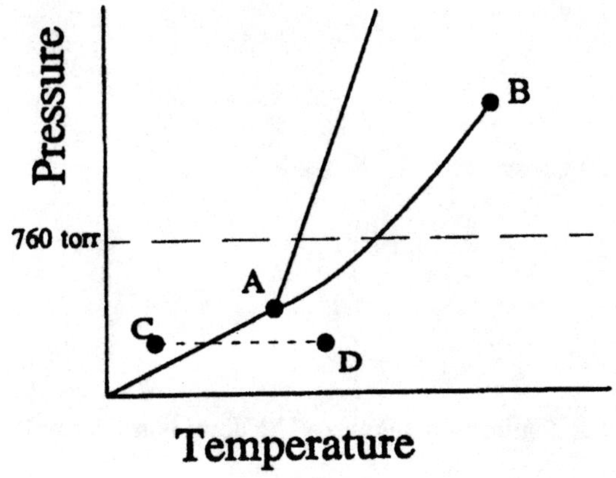

a. increasing temperature with a phase change from solid to liquid
b. increasing temperature with a phase change from solid to vapor
c. increasing temperature with a phase change from liquid to vapor
d. increasing temperature with no phase change

14. Examine the following phase diagram and determine what phase exists at point F.
E

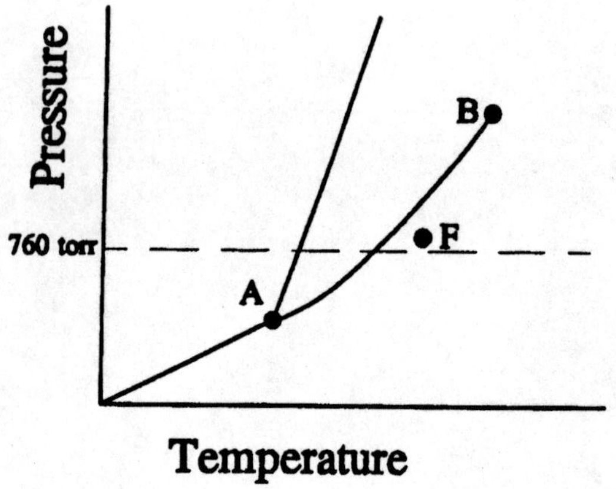

a. vapor + liquid
b. vapor
c. liquid
d. solid

15. Neon condenses due to
E

a. dipole-dipole forces
b. London dispersion forces
c. hydrogen bonding
d. covalent bonding

16. Ammonia's unusually high melting point is the result of
E

a. dipole-dipole forces
b. London dispersion forces
c. hydrogen bonding
d. covalent bonding

17. Octane is a component of fuel used in internal combustion engines. The dominant intermolecular forces in octane are
E

a. dipole-dipole forces
b. London dispersion forces
c. hydrogen bonding
d. covalent bonding

18. In hydrogen iodide ________________ are the most important intermolecular forces.
E

a. dipole-dipole forces
b. London dispersion forces
c. hydrogen bonding
d. covalent bonding

19. When the electron cloud of a molecule is easily distorted, the molecule has a high ____________.
E

a. polarity
b. polarizability
c. dipole moment
d. van der Waals radius

20. Which of the following atoms should have the greatest polarizability?
M

a. F
b. Br
c. Po
d. Pb

21. Which of the following atoms should have the smallest polarizability?
M

a. Si
b. S
c. Te
d. Bi

22. The strongest intermolecular interactions between pentane (C_5H_{12}) molecules arise from
E

a. dipole-dipole forces
b. London dispersion forces
c. hydrogen bonding
d. ion-dipole interactions

23. The strongest intermolecular interactions between ethyl alcohol (CH_3CH_2OH) molecules arise from
E

a. dipole-dipole forces
b. London dispersion forces
c. hydrogen bonding
d. ion-dipole interactions

24. The strongest intermolecular interactions between hydrogen sulfide (H_2S) molecules arise from
M

a. dipole-dipole forces
b. London dispersion forces
c. hydrogen bonding
d. ion-dipole interactions

25. The strongest intermolecular interactions between hydrogen fluoride (HF) molecules arise from
E

a. dipole-dipole forces
b. London dispersion forces
c. hydrogen bonding
d. ion-dipole interactions

26. Which of the following will form hydrogen bonds between molecules?
E

a. $(CH_3)_3N$
b. $CH_3—O—CH_3$
c. $CH_3CH_2—OH$
d. $CH_3CH_2—F$

27. Which of the following will not form hydrogen bonds?
E

a. 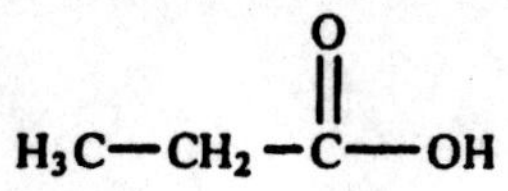

b. 

c. 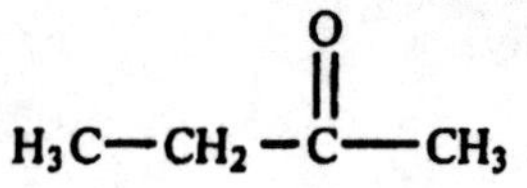

d. 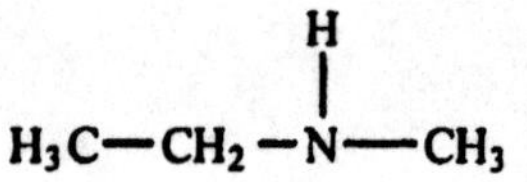

28. Which of the following pairs is arranged with the particle of higher polarizability listed first?
M

a. Se^{2-}, S^{2-}
b. I, I^-
c. Mg^{2+}, Mg
d. Br, I

29. Which of the following pairs is arranged with the particle of higher polarizability listed first?
M

a. CCl_4, CI_4
b. H_2O, H_2Se
c. C_6H_{14}, C_4H_{10}
d. NH_3, NF_3

30. Which of the following should have the highest boiling point?
E

a. CF_4
b. CCl_4
c. CBr_4
d. CI_4

31. Which of the following should have the lowest boiling point?
E

a. C_5H_{12}
b. C_6H_{14}
c. C_8H_{18}
d. $C_{10}H_{22}$

32. Which of the following has a boiling point which does not fit the general trend?
M

a. NH_3
b. PH_3
c. AsH_3
d. SbH_3

33. Select the pair of compounds in which the substance with the higher vapor pressure at a given
M temperature is listed first.

a. C_7H_{16}, C_5H_{12}
b. CCl_4, CBr_4
c. H_2O, H_2S
d. CH_3CH_2OH, $CH_3—O—CH_3$

34. Select the pair of compounds in which the substance with the lower vapor pressure at a given temperature is
M listed first.

a. $H_3C-\overset{\overset{\displaystyle O}{\|}}{C}-OH$, $CH_3CH_2CH_2OH$

b. PH_3 , NH_3

c. CF_4 , CBr_4

d. C_3H_8 , C_4H_{10}

35. Comparing the energies of the following intermolecular forces on a kJ/mol basis, which would
E normally have the highest energy (i.e. be the strongest force)?

a. ion-induced dipole
b. dipole-induced dipole
c. ion-dipole
d. dipole-dipole
e. dispersion

36. Which of the following should have the highest surface tension at a given temperature?
E

a. CF_4
b. CCl_4
c. CBr_4
d. CI_4

37. Which of the following should have the highest surface tension at a given temperature?
M

a. $H_3C-CH_2-CH_2-OH$

b. $H_3C-CH_2-O-CH_3$

c. $H_3C-CH_2-C(=O)-H$

d. $H_3C-CH_2-C(=O)-OH$

38. The energy needed to increase the surface area of a liquid by one square meter is
M

a. capillary action
b. surface tension
c. viscosity
d. cohesion

39. When the adhesive forces between a liquid and the walls of a capillary tube are greater than the cohesive forces within the liquid
E

a. the liquid level in a capillary tube will rise above the surrounding liquid and the surface in the capillary tube will have a convex meniscus
b. the liquid level in a capillary tube will rise above the surrounding liquid and the surface in the capillary tube will have a concave meniscus
c. the liquid level in a capillary tube will drop below the surrounding liquid and the surface in the capillary tube will have a convex meniscus
d. the liquid level in a capillary tube will drop below the surrounding liquid and the surface in the capillary tube will have a concave meniscus

40. The resistance of a liquid to flow is
E

a. surface tension
b. capillary action
c. viscosity
d. adhesion

41. Which of the following factors contributes to a low viscosity for a liquid?
E

a. low temperature
b. spherical molecular shape
c. hydrogen bonding
d. high molecular weight

42. Which of the following pairs of molecules is arranged so that the one with higher viscosity is listed first?
M

a. $CH_3CH_2CH_2CH_2CH_2CH_3$, $CH_3CH_2CH_2CH_2CH_2CH_2CH_2CH_3$

b. $H_3C{-}CH_2{\cdot}CH_2{-}OH$, $H_3C{-}CH_2{-}O{-}CH_3$

c. 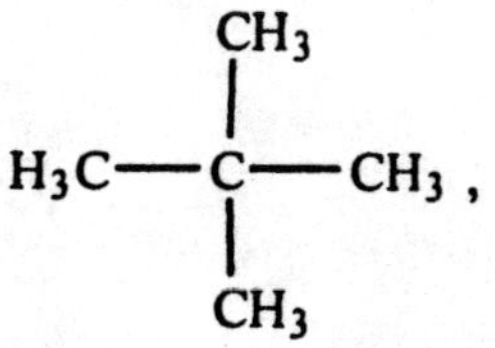 , $CH_3CH_2CH_2CH_2CH_3$

d. $H_3C{-}CH_2{\cdot}CH_2{-}OH$, $HO{-}CH_2{\cdot}CH_2{-}OH$

43. Which of the following liquid substances would you expect to have the lowest surface tension?
E

a. Pb
b. CH_3OCH_3
c. $HOCH_2CH_2OH$
d. H_2O
e. CH_3CH_2OH

44. Which of the following liquids is likely to have the highest surface tension?
E

a. Br_2
b. C_8H_{18}
c. CH_3OCH_3
d. Pb
e. CH_3OH

45. A metal in the simple cubic lattice will have _______________ atom(s) per unit cell.
M

a. 1
b. 2
c. 3
d. 4

46. A metal such as chromium in the body-centered cubic lattice will have _______________ atom(s) per unit cell.
M

a. 1
b. 2
c. 3
d. 4

47. A metal such as chromium in the face-centered cubic lattice will have _______________ atom(s) per unit cell.
M

a. 1
b. 2
c. 3
d. 4

48. Polonium crystallizes in the simple cubic lattice. What is the coordination number for Po?
E

a. 4
b. 6
c. 8
d. 12

49. Iron crystallizes in the body-centered cubic lattice. What is the coordination number for Fe?
E

a. 4
b. 6
c. 8
d. 12

50. Lead crystallizes in the face-centered cubic lattice. What is the coordination number for Pb?
M

a. 4
b. 6
c. 8
d. 12

51. Which one of the following statements about unit cells and packing in solids is incorrect?
H

a. In unit cells of solid crystals, every face of the cell must have an opposite face which is equal and parallel to it.
b. The faces of unit cells must meet at angles of 90°.
c. The coordination number of atoms in a close packed metal is 12.
d. The packing efficiency in fcc structures is higher than in bcc structures.
e. The packing efficiency in fcc and hcp structures is the same.

52. Which of the following statements concerning a face-centered cubic unit cell and the corresponding
M lattice, made up of identical atoms, is incorrect?

a. The coordination number of the atoms in the lattice is 8
b. The packing in this lattice is more efficient than for a body-centered cubic system
c. If the atoms have radius r, then the length of the cube edge is $\sqrt{8} \times r$
d. There are four atoms per unit cell in this type of packing
e. The packing efficiency in this lattice and hexagonal close packing are the same

53. Which of the following statements about the packing of monatomic solids with different unit cells
M is incorrect?

a. The coordination number of atoms in hcp and fcc structures is 12
b. The coordination number of atoms in simple cubic structures is 6
c. The coordination number of atoms in bcc structures is 8
d. A bcc structure has a higher packing efficiency than a simple cubic structure
e. A bcc structure has a higher packing efficiency than a fcc structure

54. A cubic unit cell has an edge length of 400. pm. The length of its body diagonal (internal diagonal) in pm is therefore
H

a. 512
b. 566
c. 631
d. 693
e. 724

55. Which one of the following substances does not exist in the indicated solid type?
E

a. graphite - network
b. Na - metallic
c. SiO_2 - molecular
d. NaCl - ionic
e. diamond - network

56. When liquid bromine is cooled to form a solid, which of the following types of solid would it form?
E

a. atomic
b. metallic
c. molecular
d. ionic
e. covalent network

57. For the solid forms of the following elements, which one is most likely to be of the molecular type?
E

a. Xe
b. C
c. Pb
d. S
e. Cr

58. The coordination number of sodium and chloride ions in the NaCl lattice, are, respectively:
E

a. 10 and 10
b. 8 and 8
c. 6 and 6
d. 4 and 4
e. none of the above

59. In an ionic solid MX consisting of the monatomic ions, M^+ and X^-, the coordination number of M^+ is:
H

a. 1
b. 2
c. 6
d. 8
e. impossible to predict without knowing the crystal structure of MX

60. A temperature increase causes _______________ in the conductivity of a semiconductor.
M

a. a decrease
b. an increase
c. a modulation
d. no change

61. A temperature increase causes _______________ in the conductivity of a conductor.
M

a. a decrease
b. an increase
c. a modulation
d. no change

62. The energy gap between the conduction band and the valence band is large for
M

a. conductors
b. semiconductors
c. superconductors
d. insulators

63. The highest temperature at which superconductivity has been achieved is approximately
M

a. 4 K
b. 30 K
c. 70 K
d. 100 K
e. 130 K

64. When silicon is doped with an element from group 3A(13), the device/material produced is a/an
M

a. intrinsic semiconductor
b. p-type semiconductor
c. n-type semiconductor
d. p-n junction
e. transistor

65. What word best describes the type of liquid crystal
M represented alongside?

a. nematic
b. cholesteric
c. smectic
d. isotropic
e. elastic

66. Which of the following statements about ceramics is incorrect?
M

a. Silicon carbide has a diamond-like structure.
b. Boron nitride can exist in both diamond-like and graphite-like forms.
c. Silicon carbide can be prepared by direct reaction of silicon and carbon
d. Superconducting ceramics present manufacturing difficulties owing to their brittleness.
e. Superconducting ceramic compounds usually incorporate cobalt in a key role.

Short Answer Questions

67. E Mercury melts at -39°C and boils at 357°C. Draw a diagram of the heating curve of mercury. Label all lines and axes, and clearly indicate the melting and boiling points on your diagram.

68. E Consider the phase diagram shown alongside.

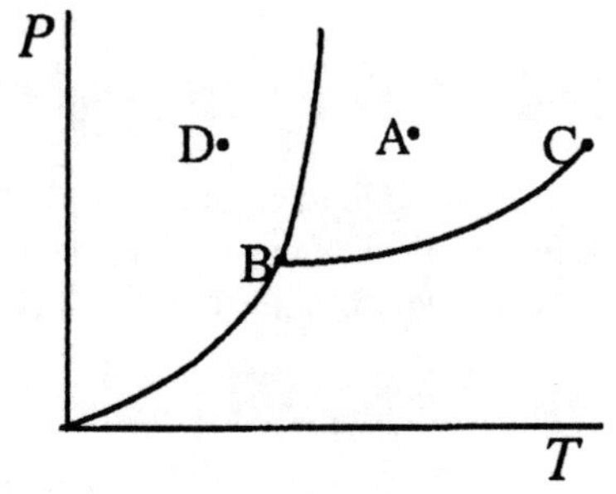

a. What phase(s) is/are present at point A?
b. What phase(s) is/are present at point B?
c. Name point C and explain its significance.
d. Starting at D, if the pressure is lowered while the temperature remains constant, describe what will happen.

69. E Draw a fully labeled phase diagram (*P* versus *T*) of a substance whose solid phase can melt due to applied pressure (i.e., solid is less dense than liquid). Clearly label the triple point and the critical temperature on your diagram.

70. M Liquid ammonia boils at -33.4°C and has a heat of vaporization of 23.5 kJ/mol. Calculate its vapor pressure at -50.0°C.

71. M Chlorine trifluoride is used in processing nuclear reactor fuel. It has a vapor pressure of 29.1 torr at -47.0°C and its heat of vaporization is 30.61 kJ/mol. At what temperature would its vapor pressure be 107.7 torr?

72. M The vapor pressure of 1-butene is 1.268 atm at 273.15 K and its heat of vaporization is 22.9 kJ/mol. What is the normal boiling point of 1-butene?

73. E
a. State the essential requirements for hydrogen bonding to be important in a compound.
b. List four properties of water which are significantly influenced by the presence of hydrogen bonding.

74. H
a. Explain what is meant by the term "unit cell".
b. Copper metal has a face-centered cubic unit cell. The edge length of the unit cell is 361 pm, and the atomic weight of copper is 63.55 amu. Calculate the density of the copper in g/cm^3. (1 amu = 1.661×10^{-24} g.)
c. From the data in (b), calculate the radius of a copper atom in pm.

75. H Strontium metal crystallizes in a cubic unit cell which has an edge length of 612 pm. If the mass of an atom of Sr is 87.62 amu, and the density of Sr metal is 2.54 g/cm^3, calculate the number of atoms per unit cell. (1 amu = 1.661×10^{-24} g.)

76. H Iron has a body-centered cubic unit cell, and a density of 7.87 g/cm^3. Calculate the edge length of the unit cell, in pm. (The atomic mass of iron is 55.85 amu. Also, 1 amu = 1.661×10^{-24} g.)

77. M Assuming that atoms are spherical, calculate the fraction of space which is occupied by atoms (i.e. the packing efficiency) in a metal with a simple cubic unit cell.

78. H Assuming that atoms are spherical, calculate the fraction of space which is occupied by atoms (i.e. the packing efficiency) in a metal with a face-centered cubic unit cell.

79. E
a. Name the two unit cells which occur in close packing of identical atoms.
b. Briefly explain how the two types of close-packed lattices of identical atoms differ, in terms of atomic arrangements.

80. Of the five major types of crystalline solid, which would you expect each of the following to form?
M (e.g. H_2O : molecular)

a. Sn
b. Si
c. KCl
d. Xe
e. F_2

81. The density of solid sodium chloride, NaCl, is 2.17 g/cm^3. Use your knowledge of the sodium
H chloride lattice to calculate the spacing between Na^+ and Cl^- nearest neighbors, in cm. (Atomic masses (amu) are: Na, 22.99; Cl, 35.45. Also, 1 amu = 1.661×10^{-24} g)

82. How do the electrical properties of semiconductors differ from those of metals?
M

83. Use molecular orbital band diagrams to explain why metals are good conductors but semiconductors
M are not.

84. Germanium is a semiconductor. With the aid of diagrams showing bands of molecular orbital,
M explain why it is a poor conductor and how doping it with phosphorus increases its conductivity.

85. List the three common classes of liquid crystals in order of decreasing degree of order.
M

True/False Questions

86. The maximum number of phases of a single substance which can coexist in equilibrium is two.
M

87. The energy of a hydrogen bond is greater than that of a typical covalent bond.
M

88. A single water molecule can participate in at most two hydrogen bonds at any instant.
M

89. The surface tension of water is lowered when a detergent is present in solution.
E

90. Only molecules which do not have dipole moments can experience dispersion forces.
M

91. In the packing of identical atoms with cubic unit cells, the packing efficiency increases as the
M coordination number increases.

92. In cubic closest packing, the unit cell is body-centered cubic.
E

93. Hexagonal close packing of identical atoms occurs when close-packed layers are stacked in an
M *abcabc....* arrangement.

94. In metals, the conduction bands and valence bands of the molecular orbitals are separated by a large
E energy gap.

95. In a transistor, the current through one semiconductor junction controls the current through a
E neighboring junction.

96. Liquid crystal displays are most commonly constructed of smectic type liquid crystals.
M

97. Smectic liquid crystals are more highly ordered than either nematic or cholesteric liquid crystals.
M

98. Ceramic superconductors often contain copper in unusual oxidation states.
E

Intermolecular Forces: Liquids, Solids, and Phase Changes
Chapter 12
Answer Key

1. d
2. b
3. b
4. c
5. a
6. b
7. d
8. a
9. d
10. d
11. c
12. c
13. b
14. b
15. b
16. c
17. b
18. a
19. b
20. d
21. b
22. b
23. c
24. a
25. c
26. c
27. c
28. a
29. c
30. d
31. a
32. a
33. b
34. a
35. c
36. d
37. d
38. b
39. b
40. c
41. b
42. b
43. b
44. d
45. a
46. b
47. d
48. b
49. c
50. d
51. b
52. a
53. e
54. d
55. c
56. c
57. d
58. c
59. e
60. b
61. a
62. d
63. e
64. b
65. a
66. e

67.

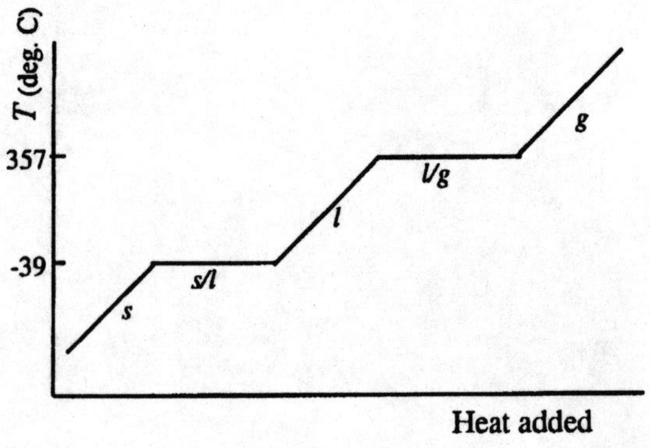

68. a. liquid
    b. solid, liquid and gas
    c. C is the critical point. Above the critical temperature the substance cannot be liquefied, regardless of the applied pressure.
    d. D is in the region of the solid phase. As pressure is lowered, the line dividing the solid and gas phases will be reached. At this point the solid will sublime.

69.

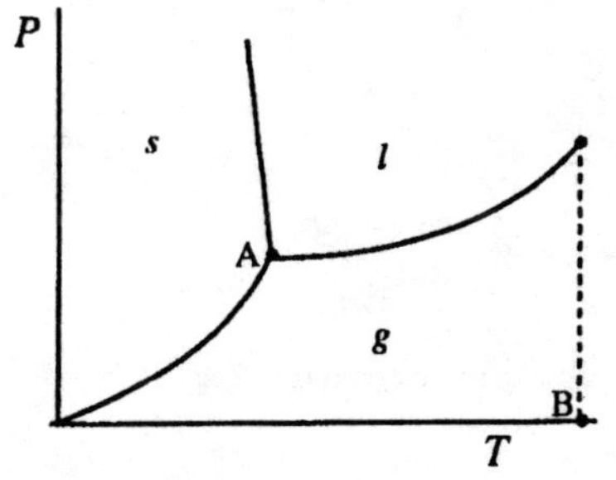

A is the triple point and B is the critical temperature.

70. 316 torr

71. -27.2°C

72. 266.9 K

73. a. The compound must contain hydrogen bonded to an atom which is small, electronegative and has lone pairs of electrons. The atoms which meet these criteria are N, O and F.
 b. High specific heat capacity; high heat of fusion; high surface tension; low density of ice compared to liquid water; other correct answers also possible.

74. a. A unit cell is the smallest repeating unit in a crystal lattice. Shifting of the unit cell along any of three directions must be capable of creating the entire lattice, without need to rotate the cell.
 b. 8.97 g/cm^3
 c. 128 pm

75. 4

76. 287 pm

77. 0.524

78. 0.740

79. a. face-centered cubic; hexagonal
 b. Cubic closest packing, which has the fcc unit cell, arises when close packed planes of atoms are stacked in an *abcabc...* arrangement; i.e. the fourth layer is above the first layer. Hexagonal close packing arises when the planes are stacked in an *abab....* arrangement; i.e. the third layer is above the first.

80. a. metallic
 b. network covalent
 c. ionic
 d. atomic
 e. molecular

81. 2.82×10^{-8} cm

82. Metals are good conductors whose conductivity decreases as the temperature rises. Semiconductors are poor conductors, but their conductivity increases with temperature.

83.

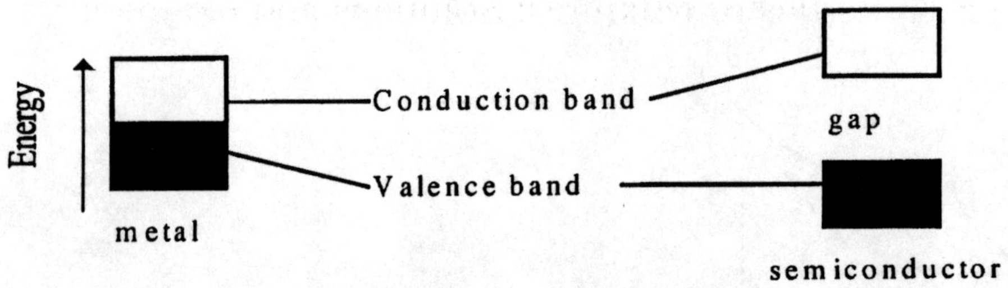

In metals, the conduction and valence bands overlap. Electrons can freely spill into the conduction band from the valence band, resulting in good conductivity. In semiconductors, the moderate gap between the bands results in few electrons being available to carry current, and conductivity is poor.

84.

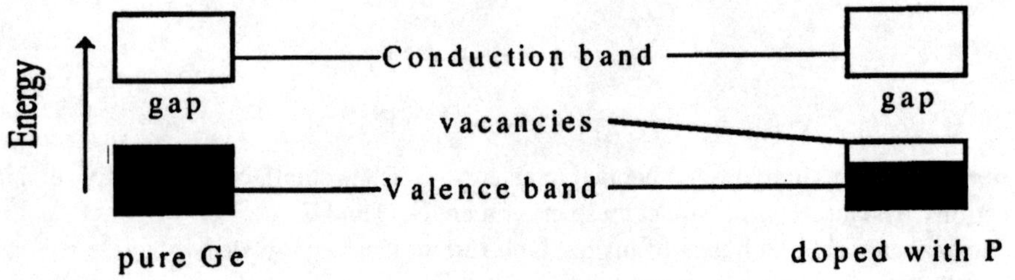

In pure Ge, very few electrons have sufficient energy to move into the conduction band, and conductivity is poor. Since P has only three valence electrons to the four of Ge, there will be some vacancies in the valence band when Ge is doped with P. Movement of these holes results in current flow; it will be a p-type semiconductor.

85. smectic > cholesteric > nematic

86. F

87. F

88. F

89. T

90. F

91. T

92. F

93. F

94. F

95. T

96. F

97. T

98. T

The Properties of Mixtures: Solutions and Colloids
Chapter 13

Multiple Choice Questions

1. Which of the following sets of conditions could exist when two liquids which are completely miscible in one
E another are mixed?

a. $\Delta H_{soln} > 0$, entropy decreases
b. $\Delta H_{soln} \approx 0$, entropy decreases
c. $\Delta H_{soln} \approx 0$, entropy change ≈ 0
d. $\Delta H_{soln} \approx 0$, entropy increases

2. If a solute dissolves in an endothermic process
M

a. H bonds must exist between solvent and solute.
b. strong ion-dipole forces must exist in the solution.
c. the entropy of the solution must be greater than that of its pure components.
d. the solute must be a gas.

3. When two pure substances are mixed to form a solution
M

a. heat is released.
b. heat is absorbed.
c. there is an increase in entropy.
d. there is a decrease in entropy.

4. A solution of sucrose (sugar) in water is in equilibrium with solid sucrose. If more solid sucrose is
M now added, with stirring,

a. the concentration of the solution will increase.
b. the concentration of the solution will decrease.
c. the concentration of the solution will remain the same.
d. a supersaturated solution will be produced.

5. Which of the following pairs of ions is arranged so that the ion with the smaller charge density is listed first?
M

a. K^+, Rb^+
b. Cl^-, K^+
c. Cl^-, Br^-
d. Ca^{2+}, Ba^{2+}

6. Which of the following pairs of ions is arranged so that the ion with the larger (i.e. more negative) heat of
M hydration is listed first?

a. Br^-, K^+
b. Mg^{2+}, Sr^{2+}
c. Ca^{2+}, Sc^{3+}
d. Na^+, Li^+

7. A solution of potassium hydroxide is in equilibrium with undissolved solute at 45°C. What will happen if the temperature is raised to 50°C? (ΔH_{soln} = -57.6 kJ/mol)
M

a. The mass of dissolved KOH will increase.
b. The mass of dissolved KOH will decrease.
c. The mass of dissolved KOH will be unchanged.
d. The mass of water in the solution will increase.

8. The Henry's Law constant *(k)* for carbon monoxide in water at 25°C is 9.71×10^{-4} mol/(L·atm). How many grams of CO will dissolve in 1.00 L of water if the partial pressure of CO is 2.75 atm?
M

a. 3.53×10^{-4} g
b. 2.67×10^{-3} g
c. 9.89×10^{-3} g
d. 7.48×10^{-2} g

9. Methane has a Henry's Law constant *(k)* of 9.88×10^{-2} mol/(L·atm) when dissolved in benzene at 25°C. How many grams of CH_4 will dissolve in 3.00 L of benzene if the partial pressure of CH_4 is 1.48 atm?
M

a. 0.0667 g
b. 0.146 g
c. 2.34 g
d. 7.02 g

10. For a given solution, which of the following concentration values will change as temperature changes?
M

a. mass percent
b. molality
c. mole fraction
d. molarity

11. Potassium fluoride is used for frosting glass. Calculate the molarjty of a solution prepared by dissolving 78.6 g of KF in enough water to produce 225 mL of solution.
E

a. 0.304 *M*
b. 0.349 *M*
c. 3.29 *M*
d. 6.01 *M*

12. Potassium hydrogen phosphate is used in the preparation of non-dairy powdered creamers. Calculate the molarity of a solution prepared by dissolving 238 g of K_2HPO_4 in enough water to produce 275 mL of solution.
E

a. 0.732 *M*
b. 0.865 *M*
c. 2.66 *M*
d. 4.97 *M*

13. Calculate the molarity of a solution prepared by diluting 1.85 L of 6.5 *M* KOH to 11.0 L.
E

a. 0.28 *M*
b. 0.91 *M*
c. 1.1 *M*
d. 3.1 *M*

14. What volume of concentrated (14.7 M) phosphoric acid is needed to prepare 25.0 L of 3.0 M H_3PO_4?
E

a. 0.20 L
b. 0.57 L
c. 1.8 L
d. 5.1 L

15. Saccharin, one of the first non-nutritive sweeteners used in soft-drinks, is 500 times sweeter than sugar in
E dilute aqueous solutions. The solubility of saccharin is 1.00 gram per 290 mL of solution. What is the molarity of a saturated saccharin solution? $\mathcal{M}_{saccharin} = 183.2$ g/mol

a. 0.0188 M
b. 0.632 M
c. 1.58 M
d. 3.45 M

16. Which of the following statements describes the correct method of preparation of 1.00 L of a 2.0 M urea
E solution? $\mathcal{M}_{urea} = 60.06$ g/mol

a. Dissolve 120 g of urea in 1.00 kg of distilled water.
b. Dissolve 120 g of urea in 880 g of distilled water.
c. Dissolve 120 g of urea in enough distilled water to produce 1.00 L of solution.
d. Dissolve 120 g of urea in 1.00 liter of distilled water.

17. Copper(II) bromide is used as a wood preservative. What mass of $CuBr_2$ is needed to prepare 750.0 mL of a
E 1.25 M solution?

a. 134 g
b. 209 g
c. 372 g
d. 938 g

18. What is the molality of a solution prepared by dissolving 86.9 g of diethyl ether, $C_4H_{10}O$, in 425 g of
E benzene, C_6H_6?

a. 0.362 m
b. 0.498 m
c. 2.01 m
d. 2.76 m

19. Calcium nitrite is used as a corrosion inhibitor in lubricants. What is the molality of a solution prepared by
E dissolving 18.5 g of calcium nitrite in 83.5 g of distilled water?

a. 0.0342 m
b. 0.0855 m
c. 0.222 m
d. 1.68 m

20. Cadmium bromide is used in photography and lithography. Calculate the molality of a solution prepared by
E dissolving 45.38 g of $CdBr_2$ in 375.0 g of water.

a. 0.03035 m
b. 0.01600 m
c. 0.1210 m
d. 0.4446 m

21. H Isoamyl salicylate ($\mathcal{M}$ = 208.25 g/mol) has a pleasant aroma and is used in perfumes and soaps. Which of the following combinations gives a 0.75 *m* solution of isoamyl salicylate in ethyl alcohol (*d* = 0.7893 g/mL)?

a. 117.2 g isoamyl salicylate in 950.0 mL of ethyl alcohol
b. 117.2 g isoamyl salicylate in 750.0 mL of ethyl alcohol
c. 117.2 g isoamyl salicylate in 750.0 mL of solution
d. 117.2 g isoamyl salicylate in 592.0 g of ethyl alcohol

22. H The chemist, Anna Lytic, must prepare 1.00 kg of 15.0% (w/w) acetic acid using a stock solution which is 36.0% (w/w) acetic acid (*d* = 1.045 g/mL). Which of the following combinations will give her the solution she wants?

a. 417 mL of 36% acetic acid in 583 mL of distilled water
b. 417 g of 36% acetic acid in 583 g of distilled water
c. 360 mL of 36% acetic acid in 640 mL of distilled water
d. 360 g of 36% acetic acid in 640 g of distilled water

23. M The solubility of the oxidizing agent potassium permanganate is 7.1 g per 100.0 g of water at 25°C. What is the mole fraction of potassium permanganate in this solution?

a. 0.0080
b. 0.0086
c. 0.45
d. 0.48

24. M Aqueous ammonia is commercially available in a solution that is 28% (w/w) ammonia. What is the mole fraction of ammonia in such a solution?

a. 0.017
b. 0.023
c. 0.24
d. 0.29

25. H Sodium hydroxide is a common ingredient in drain cleaners such as Drano. The mole fraction of sodium hydroxide in a saturated aqueous solution is 0.310. What is the molality of the solution?

a. 0.310 *m*
b. 0.690 *m*
c. 12.4 *m*
d. 25.0 *m*

26. H The mole fraction of potassium nitrate in an aqueous solution is 0.0194. The solution's density is 1.0627 g/mL. Calculate the molarity of the solution.

a. 0.0194 *M*
b. 0.981 *M*
c. 1.05 *M*
d. 1.96 *M*

27. A 0.89% (w/v) sodium chloride solution is referred to as physiological saline solution because it has the same
M concentration of salts as human blood. What is the molarity of a physiological saline solution?

a. 0.0028 *M*
b. 0.015 *M*
c. 0.15 *M*
d. 0.30 *M*

28. A 2.0% (w/v) solution of sodium hydrogen citrate, $Na_2C_6H_6O_7$, which also contains 2.5% (w/v) of dextrose,
M $C_6H_{12}O_6$, is used as an anticoagulant for blood which is to be used for transfusions. What is the molarity of the sodium hydrogen citrate in the solution?

a. 0.085 *M*
b. 0.19 *M*
c. 0.53 *M*
d. 1.2 *M*

29. Procaine hydrochloride ($\mathcal{M}$ = 272.77 g/mol) is used as a local anesthetic. Calculate the molarity of a 4.666 *m*
H solution which has a density of 1.1066 g/mL.

a. 2.272 *M*
b. 4.056 *M*
c. 4.216 *M*
d. 4.666 *M*

30. The concentration of iodine in sea water is 60. parts per billion by mass. If one assumes that the iodine exists
M in the form of iodide anions, what is the molarity of iodide in sea water? (The density of sea water is 1.025 g/mL.)

a. 4.8×10^{-10} *M*
b. 4.8×10^{-7} *M*
c. 4.7×10^{-4} *M*
d. 4.7×10^{-1} *M*

31. Children under the age of six with more than 0.10 ppm of lead in their blood can suffer a reduction in I.Q. or
M have behavior problems. What is the molality of a solution which contains 0.10 ppm of lead?

a. 4.8×10^{-10} *m*
b. 4.8×10^{-7} *m*
c. 4.8×10^{-4} *m*
d. 4.8×10^{-1} *m*

32. Colligative properties depend on
E

a. the chemical properties of the solute.
b. the chemical properties of the solvent.
c. the number of particles dissolved.
d. the molar mass of the solute.

33. Raoult's Law relates the vapor pressure of the solvent above the solution to its mole fraction in the solution. Which of the following is an accurate statement?
E

a. Raoult's Law applies exactly to all solutions.
b. Raoult's Law works best when applied to concentrated solutions.
c. Raoult's Law works best when applied to dilute solutions.
d. Raoult's Law applies only to non-ideal solutions.

34. Select the aqueous solution which should have the highest boiling point.
M

a. 1.0 *M* KNO_3
b. 0.75 *M* NaCl
c. 0.75 *M* $CuCl_2$
d. 2.0 *M* $C_{12}H_{22}O_{11}$ (sucrose)

35. Which of the following aqueous solutions will show the greatest freezing point depression?
M

a. 0.75 *M* $(NH_4)_3PO_4$
b. 1.0 *M* $CaSO_4$
c. 1 0 *M* $LiClO_4$
d. 1.5 *M* CH_3OH, methyl alcohol

36. Which of the following aqueous solutions should demonstrate the most ideal behavior?
M

a. 0.1 *M* K_2SO_4
b. 0.1 *M* $CaCl_2$
c. 0.1 *M* NaCl
d. 0.1 *M* $MgSO_4$

37. Which of the following aqueous solutions should demonstrate the most non-ideal behavior?
M

a. 0.1 *M* NaI
b. 0.2 *M* $CuSO_4$
c. 0.5 *M* KBr
d. 2.0 *M* $CuCl_2$

38. Select the strongest electrolyte from the following set.
M

a. CH_3CH_2OH, ethanol
b. $LiNO_3$
c. $C_6H_{12}O_6$, glucose
d. CCl_4

39. Select the weakest electrolyte from the following set.
M

a. Na_2SO_4
b. KCl
c. CH_3CH_2COOH, propionic acid
d. $CaCl_2$

40. How many moles of sulfate ions are present in 1.0 L of 0.5 M Li_2SO_4?
E

a. 0.5 mol
b. 1.0 mol
c. 1.5 mol
d. 3.0 mol

41. How many moles of bromide ions are present in 750.0 mL of 1.35 M $MgBr_2$?
E

a. 0.506 mol
b. 1.01 mol
c. 2.03 mol
d. 3.04 mol

42. How many moles of solute particles are present in 100.0 mL of 2.50 M $(NH_4)_3PO_4$?
M

a. 0.250 mol
b. 0.500 mol
c. 0.750 mol
d. 1.00 mol

43. Two aqueous are prepared: 1.00 m Na_2CO_3 and 1.00 m LiCl. Which of the following statements is true?
M

a. The Na_2CO_3 solution has a higher osmotic pressure and higher vapor pressure than the LiCl solution.
b. The Na_2CO_3 solution has a higher osmotic pressure and higher boiling point than the LiCl solution.
c. The Na_2CO_3 solution has a lower osmotic pressure and lower vapor pressure than the LiCl solution.
d. The Na_2CO_3 solution has a lower osmotic pressure and higher boiling point than the LiCl solution.

44. Two aqueous solutions are prepared: 2.0 m $Cu(NO_3)_2$ and 2.0 m NaBr. Which of the following statements is true?
M

a. The $Cu(NO_3)_2$ solution has a higher vapor pressure and lower freezing point than the NaBr solution.
b. The $Cu(NO_3)_2$ solution has a higher vapor pressure and higher freezing point than the NaBr solution.
c. The $Cu(NO_3)_2$ solution has a lower vapor pressure and lower freezing point than the NaBr solution.
d. The $Cu(NO_3)_2$ solution has a lower vapor pressure and higher freezing point than the NaBr solution.

45. Which of the following aqueous solutions will have the lowest osmotic pressure?
M

a. 0.10 m KOH
b. 0.10 m RbCl
c. 0.05 m $CaSO_4$
d. 0.05 m $BaCl_2$

46. Which of the following aqueous solutions will have the lowest freezing point?
M

a. 0.5 m $C_{12}H_{22}O_{11}$ (sucrose)
b. 0.5 m $Ca(NO_3)_2$
c. 0.5 m $NiSO_4$
d. 0.5 m Li_3PO_4

47. Calculate the vapor pressure of a solution prepared by dissolving 0.50 mol of a non-volatile solute in 275 g of hexane ($\mathcal{M}$= 86.18 g/mol) at 49.6°C. $P°_{hexane}$ = 400.0 torr @ 49.6°C
M

a. 54 torr
b. 154 torr
c. 246 torr
d. 346 torr

48. Safrole is used as a topical antiseptic. Calculate the vapor pressure of a solution prepared by dissolving 0.75 mol of safrole in 950 g of ethanol ($\mathcal{M}$= 46.07 g/mol). $P°_{ethanol}$ = 50.0 torr at 25°C
M

a. 1.8 torr
b. 11 torr
c. 40 torr
d. 48 torr

49. Diethyl ether has a vapor pressure of 400.0 torr at 18 °C. When a sample of benzoic acid is dissolved in ether, the vapor pressure of the solution is 342 torr. What is the mole fraction of benzoic acid in the solution?
M

a. 0.0169
b. 0.0197
c. 0.145
d. 0.855

50. Determine the freezing point of a solution which contains 0.31 mol of sucrose in 175 g of water. K_f = 1.86°C/*m*
E

a. 3.3°C
b. 1.1°C
c. -1.1°C
d. -3.3°C

51. Benzaldehyde ($\mathcal{M}$= 106.1 g/mol), also known as oil of almonds, is used in the manufacture of dyes and perfumes and in flavorings. What would be the freezing point of a solution prepared by dissolving 75.00 g of benzaldehyde in 850.0 g of ethanol? K_f = 1.99°C/*m*, freezing point of pure ethanol = -117.3°C
E

a. -117.5°C
b. -118.7°C
c. -119.0°C
d. -120.6°C

52. Carbon tetrachloride, once widely used in fire extinguishers and as a dry cleaning fluid, has been found to cause liver damage to those exposed to its vapors over long periods of time. What is the boiling point of a solution prepared by dissolving 375 g of sulfur (S_8, $\mathcal{M}$= 256.5 g/mol) in 1250 g of CCl_4? K_b = 5.05°C/*m*, boiling point of pure CCl_4 = 76.7°C
E

a. 70.8°C
b. 75.2°C
c. 78.2°C
d. 82.6°C

53.
E

Barbiturates are synthetic drugs used as sedatives and hypnotics. Barbital ($\mathcal{M}$ = 184.2 g/mol) is one of the simplest of these drugs. What is the boiling point of a solution prepared by dissolving 42.5 g of barbital in 825 g of acetic acid? K_b = 3.07°C/*m*, boiling point of pure acetic acid = 1 17.9°C

a. 117.0°C
b. 117.7°C
c. 118.1°C
d. 118.8°C

54.
E

Dimethylglyoxime, DMG, is an organic compound used to test for aqueous nickel(II) ions. A solution prepared by dissolving 65.0 g of DMG in 375 g of ethanol boils at 80.3°C. What is the molar mass of DMG? K_b = 1.22°C/*m*, boiling point of pure ethanol = 78.5°C

a. 44.1 g/mol
b. 65.8 g/mol
c. 117 g/mol
d. 553 g/mol

55.
M

Hexachlorophene is used as a disinfectant in germicidal soaps. What mass of hexachlorophene ($\mathcal{M}$= 406.9 g/mol) must be added to 125 g of chloroform to give a solution with a boiling point of 62.60°C? K_b = 3.63°C/*m*, boiling point of pure chloroform = 61.70°C

a. 12.6 g
b. 17.2 g
c. 31.0 g
d. 101 g

56.
M

Cinnamaldehyde ($\mathcal{M}$= 132.15 g/mol) is used as a flavoring agent. What mass of cinnamaldehyde must be added to 175 g of ethanol to give a solution whose boiling point is 82.7°C? K_b = 1.22°C/*m*, boiling point of pure ethanol = 78.5°C

a. 62.4 g
b. 67.8 g
c. 76.2 g
d. 79.6 g

57.
M

Calculate the freezing point of a solution made by dissolving 3.50 g of potassium chloride ($\mathcal{M}$= 74.55 g/mol) in 100.0 g of water. Assume ideal behavior for the solution; K_f= 1.86°C/*m*

a. -1.7°C
b. -0.9°C
c. 0.9°C
d. 1.7°C

58.
M

A 0.100 *m* K_2SO_4 solution has a freezing point of -0.43°C. What is the van't Hoff factor for this solution? K_f = 1.86°C/*m*

a. 0.77
b. 1.0
c. 2.3
d. 3.0

59. M A 0.100 *m* $MgSO_4$ solution has a freezing point of -0.23°C. What is the van't Hoff factor for this solution? $K_f = 1.86°C/m$

a. 0.62
b. 1.0
c. 1.2
d. 2.0

60. E Human blood has a molar concentration of solutes of 0.30 *M*. What is the osmotic pressure of blood at 25°C?

a. 0.012 atm
b. 0.62 atm
c. 6.8 atm
d. 7.3 atm

61. M Lysine is an amino acid that is an essential part of nutrition but which is not synthesized by the human body. What is the molar mass of lysine if 750.0 mL of a solution containing 8.60 g of lysine has an osmotic pressure of 1.918 atm? T = 25.0°C

a. 110 g/mol
b. 146 g/mol
c. 1340 g/mol
d. 1780 g/mol

62. E Which one of the following pairs of dispersed phases and dispersing media can never form a colloid?

a. solid and gas
b. liquid and gas
c. solid and solid
d. liquid and liquid
e. gas and gas

63. E An emulsion is a dispersion consisting of a

a. solid in a liquid.
b. liquid in a liquid.
c. gas in a liquid.
d. liquid in a solid.
e. gas in a solid.

Short Answer Questions

64. E The density of pure water at 25°C is 0.997 g/mL. Considering water as being both solvent and solute, calculate its molarity.

65. M A 50.0 % (w/w) solution of sulfuric acid (H_2SO_4) in water has a density of 1.395 g/mL. What is its molarity?

66. M A 7.112 *M* solution of sulfuric acid (H_2SO_4) in water has a density of 1.395 g/mL. What is its molality?

67. M Octane and nonane are liquids which are components of gasoline. Their vapor pressures at 25°C are 13.9 torr and 4.7 torr, respectively. What is the vapor pressure of a mixture consisting of 1 mole of each of these compounds?

68. If a 100-mL sample of a volatile liquid such as diethyl ether is introduced into a 250-mL flask which
M is immediately sealed, the pressure inside will increase above atmospheric pressure. Explain.

69. If the shell of a raw egg is carefully dissolved away, and the egg in its flexible membrane is then
M placed in distilled water, the egg's volume will expand. Explain.

70. 1.00 L of an aqueous solution contains 1.52 g of a compound used in antifreeze. If the osmotic
M pressure of this solution at 20.0°C is 448 torr, calculate the molar mass of the antifreeze compound.

71. Predict the van't Hoff factor (*i*) for each of the following solutes:
E

a. $CaCl_2$
b. $Fe_2(SO_4)_3$
c. NH_4Cl

72. A 0.137 m solution of sodium ferrocyanide ($NaFe(CN)_6.10H_2O$) in water has a freezing point
M depression of 0.755°C. Calculate the van't Hoff factor (*i*) under these conditions. (K_f= 1.86°C/*m*)

73. Classify the following as either solutions or colloids. If a colloid, name the type of colloid and
M identify both the dispersed and the dispersing phases.

a. glucose in water
b. smoke in air
c. carbon dioxide in air
d. milk

True/False Questions

74. In general, water is a good solvent for both polar and non-polar compounds.
E

75. Gases with high boiling points tend to be more soluble in water than ones with low boiling points.
M

76. The heat of solution is the total enthalpy change when a solution is formed from the separated solute
M and solvent.

77. Heats of solution may be either positive or negative.
M

78. Heats of hydration may be either positive or negative.
H

79. The larger the ion, the greater in magnitude will be its heat of hydration.
E

80. The solubility of gases in water increases with increase in the pressure of the gas.
E

81. If the density of a solution is less than 1.0 g/mL, its molarity will be greater than its molality.
M

82. Raoult's Law relates the vapor pressure of a solvent above a solution to the mole fraction of the
M solvent and the vapor pressure of the pure solvent.

83. **Ideal solutions do not conform to Raoult's Law.**
E

84. **Electrolyte solutions generally behave less ideally as the solute concentration increases.**
E

85. **Colloidal particles may be either solids, liquids or gases.**
M

The Properties of Mixtures
Chapter 13
Answer Key

1.	d	22.	b	43.	b
2.	c	23.	a	44.	c
3.	c	24.	d	45.	c
4.	c	25.	d	46.	d
5.	b	26.	c	47.	d
6.	b	27.	c	48.	d
7.	b	28.	a	49.	c
8.	d	29.	a	50.	d
9.	d	30.	b	51.	c
10.	d	31.	b	52.	d
11.	d	32.	c	53.	d
12.	d	33.	c	54.	c
13.	c	34.	c	55.	a
14.	d	35.	a	56.	d
15.	a	36.	c	57.	a
16.	c	37.	d	58.	c
17.	b	38.	b	59.	c
18.	d	39.	c	60.	d
19.	d	40.	a	61.	b
20.	d	41.	c	62.	e
21.	a	42.	d	63.	b

64. 55.3 *M*

65. 7.112 *M*

66. 10.20 *m*

67. 9.3 torr

68. The pressure inside the flask will initially be the same as atmospheric pressure, and the gas present will be air with some ether vapor. Initially, the gas above the liquid will not be saturated with ether vapor. More liquid will evaporate until equilibrium is reached, leading to an increase in pressure.

69. The osmotic pressure of the solution inside the membrane is greater than that of the pure water outside the membrane. By osmosis, water molecules will enter the egg, leading to an increase in its volume.

70. 62.1 g/mol

71.
   a. 3
   b. 4
   c. 2

72. 2.96

73.
   a. solution
   b. colloid; aerosol; solid in gas
   c. solution
   d. colloid; emulsion; liquid in liquid

74. F

75. T

76. T

77. T

78. F

79. F

80. T

81. F

82. T

83. F

84. T

85. T

Periodic Patterns in the Main-Group Elements: Bonding, Structure, and Reactivity
Chapter 14

Multiple Choice Questions

1. Although the periodic table is organized according to the atomic numbers of the elements, chemists are more
M interested in the arrangement of the electrons for their studies. Which of the following statements about the electron configurations and their quantum numbers is correct?

a. The size of an atom is associated with the angular momentum quantum number.
b. The valence electrons of atoms in a particular group have the same principal and angular momentum quantum numbers.
c. The valence electrons of atoms in a particular group have the same angular momentum quantum number but have different principal quantum numbers.
d. Quantum numbers for the electrons tell us little about the relative energies of the electrons.

2. Which of the following statements about the effective nuclear charge, Z_{eff}, is correct?
M

a. Z_{eff} increases with the size of the atom.
b. Z_{eff} decreases across a period and increases down a group.
c. Z_{eff} increases across a period and is relatively constant down a group.
d. Z_{eff} increases as the value of the principal quantum number increases.

3. The largest ionization energies are found in the ___________ ___________ region of the periodic table.
E

a. upper left
b. upper right
c. lower left
d. lower right

4. The largest electronegativities are found in the ___________ ___________ region of the periodic table.
E

a. upper left
b. upper right
c. lower left
d. lower right

5. Which of the following elements has the lowest electronegativity?
E

a. Al
b. S
c. Ba
d. In

6. Bromine will form compounds with each of the other elements in Period 4 of the periodic table. How does the
E type of bonding in the compounds change as one moves from potassium bromide to selenium bromide?

a. polar covalent to ionic
b. ionic to polar covalent
c. polar covalent to non-polar covalent
d. coordinate covalent to polar covalent

7. M The atomic radius of sodium is 186 pm and of chlorine is 100 pm. The ionic radius for Na^+ is 102 pm and for Cl^- is 181 pm. In going from Na to Cl in Period 3, why does the atomic radius decrease while the ionic radius increases?

a. The inner electrons in the sodium cation shield its valence electrons more effectively than the inner electrons in the chloride anion do.
b. The inner electrons shield the valence electrons more effectively in the chlorine atom than in the chloride anion.
c. The outermost electrons in chloride experience a smaller effective nuclear charge than those in the sodium cation do.
d. The outermost electrons in chloride experience a larger effective nuclear charge than those in the sodium cation do.

8. E Select the element with the lowest first ionization energy.

a. Se
b. S
c. Sr
d. Sn

9. M Select the element with the highest first ionization energy.

a. Mg
b. Ca
c. Ba
d. Ra

10. M Which of the following pairs of elements will form the longest single bond?

a. C, F
b. C, N
c. C, S
d. C, O

11. E Which of the following ions has the greatest radius?

a. Se^{2-}
b. Br^-
c. Rb^+
d. Sr^{2+}

12. M Which of the following atoms has the smallest volume?

a. Ba
b. Cs
c. Sr
d. I

13. E Which of the following oxides will give the most basic solution when dissolved in water?

a. SO_2
b. CO_2
c. K_2O
d. P_4O_{10}

14. Which of the following oxides will give the most acidic solution when dissolved in water?
E

a. MgO
b. Al_2O_3
c. Cl_2O
d. SrO

15. Which of the following will have the highest boiling point?
E

a. O_2
b. Cl_2
c. Br_2
d. I_2

16. Which of the following pure substances will not participate in hydrogen bonding?
M

a. CH_3NH_2
b. $CH_3CH_2OCH_2CH_3$
c. CH_3CH_2OH
d. HF

17. What are the products of the following reaction of strontium hydride and water?
M

$SrH_2(s) + H_2O(l) \rightarrow$

a. $Sr^{2+}(aq) + H_2(g) + O_2(g)$
b. $Sr^{2+}(aq) + H_2(g) + OH^-(aq)$
c. $Sr(s) + H_2(g) + OH^-(aq)$
d. $Sr(s) + H_3O^+(aq)$

18. What are the products of the following reaction of potassium hydride and water?
M

$KH(s) + H_2O(l) \rightarrow$

a. $K(s) + H_2(g) + OH^-(aq)$
b. $K(s) + H_2(g) + O_2(g)$
c. $K^+(aq) + H_2(g) + OH^-(aq)$
d. $K^+(aq) + H_2(g) + O_2(g)$

19. Hydrogen forms metallic (interstitial) hydrides with the *d* and *f* transition elements. Which of the following statements is correct?
M

a. These substances have distinct stoichiometric formulas like ionic hydrides.
b. Hydrogen forms bonds with the metals by donating its electron to the valence band of the metal.
c. Hydrogen molecules and atoms occupy holes within the crystal structure of the metal.
d. These substances are useful catalysts.

20. Which one of the following elements is likely to exhibit the most violent and rapid reaction with water?
M

a. Na
b. Rb
c. Mg
d. Sr
e. Cl_2

21. Which of the following bonds should have the greatest ionic character?
M

a. O–F
b. C–F
c. B–F
d. N–F

22. Which element forms compounds which are used to treat individuals suffering from
M manic-depressive disorders?

a. fluorine
b. lithium
c. boron
d. beryllium

23. Which element forms compounds which are used as coatings for other substances?
M

a. boron
b. beryllium
c. fluorine
d. nitrogen

24. Which element forms compounds which are involved in smog and acid rain?
M

a. carbon
b. fluorine
c. nitrogen
d. boron

25. Predict the products for the following set of reactants.
M

$Li(s) + H_2O(l) \rightarrow$

a. $Li^+(aq) + H_2(g) + O_2(g)$
b. $Li^+(aq) + H_2(g) + OH^-(aq)$
c. $LiH(s) + O_2(g)$
d. $Li^+(aq) + H_2O_2(aq)$

26. Predict the products for the following set of reactants.
M

$K_2O(s) + H_2O(l) \rightarrow$

a. $K^+(aq) + OH^-(aq) + H_2(g)$
b. $K^+(aq) + OH^-(aq)$
c. $K^+(aq) + H_2(g) + O_2(g)$
d. $KH(s) + O_2(g)$

27. Most of the alkali metal salts are soluble in water while many alkaline earth salts have very low solubilities.
M Why is this so?

a. The alkali metal cations are smaller than the alkaline earth cations and are more easily hydrated.
b. The alkali metals have lower ionization energies than alkaline earth elements.
c. The alkaline earth salts have much greater lattice energies than the alkali metal salts.
d. The alkaline earth metals have greater heats of atomization than the alkali metals.

28. H Unlike the remainder of the Group 1A(1) elements, lithium forms many salts that have some covalent bond character. What is a reason for this behavior?

a. The high first ionization energy of lithium makes sharing the electron easier than transferring it.
b. The high charge density on the lithium cation deforms nearby polarizable electron clouds.
c. The atomic radius of lithium enables it to share its valence electron effectively.
d. Since lithium has only 1 electron in its 2s orbital, it can accept an electron from another element.

29. M In which of the following ways is lithium different from the other alkali metals?

a. Its salts are much more soluble in water than those of the other alkali metals.
b. It has an unusually high density.
c. It forms molecular compounds with the hydrocarbon groups of organic halides.
d. Its ionization energy is lower than expected.

30. Predict the products for the following set of reactants.

$Li(s) + CH_3Cl(g) \rightarrow$

a. $CH_3CH_3(g) + LiCl(s)$
b. $CH_4(g) + LiCl(s)$
c. $CH_3Li(s) + LiCl(s)$
d. $CH_3Li(s) + Cl_2(g)$

31. E The elements from Groups 1A(1) and 2A(2) are

a. strong oxidizing agents
b. strong reducing agents
c. strong acids
d. strong bases

32. M Predict the products for the reaction of the following set of reactants.

$CaO(s) + H_2O(l) \rightarrow$

a. $Ca^{2+}(aq) + OH^-(aq) + H_2(g)$
b. $Ca^{2+}(aq) + H_2(g) + O_2(g)$
c. $Ca^{2+}(aq) + H_3O^+(aq)$
d. $Ca(OH)_2(s)$

33. M Predict the products for the reaction of the following set of reactants.

$Sr(s) + H_2O(l)$

a. $SrO(s) + H_2(g)$
b. $Sr^{2+}(aq) + H_2(g) + O_2(g)$
c. $Sr^{2+}(aq) + H_3O^+(aq)$
d. $Sr(OH)_2(s) + H_2(g)$

34. Predict the products for the reaction of the following set of reactants.
E

$BaO(s) + CO_2(g) \rightarrow$

a. $BaCO_3(s)$
b. $Ba(s) + CO(g) + O_2(g)$
c. $BaO_2(s) + CO(g)$
d. $BaC_2(s) + O_2(g)$

35. Predict the products for the reaction of the following set of reactants.
E

$Mg(s) + Cl_2(g) \rightarrow$

a. $MgCl(s)$
b. $MgCl_2(s)$
c. $MgCl(l)$
d. $MgCl_2(l)$

36. Predict the products for the reaction of the following set of reactants.
M

$MgCO_3(s)$ + heat $\rightarrow$

a. $Mg(s) + CO(g)$
b. $Mg(s) + CO_2(g)$
c. $MgO(s) + CO(g)$
d. $MgO(s) + CO_2(g)$

37. Predict the products for the reaction of the following set of reactants.
E

$Ca(s) + H_2(g) \rightarrow$

a. $CaH(s)$
b. $CaH_2(s)$
c. $Ca_2H_3(s)$
d. $Ca_3H_2(s)$

38. Magnesium oxide, MgO, is an important industrial material. Which of the following is one of its uses?
H

a. antacid
b. furnace bricks
c. toothpaste abrasive
d. construction material

39. Thallium can form two oxides, Tl_2O and Tl_2O_3. Which will be the more basic substance?
H

a. Tl_2O
b. Tl_2O_3
c. They have the same strength as bases.
d. More information is needed to make an accurate prediction.

40. Which of the following oxides will be the most acidic?
H

a. Al_2O_3
b. Ga_2O_3
c. In_2O_3
d. Tl_2O_3

41. Which of the following hydroxides will be the most basic?
H

a. $B(OH)_3$
b. $Al(OH)_3$
c. $In(OH)_3$
d. TlOH

42. The most basic oxides contain elements from the ________________ ____________ region of the periodic table.
M

a. upper right
b. upper left
c. lower right
d. lower left

43. The basic character of the oxides of an element
M

a. increases as the oxidation number of the central element increases.
b. increases as the oxidation number of the central element decreases.
c. is unaffected by the oxidation number of the central element.

44. Predict the products for the reaction of the following set of reactants.
H

$(CH_3)_3Ga(g) + AsH_3(g) \rightarrow$

a. $(CH_3)_3As(g) + GaH_3(g)$
b. $CH_4(g) + GaAs(s)$
c. $CH_3CH_3(g) + GaAs(s)$
d. $CH_4(g) + H_2GaAs(s)$

45. Boron has 3 valence electrons. Which of the following processes is involved in boron's achieving a complete outer shell?
H

a. Formation of a B^{3+} cation
b. Formation of bridge bonds
c. Formation of π-bonds using its *d*-orbitals
d. Formation of π-bonds using sp^3-orbitals

46. Which of the following elements exists in allotropic forms?
M

a. silicon
b. germanium
c. tin
d. lead

47. When elements from a group exhibit more than one oxidation state,
H

a. the higher oxidation state is more important as one goes down a group.
b. the lower oxidation state is more important as one goes down a group.
c. both oxidation states are equally important throughout the group.
d. the oxidation state will be affected by the elements on either side of it in the period.

48. One important feature of the chemistry of carbon is
M

a. its ability to form multiple bonds with hydrogen.
b. its ability to catenate.
c. its large radius that allows other atoms to fit easily around it.
d. its low electronegativity that allows it to ionize easily.

49. The polymers containing silicon differ from polymers of carbon in which of the following ways?
M

a. Silicon-based polymers are larger molecules than carbon-based polymers.
b. Silicon-based polymers generally have a repeating silicon-oxygen link while carbon-based polymers can have carbon-carbon links.
c. Silicon-based polymers generally have inorganic elements attached to the chain while carbon-based polymers generally have organic groups attached.
d. Silicon-based polymers tend to be rigid while carbon-based polymers are generally flexible.

50. Silicon halides have stronger bonds than corresponding carbon halides. Which of the following is a possible
H explanation of this phenomenon?

a. The larger silicon atoms permit better overlap of its atomic orbitals with those of the halogens than the smaller carbons atoms do.
b. The large electronegativity difference between silicon and the halogens makes their bonds stronger than those of carbon.
c. Silicon has the ability to form a partial double bond with a halogen through the overlap of its *d*-orbital with a *p*-orbital of the halogen.
d. Silicon has a larger effective nuclear charge than carbon which allows it to bond more strongly to the more negative halogens than carbon.

51. Certain Period 2 elements exhibit behaviors similar to Period 3 elements immediately below and to the right.
H One of these interesting diagonal relationships occurs between lithium and magnesium. Which of the following is one of their similarities?

a. They both form insoluble carbonate salts.
b. They both form organic compounds with polar covalent bonds from the metal to hydrocarbon group.
c. Their first ionization energies are almost equal.
d. Their densities are very similar.

52. Certain Period 2 elements exhibit behaviors similar to Period 3 elements immediately below and to the right.
M One of these interesting diagonal relationships occurs between beryllium and aluminum. Which of the following is one of their differences?

a. Some aluminum compounds and all beryllium compounds show significant covalent character in the gas phase.
b. Beryllium forms bridge bonds in its hydrides while aluminum does not.
c. Both form oxides that are impervious to reaction with water.
d. The cations for both strongly polarize nearby electron clouds.

53. H Certain Period 2 elements exhibit behaviors similar to Period 3 elements immediately below and to the right. One of the interesting diagonal relationships occurs between boron and silicon. Which of the following is one of their similarities?

a. Both exhibit electrical properties of a conductor.
b. The oxoanions of both elements occur in extended ionic networks.
c. Both elements form compounds--boranes and silanes--that are good oxidizing agents.
d. Boric acid and silicic acid occur in layers with widespread hydrogen bonding.

54. H Silicon carbide is an important industrial chemical. Which of the following statements about silicon carbide is true?

a. It will react with water to form acetylene.
b. It is used in preparation of glass which can be used at high temperature.
c. It can be doped to form a high temperature semiconductor.
d. It is a naturally occurring mineral source of silicon.

55. H Tin reacts to form several organic compounds. Which of the following is a use for organotin compounds?

a. They are starting materials for polymers.
b. They can be used as fungicides.
c. They can be used to prepare semiconductors.
d. They can be used as solvents for reactions involving polar organic compounds.

56. H Which of the following oxides is most basic?

a. As_2O_3
b. P_4O_{10}
c. Sb_2O_3
d. Sb_2O_5

57. H Predict the products for the following set of reactants.

$$Ca_3As_2(s) + H_2O(l) \rightarrow$$

a. $As^{3+}(aq) + Ca(OH)_2(aq)$
b. $As(OH)_3(s) + Ca^{2+}(aq) + H_2(g)$
c. $As(OH)_3(s) + Ca(OH)_2(aq)$
d. $AsH_3(g) + Ca(OH)_2(aq)$

58. H Predict the products for the following set of reactants.

$$PCl_3(l) + H_2O(l) \rightarrow$$

a. $H_3PO_3(aq) + HCl(aq)$
b. $H_3PO_4(aq) + Cl_2(g)$
c. $PH_3(g) + HCl(aq) + O_2(g)$
d. $P_2O_5(s) + HCl(aq)$

59. H Which of the following would you predict to have the greatest thermal stability?

a. AsF_3
b. $AsCl_3$
c. $AsBr_3$
d. AsI_3

60. Dinitrogen monoxide, N_2O, is
M

a. a brown poisonous gas that is one of the chemicals involved in the production of photochemical smog.
b. a colorless gas used in the production of nitric acid.
c. a colorless gas used as a propellant in canned whipped cream.
d. a colorless gas that disproportionates into nitrogen and oxygen.

61. Sodium tripolyphosphate is useful
M

a. as a starting material in the synthesis of organic phosphorus compounds.
b. as a water-softening agent.
c. as an emulsifier in making of processed cheese.
d. as a radiator corrosion inhibitor.

62. The nitrate anion is
M

a. a strong oxidizing agent.
b. a strong reducing agent.
c. a strong acid.
d. a strong base.

63. Predict the products for the following set of reactants.
H

$Bi(s) + Cl_2(g) \rightarrow$

a. $BiCl$
b. $BiCl_2$
c. $BiCl_3$
d. $BiCl_5$

64. Which of the following has the most allotropes?
M

a. carbon
b. sulfur
c. oxygen
d. selenium

65. Which of the following is an accurate comparison of the properties of nitrogen and oxygen?
M

a. Both oxygen and nitrogen form anions easily.
b. Oxygen is a strong oxidizing agent while nitrogen is a strong reducing agent.
c. Nitrogen uses more oxidation states in its compounds than oxygen does.
d. Nitrogen is more reactive than oxygen.

66. Predict the products formed from the following set of reactants.
M

$FeSe(s) + HCl(aq) \rightarrow$

a. $Se(s) + FeCl_2(aq)$
b. $SeCl_2(s) + Fe(s) + H_2(g)$
c. $H_2Se(g) + FeCl_2(aq)$
d. $H_2Se(g) + Fe(s) + Cl_2(g)$

67. Sulfur hexafluoride, SF_6, is
M

a. a reactive gas that decomposes when exposed to moisture.
b. a reactive gas used to fluorinate organic compounds.
c. an inert gas that is used as an electrical insulator.
d. a reactant in the process of manufacturing non-stick coatings.

68. Hydrogen peroxide, H_2O_2, is
H

a. used in the production of polymers.
b. used as a drying agent.
c. colored liquid with a low boiling point.
d. an acid.

69. The chemical that ranks first in production among all industrial chemicals is
M

a. NH_3, ammonia.
b. H_3PO_4, phosphoric acid.
c. NaOH, sodium hydroxide.
d. H_2SO_4, sulfuric acid.

70. The halogens are
E

a. strong oxidizing agents.
b. strong reducing agents.
c. strong acids.
d. strong bases.

71. The strongest oxidizing agents are found in the _________ _________ region of the periodic table while the
E strongest reducing agents are found in the ________ ________ region of the periodic table.

a. upper left, lower right
b. upper right, lower left
c. lower left, upper right
d. lower right, upper left

72. Predict the products for the following set of reactants.
E

$$Cs(s) + Br_2(l) \rightarrow$$

a. $CsBr(s)$
b. $CsBr_2(s)$
c. $CsBr(l)$
d. $CsBr_2(l)$

73. Predict the products for the following set of reactants.
M

$$Cl_2(g) + I^-(aq) \rightarrow$$

a. ICl
b. ICl_2
c. ICl_3
d. $I_2 + Cl^-$

74. Which of the following is the strongest acid?
H

a. $HClO_3$
b. $HBrO_3$
c. HIO_3
d. HIO_2

75. Predict the products for the following set of reactants.
M

$$H_2O(l) + ClF_5(l) \rightarrow$$

a. $HClO_2(aq) + HF(aq)$
b. $HClO_3(aq) + HF(aq)$
c. $HClO(aq) + HF(aq)$
d. $HCl(aq) + HF(aq)$

76. Sodium hypochlorite is used
M

a. in chemical analysis.
b. as an oxidizer in rocket fuels.
c. as a disinfectant.
d. in the manufacture of steel.

77. Hydrogen fluoride is used
M

a. in the manufacture of steel.
b. in the synthesis of cryolite for aluminum production.
c. as an oxidizing agent.
d. as a disinfectant.

78. Which of the following generalized formulas does not exist for interhalogen compounds? (X and Y represent
M different halogens.)

a. XY
b. XY_2
c. XY_3
d. XY_5

79. Xenon forms several compounds with oxygen and fluorine. It is the most reactive non-radioactive noble
H gas because

a. its large radius allows oxygen and fluorine to bond without being crowded.
b. it has the lowest ionization energy of these noble gases.
c. it has the highest electron affinity of these gases.
d. its effective nuclear charge is lower than the other noble gases.

80. Predict the molecular shape of XeF_3^-.
M

a. linear
b. trigonal planar
c. trigonal pyramid
d. T-shaped

Short Answer Questions

81. Name the three different classes (types) of hydride, and list some of their important characteristics.
M

82. If an alkali metal is represented by the symbol M, what is the formula of
E

a. the oxide of M?
b. the peroxide of M?
c. the superoxide of M?

83. Name the three allotropes of carbon and briefly describe their properties and structures.
E

84. Write balanced equations, showing all reactants and products, to represent
E

a. the reaction of calcium metal with water
b. the reaction of aluminum metal with oxygen gas

85. Write balanced equations, showing all reactants and products, to represent
M

a. the reaction of lithium metal with oxygen to form lithium oxide
b. the formation of ammonia in the Haber process

86. Write balanced equations, showing all reactants and products, to represent
M

a. roasting of limestone ($CaCO_3$) to give lime
b. tetraphosphorus decaoxide (P_4O_{10}) reacting with water to produce phosphoric acid

87. Consider the oxides XO_2, where X is a main-group element. Identify X in each of the following cases:
M

a. XO_2 is a natural component of the atmosphere, implicated in global warming
b. XO_2 is a toxic, brown gas and a component of photochemical smog
c. XO_2 is a colorless, toxic gas, implicated in acid rain
d. XO_2 is network solid; one form of it is the mineral quartz

88. Identify by number (1A-8A) the main-group of the periodic table to which the described element (X)
M belongs, in each of the following cases.

a. X commonly forms the ion X^{2-}
b. X reacts with water according to the equation

$$X(s) + 2H_2O(l) \rightarrow X(OH)_2(aq) + H_2(g)$$

c. X exists as molecules, X_2; its hydride has the formula HX
d. X is a metal which reacts with oxygen to produce the peroxide, X_2O_2

89. Each one of the following statements applies to an element or elements from one of the main groups
M (1A through 8A). Identify the group number appropriate to each of the statements, and write it in the left hand margin alongside that statement.

a. This group contains the element with the lowest ionization potential of any element.
b. The maximum oxidation number of any element in this group is +5.
c. An element X in this group forms a stable, unreactive gas, XF_6.

True/False Questions

90. Ionic hydrides do not have exact (stoichiometric) formulas.
E

91. The acidity of oxides of main group elements increases across a period from left to right.
E

92. The acidity of oxides of main group elements increases down a group, from top to bottom.
E

93. Potassium is a strong reducing agent.
E

94. Potassium nitrate is a strong reducing agent.
M

95. In gaseous $BeCl_2$, Be does not obey the octet rule; in solid $BeCl_2$ it does.
M

96. In the gas phase, $AlCl_3$ exists as molecules of Al_2Cl_6.
M

97. Carbon monoxide's toxicity is related to its ability to bond to iron in hemoglobin.
M

98. The Haber process is the first step in the manufacture of sulfuric acid.
M

99. Phosphoric acid (H_3PO_4) is a strong acid.
H

100. Ozone is both a pollutant and a natural component of the atmosphere.
M

101. Sulfuric acid is produced when sulfur dioxide dissolves in water.
M

102. The halogens act as oxidizing agents in most of their reactions.
M

103. Neon does not form any known compounds.
M

Periodic Patterns in the Main Group Elements: Bonding, Structure, & Reactivity
Chapter 14
Answer Key

1.	c	28.	b	55.	b
2.	c	29.	c	56.	c
3.	b	30.	c	57.	d
4.	b	31.	b	58.	a
5.	c	32.	d	59.	a
6.	b	33.	d	60.	c
7.	c	34.	a	61.	b
8.	c	35.	b	62.	a
9.	a	36.	d	63.	c
10.	c	37.	b	64.	b
11.	a	38.	b	65.	c
12.	d	39.	a	66.	c
13.	c	40.	a	67.	c
14.	c	41.	d	68.	a
15.	d	42.	d	69.	d
16.	b	43.	b	70.	a
17.	b	44.	b	71.	b
18.	c	45.	b	72.	a
19.	c	46.	c	73.	d
20.	b	47.	b	74.	a
21.	c	48.	b	75.	b
22.	b	49.	b	76.	c
23.	c	50.	c	77.	b
24.	c	51.	b	78.	b
25.	b	52.	b	79.	b
26.	b	53.	d	80.	d
27.	c	54.	c		

81. Ionic hydrides form with metals from groups 1A (1) and 2A (2). These are salt-like solids, in which hydrogen is in the -1 oxidation state. Examples: NaH, CaH_2.
Covalent hydrides form with various non-metals. These are small molecules which are often gaseous under normal conditions (e.g., CH_4, H_2O, NH_3, HCl).
Metallic (interstitial) hydrides occur with d and f block metals. These are non-stoichiometric solutions of hydrogen in the metal in which the hydrogen occupies holes in the metal lattice.

82. a. M_2O
b. M_2O_2
c. MO_2

83. Graphite is the stablest allotrope under normal conditions. It is a soft solid, a lubricant and conducts electricity. The carbon atoms are bonded in planar sheets. Diamond is extremely hard, a non-conductor and is the stablest form under very high pressures. The carbons are bonded in a 3-D network. Buckminsterfullerene (C_{60}) is composed of soccer ball-like molecules, involving 5- and 6-membered carbon rings. It is a liquid.

84. a. $Ca(s) + 2H_2O(l) \rightarrow Ca(OH)_2(aq) + H_2(g)$
b. $4Al(s) + 3O_2(g) \rightarrow 2Al_2O_3(s)$

85. a. $4Li(s) + O_2(g) \rightarrow 2Li_2O(s)$
b. $N_2(g) + 3H_2(g) \rightarrow 2NH_3(g)$

86. a. $CaCO_3(s) \rightarrow CaO(s) + CO_2(g)$
b. $P_4O_{10}(s) + 6H_2O(l) \rightarrow 4H_3PO_4(aq)$

87. a. C
b. N
c. S
d. Si

88. a. 6A (16)
b. 2A (2)
c. 7A (17)
d. 1A (1)

89. a. 1A (1)
b. 5A (5)
c. 6A (6)

90. F

91. T

92. F

93. T

94. F

95. T

96. T

97. T

98. F

99. F

100. T

101. F

102. T

103. T

Organic Compounds and the Atomic Properties of Carbon
Chapter 15

Multiple Choice Questions

1. Select the correct name for the following compound.
E

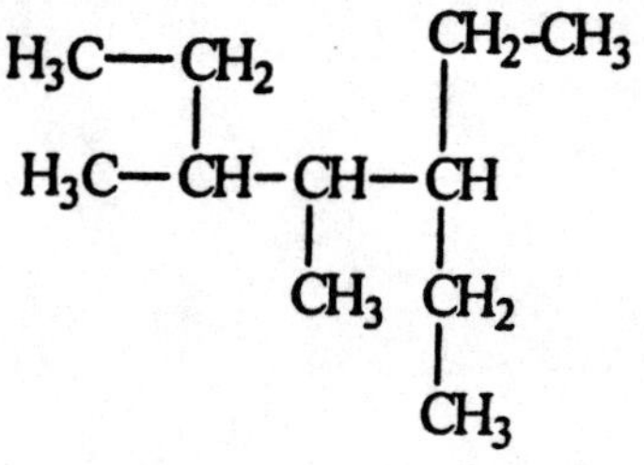

a. 1, 1, 3-triethyl-2-methylbutane
b. 1, 1-diethyl-2, 3-dimethylpentane
c. 2, 4-diethyl-3-methylhexane
d. 3-ethyl-4, 5-dimethylheptane

2. Select the correct name for the following compound.
E

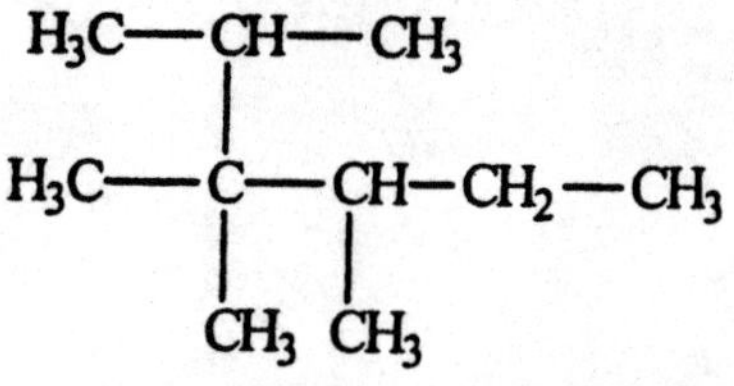

a. 2-isopropyl-2, 3, 4-trimethylbutane
b. 2-isopropyl-2, 3-dimethylpentane
c. 2, 3, 3, 4 -tetramethylhexane
d. 1, 1, 2, 2, 3-pentamethylpentane

3. Select the correct name for the following compound.
M

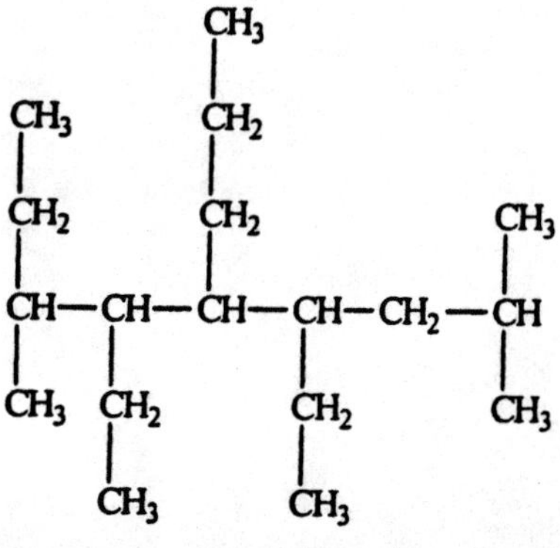

a. 1, 2-diethyl-1-methyl-3-propyl-4-isobutylhexane
b. 1, 6, 6-trimethyl-1, 2, 4-triethyl-3-propylhexane
c. 1, 1, 6-trimethyl-3, 5, 6-triethyl-4-propylhexane
d. 4, 6-diethyl-2, 7 dimethyl-5-propylnonane

4. Select the correct name for the following compound.
E

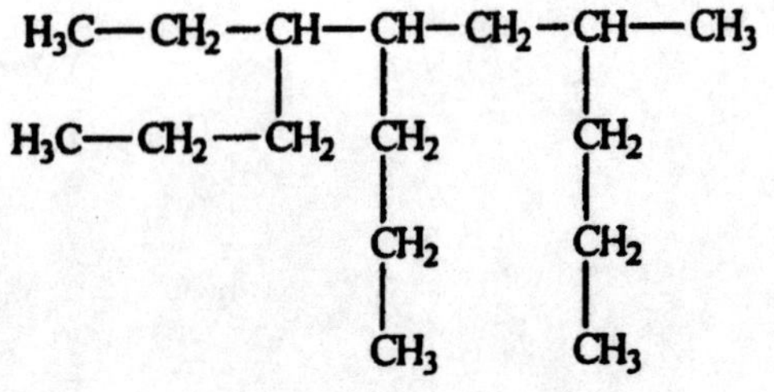

a. 2, 4, 5-tripropylheptane
b. 4-ethyl-7-methyl-5-propyldecane
c. 4-ethyl-5, 7-dipropyloctane
d. 5-ethyl-2, 4-dipropyloctane

5. Select the correct name for the following compound.
M

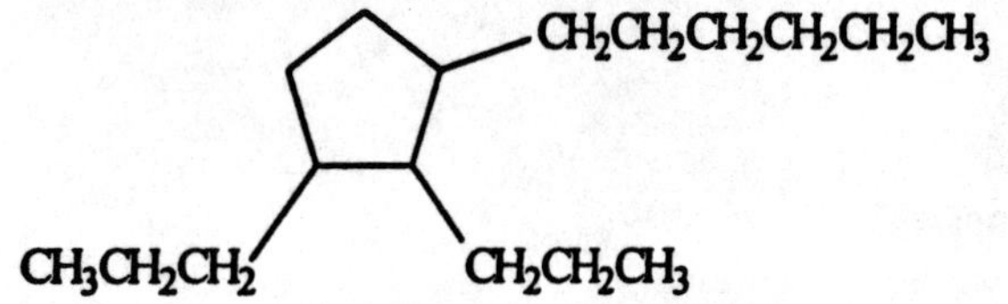

a. *ortho*-dipropylcyclopentylhexane
b. 2, 3-dipropylcyclopentylhexane
c. 2-hexyl-1, 5-dipropylcyclopentane
d. 1-hexyl-2, 3-dipropylcyclopentane

6. Select the correct name for the following compound.
E

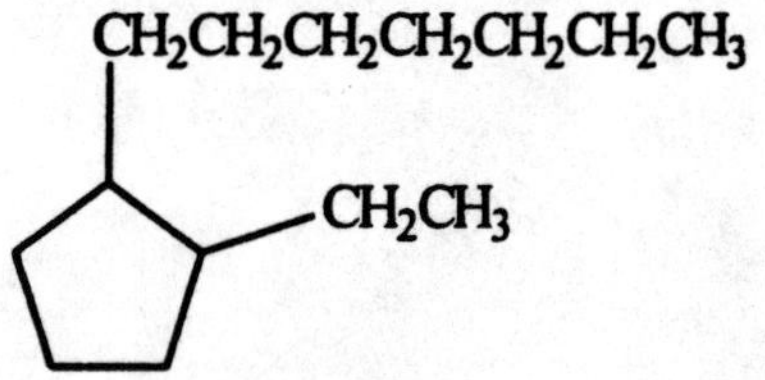

a. *ortho*-ethylheptylcyclopentane
b. *meta*-ethylheptylcyclopentane
c. 1-ethyl-2-heptylcyclopentane
d. ethylcyclopentylheptane

7. Select the correct name for the following compound.
E

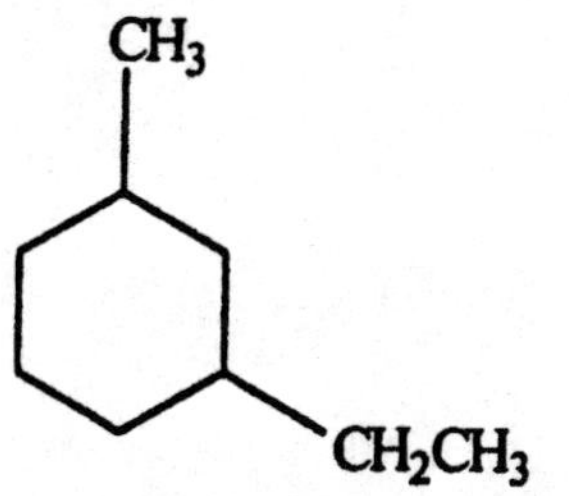

a. 1-ethyl-3-methylcyclohexane
b. 1-methyl-5-thylcyclohexane
c. *meta*-ethylmethylcyclohexane
d. *meta*-ethylmethylbenzene

8. Select the correct name for the following compound.
E

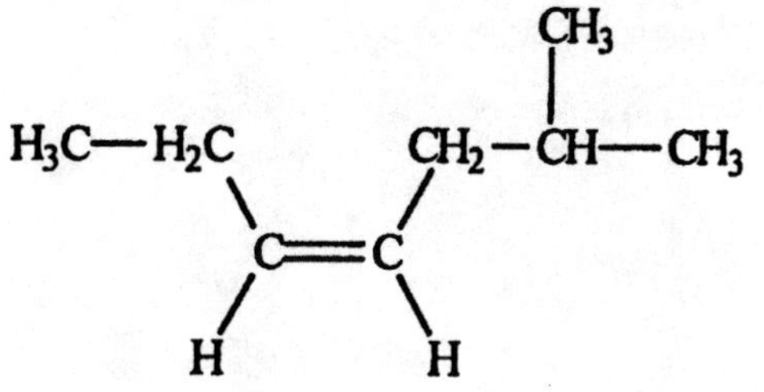

a. *cis*-2-methyl-4-heptene
b. *trans*-2-methyl-4-heptene
c. *cis*-6-methyl-3-heptene
d. *trans*-6-methyl-3-heptene

9. Select the correct name for the following compound.
M

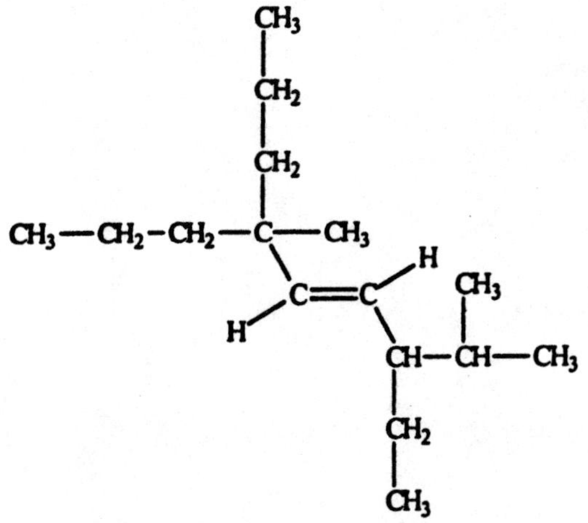

a. *cis*-3-ethyl-2,6–dimethyl-6-propyl-4-nonene
b. *trans*-3-ethyl-2,6-dimethyl-6-propyl-4-nonene
c. *cis*-7-ethyl-4,8-dimethyl-4-propyl-5-nonene
d. *trans*-7-ethyl-4,8-dimethyl-4-propyl-5-nonene

10. Select the correct name for the following compound.
M

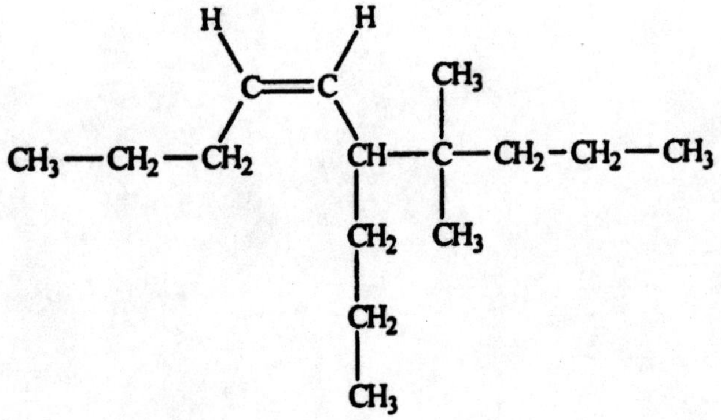

a. *cis*-2-methyl-2,3-dipropyl-4-octene
b. *cis*-7-methyl-6,7-dipropyl-4-octene
c. *cis*-4,4-dimethyl-5-propyl-6-decene
d. *cis*-7,7-dimethyl-6-propyl-4-decene

11. Select the correct name for the following compound.
M

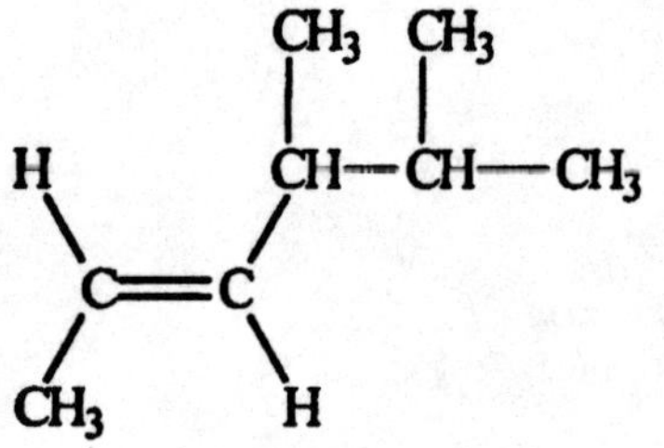

a. *cis*-2,3-dimethyl-4-hexene
b. *trans*-2,3-dimethyl-4-hexene
c. *cis*-4,5-dimethyl-2-hexene
d. *trans*-4,5-dimethyl-2-hexene

12. Select the correct name for the following compound.
M

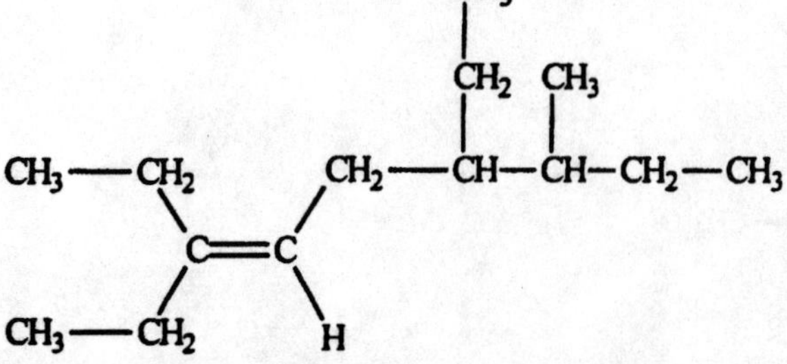

a. *cis*-2,5-diethyl-6-methyl-3-nonene
b. 2,5-diethyl-6-methyl-3-nonene
c. *cis*-4,6-diethyl-3-methyl-6-nonene
d. 4,6-diethyl-3-methyl-6-nonene

13. Select the correct name for the following compound.
E

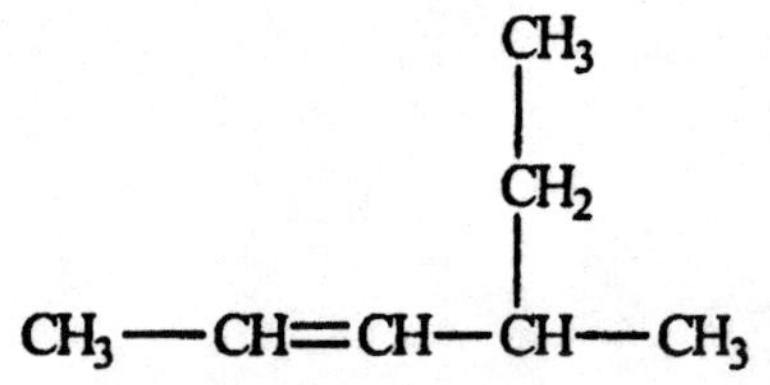

a. 2-ethyl-3-pentene
b. 4-ethyl-2-pentene
c. 3-methyl-4-hexene
d. 4-methyl-2-hexene

14. Select the correct name for the following compound.
M

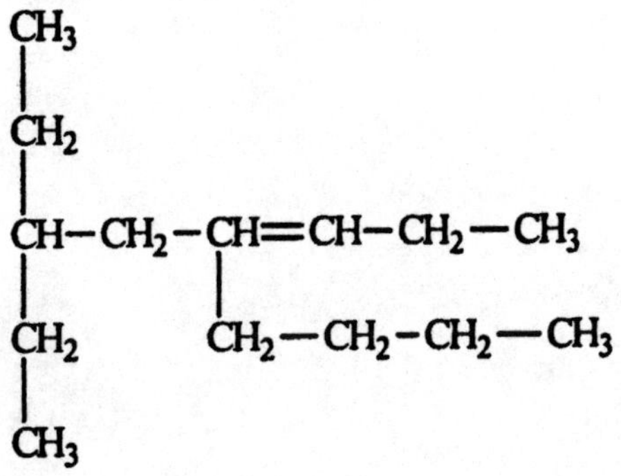

a. 1,1-diethyl-3-butyl-3-hexene
b. 5-butyl-3-ethyl-5-octene
c. 4-butyl-6-ethyl-3-octene
d. 3-ethyl-5-propyl-5-nonene

15. Select the correct name for the following compound.
M

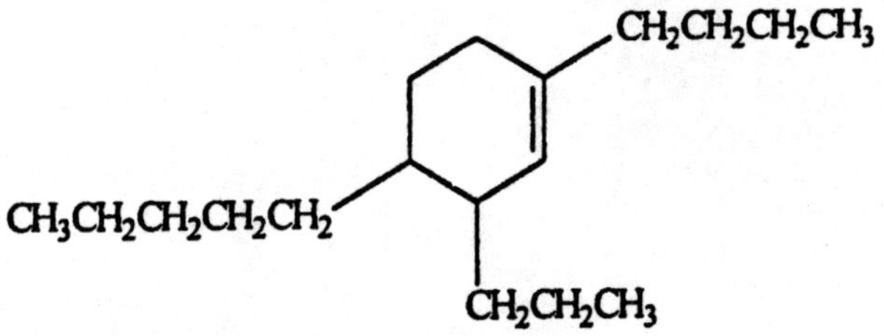

a. 1-butyl-4-pentyl-3-propylcyclohexene
b. 1-butyl-4-pentyl-5-propylcyclohexene
c. 2-butyl-5-pentyl-6-propylcyclohexene
d. 4-butyl-1-pentyl-2-propylcyclohexene

16. Select the correct name for the following compound.
M

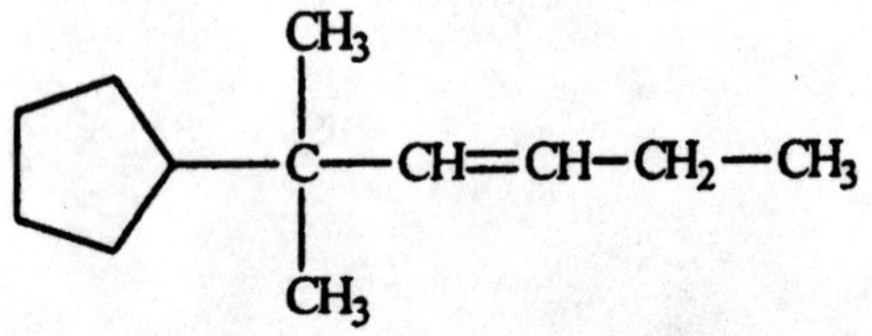

a. 1,1-dimethyl-1-cylcopentyl-2-pentane
b. 1,1-dimethyl-2-hexene-cyclopentane
c. 2-cyclopentyl-2-methyl-3-hexene
d. 5,5-dimethyl-5-cyclopentyl-3-pentane

17. Select the correct name for the following compound.
M

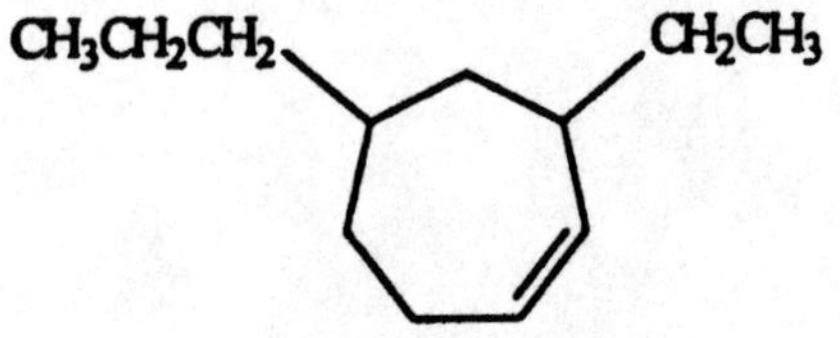

a. 2-ethyl-4-propylcycloheptene
b. 3-ethyl-5-propylcycloheptene
c. 6-ethyl-4-propylcycloheptene
d. 7-ethyl-5-propylcycloheptene

18. Select the correct name for the following compound.
E

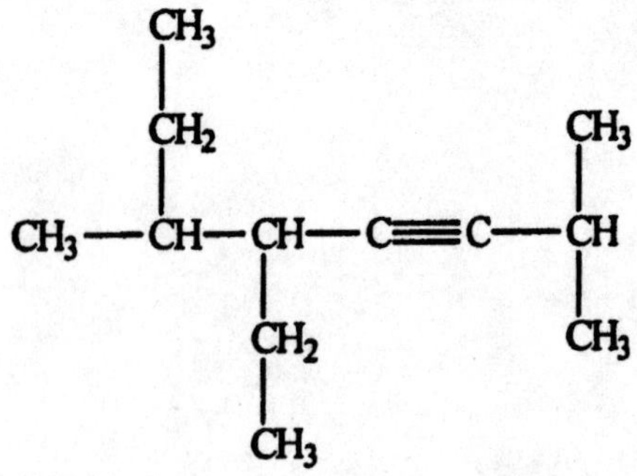

a. 4-ethyl-1,1,5-trimethyl-2-heptyne
b. 4,5-diethyl-1,1-dimethyl-2-heptyne
c. 5-ethyl-2,6-dimethyl-3-octyne
d. 3-ethyl-3,7-dimethyl-5-octyne

19. Select the correct name for the following compound.
E

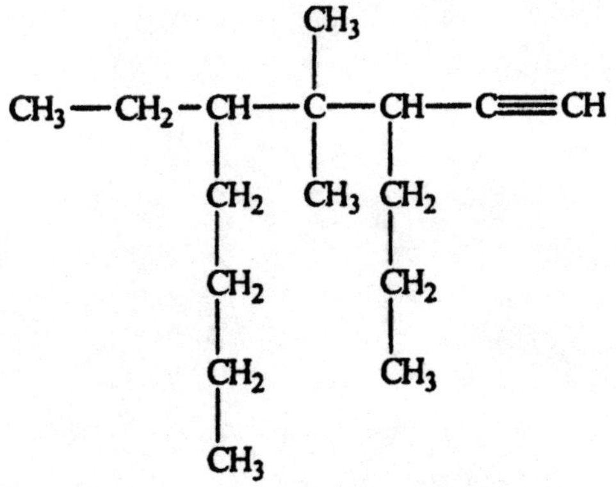

a. 5-butyl-4,4-dimethyl-3-propyl-1-heptene
b. 5-ethyl-4-4-dimethyl-3-propyl-1-nonyne
c. 5-ethyl-6-6-dimethyl-7-propyl-8-nonyne
d. 4,6-diethyl-5-5-dimethyl-4-decyne

20. Select the correct name for the following compound.
E

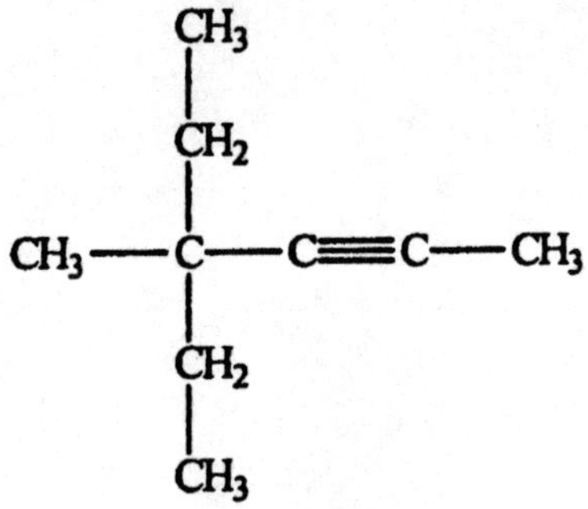

a. 4-4-diethyl-2-pentyne
b. 2,2-diethyl-3-pentyne
c. 2-ethyl-2-methyl-4-hexyne
d. 4-ethyl-4-methyl-2-hexyne

21. Select the correct name for the following compound.
M

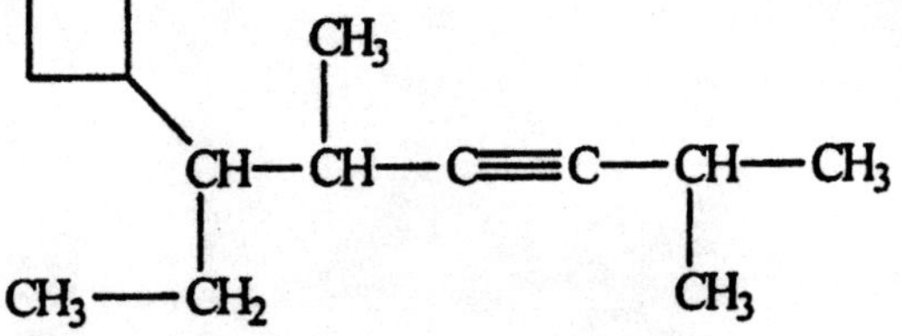

a. 2,5-dimethyl-3-octyne-6-cyclobutane
b. 4-cyclobutyl-1-isopropyl-3-methyl-1-hexyne
c. 3-cyclobutyl-4,7-dimethyl-5-octyne
d. 6-cyclobutyl-2,5-dimethyl-3-octyne

22. Select the correct name for the following compound.
E

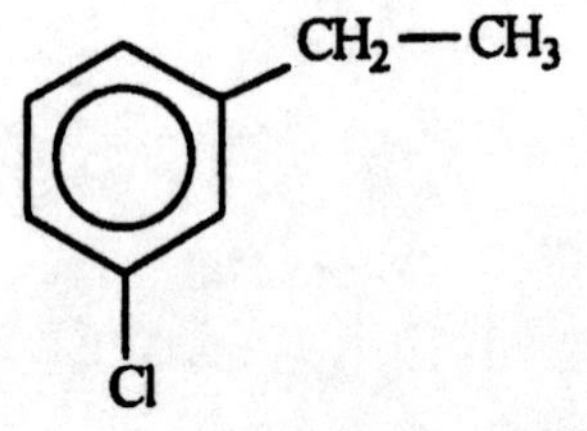

a. *meta*-chloroethylcyclohexene
b. 1-chloro-5-ethylcyclohexene
c. *meta*-chloroethylbenzene
d. 1-chloro-5-ethylbenzene

23. Select the correct name for the following compound.
M

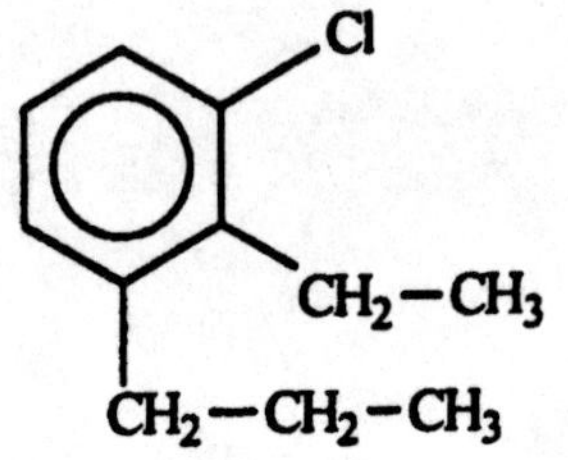

a. 1-chloro-2-ethyl-3-propylbenzene
b. 1-chloro-2-ethyl-3-propylcyclohexene
c. 1-chloro-*ortho*-ethylpropylbenzene
d. 1-chloro-*meta*-ethylpropylbenzene

24. Select the correct name for the following compound.
E

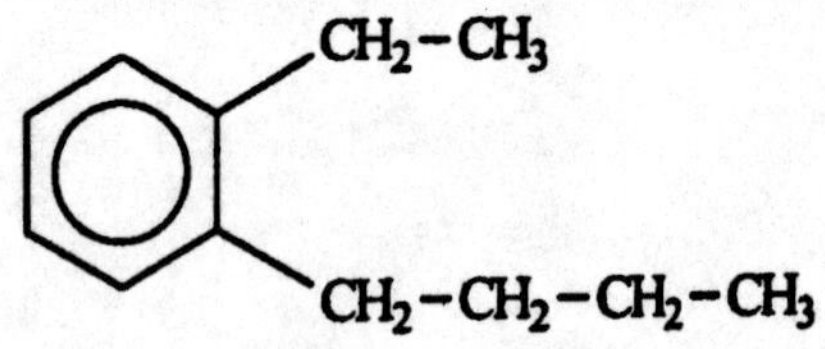

a. *ortho*-butylethylbenzene
b. *meta*-butylethylbenzene
c. *para*-butylethylbenzene
d. 1-butyl-2-ethylcyclohexene

25. Select the correct name for the following compound.
E

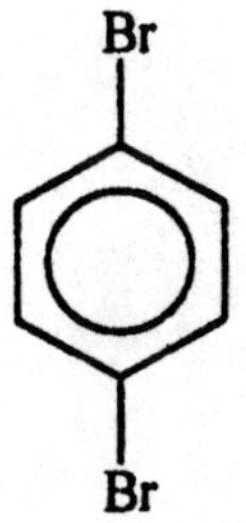

a. *ortho*-dibromobenzene
b. *meta*-dibromobenzene
c. *para*-dibromobenzene
d. 1,4-dibromocyclohexene

26. Select the correct name for the following compound.
M

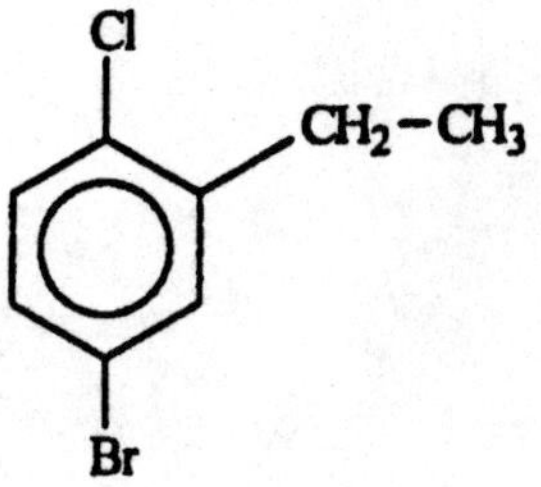

a. *para*-bromochloro-2-ethylbenzene
b. 4-bromo-1-chloro-2-ethylbenzene
c. 5-bromo-2-chloro-1-ethylbenzene
d. 1-bromo-4-chloro-3-ethylcyclohexene

27. Select the correct name for the following compound.
M

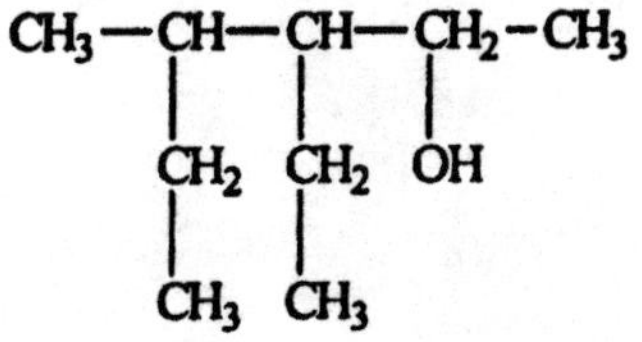

a. 3,4-diethyl-4-methyl-2-butanol
b. 2,3-diethyl-4-pentanol
c. 3,4-diethyl-2-pentanol
d. 3-ethyl-4-methyl-2-hexanol

28.
M

Select the correct name for the following compound.

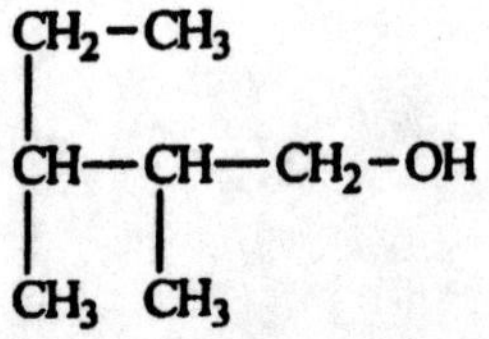

a. 3-ethyl-2,3-dimethyl-1-propanol
b. 2,3,4-trimethyl-1-butanol
c. 2,3-dimethyl-1-pentanol
d. 3,4-dimethyl-5-pentanol

29.
E

Select the compound that will be optically active.

a. 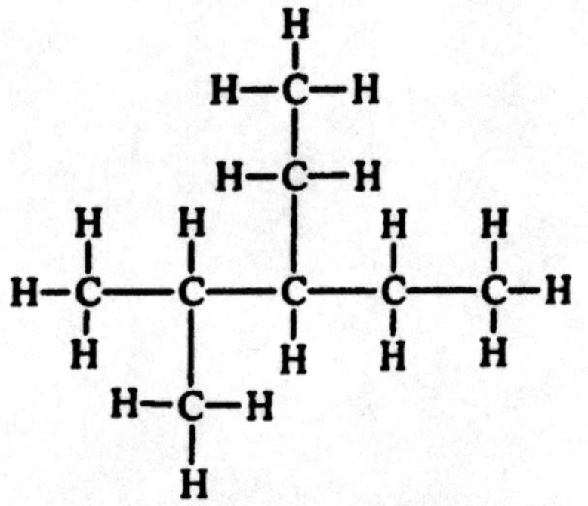

b. 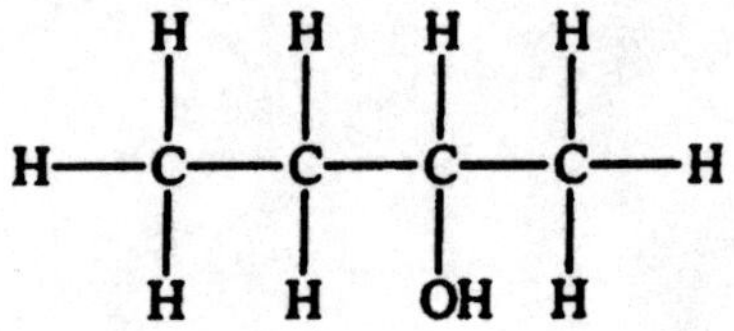

c. 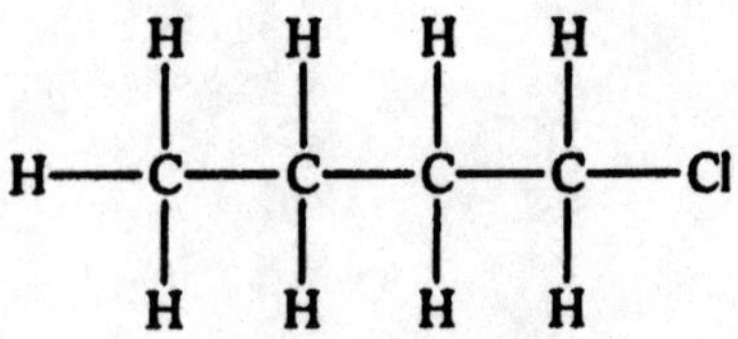

d. 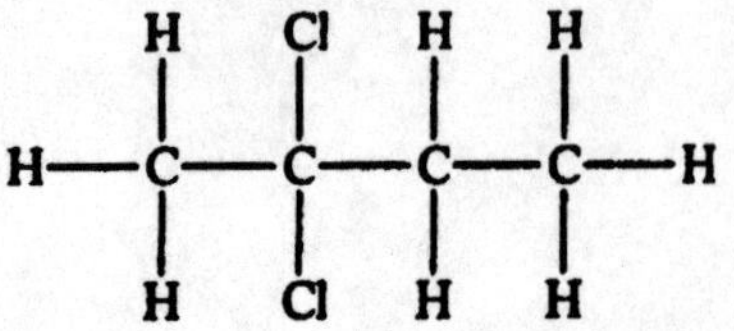

30. Select the compound that is not optically active.
E

a. 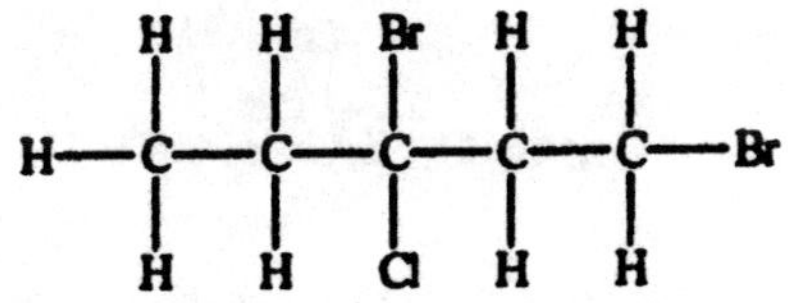

b. 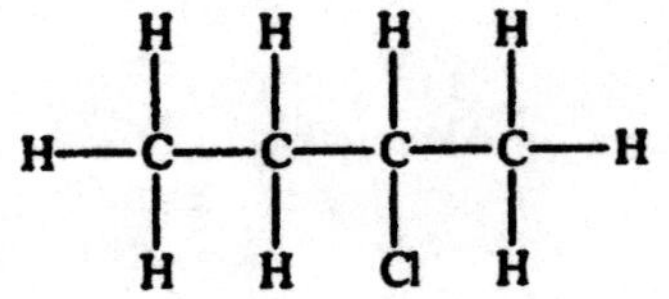

c. 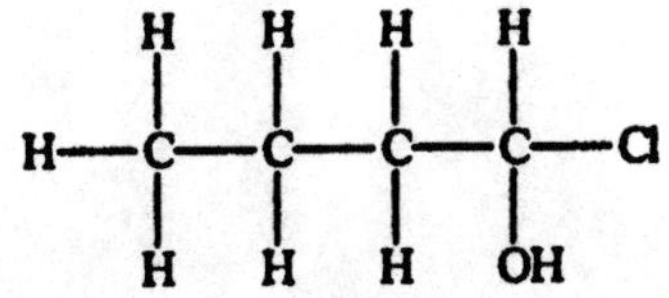

d. 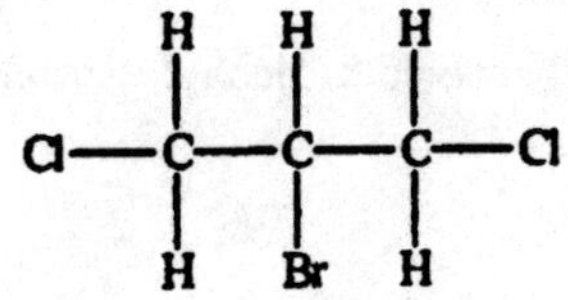

31. Select the correct type for the following reaction.
E

$$CH_3CH_2CH_2CH_2OH + HBr \rightarrow CH_3CH_2CH_2CH_2Br + H_2O$$

a. addition
b. elimination
c. substitution

32. Benzene will react with chloromethane in the presence of a catalyst to produce toluene and hydrogen chloride. Select the correct reaction type for the process.
E

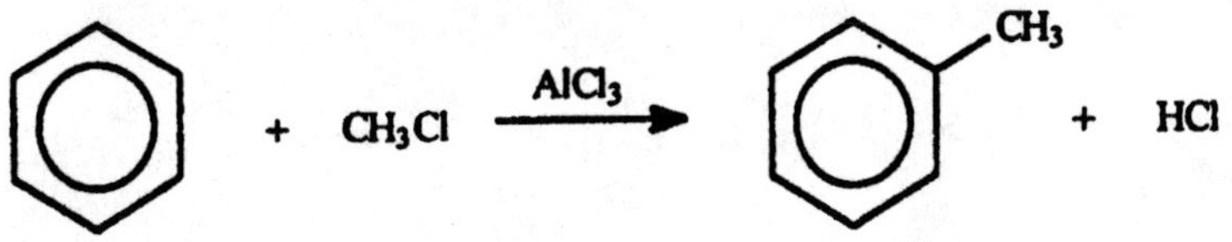

a. addition
b. elimination
c. substitution

33. Select the correct reaction type for the following process.
E

a. addition
b. elimination
c. substitution

34. The reaction of bromine with an alkene such as cyclopentene is a good laboratory test for the presence of a double bond in a compound. What type of reaction is it?
E

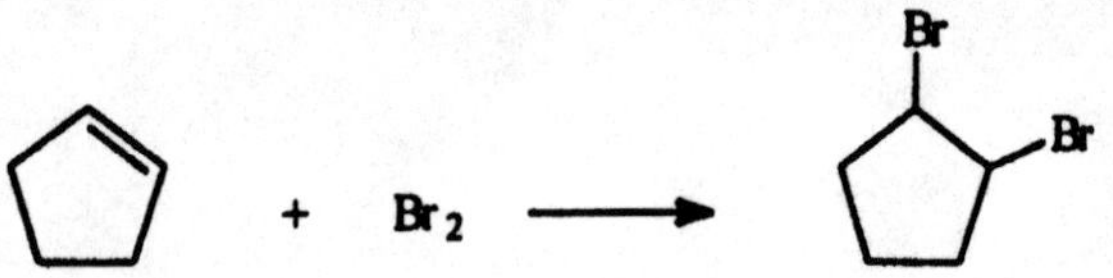

a. addition
b. elimination
c. substitution

35. 2-chloro-2,3-dimethylbutane will react with potassium hydroxide dissolved in alcohol to produce 2,3-dimethyl-2-butene. What type of reaction is this?
E

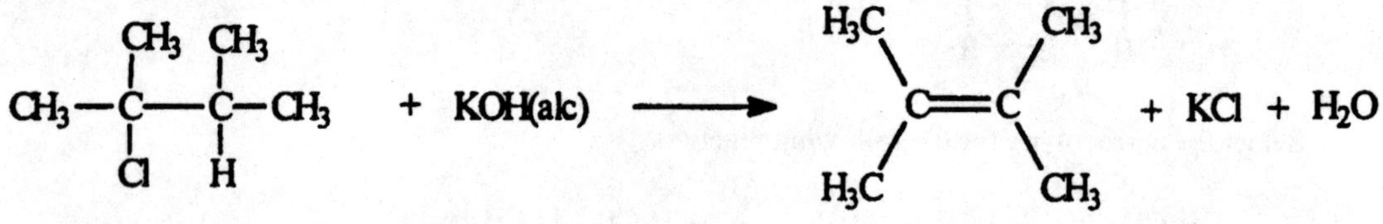

a. addition
b. elimination
c. substitution

36. Select the correct type for the following reaction.
E

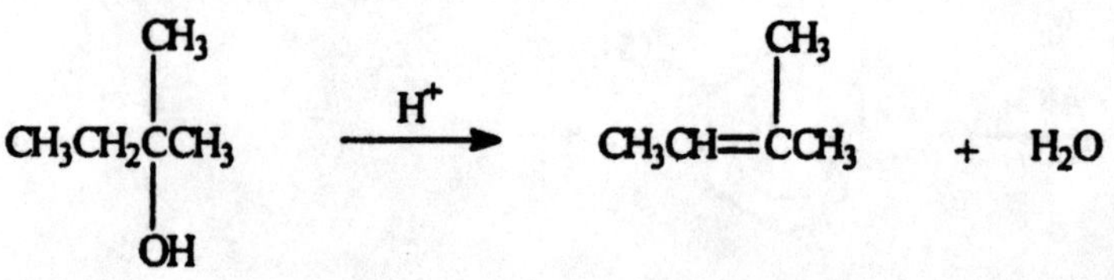

a. addition
b. elimination
c. substitution

37. Select the correct type for the following reaction.
M

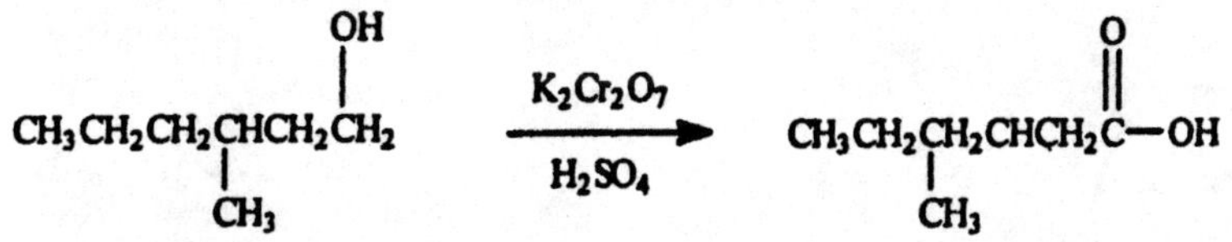

a. addition
b. elimination
c. substitution
d. oxidation

38. Select the correct type for the following reaction.
M

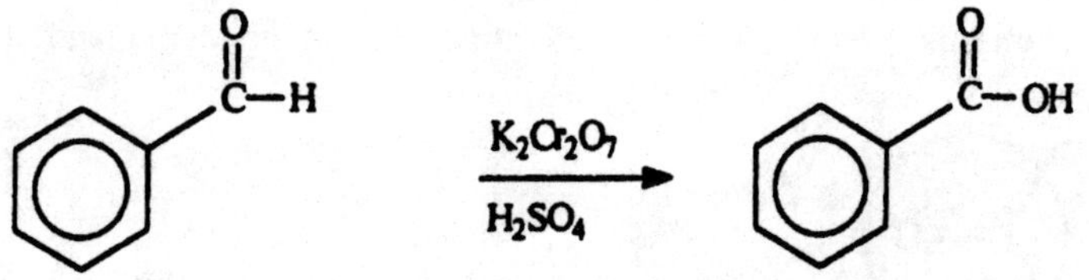

a. addition
b. elimination
c. substitution
d. oxidation

39. Identify the functional group circled.
E

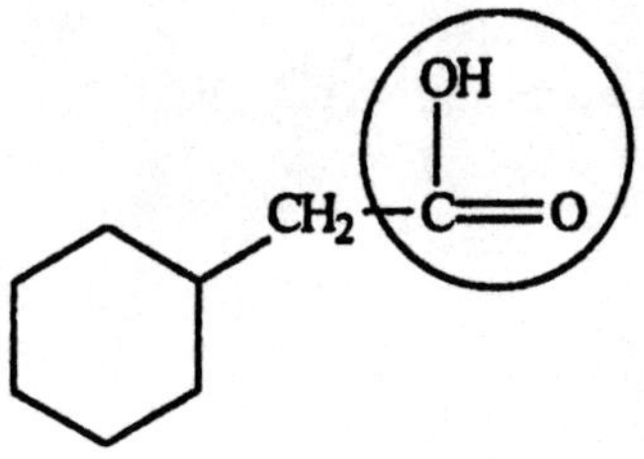

a. aldehyde
b. ketone
c. alcohol
d. carboxylic acid

40. Vanillin is used as a flavoring agent. Identify the functional group circled.
E

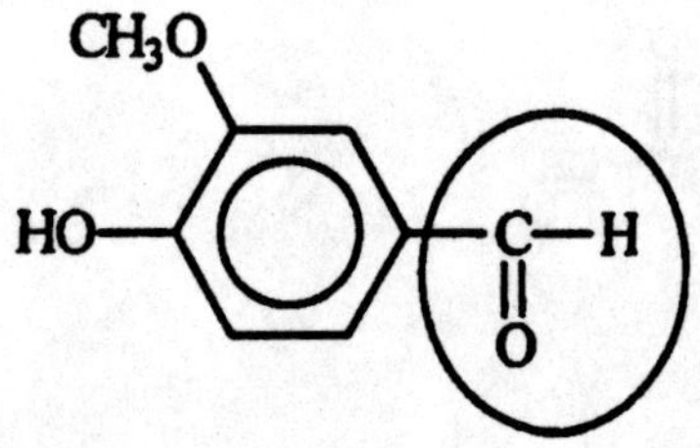

a. aldehyde
b. ketone
c. alcohol
d. carboxylic acid

41. One source of a musky odor in perfumes is civetone, a compound extracted from the scent gland of the civet cat. Identify the functional group circled.
E

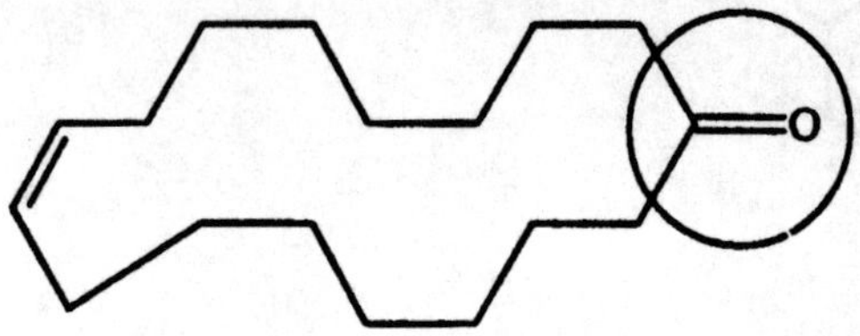

a. aldehyde
b. ketone
c. alcohol
d. carboxylic acid

42. Aspirin is an effective and widely used pain reliever. Identify the functional group circled.
M

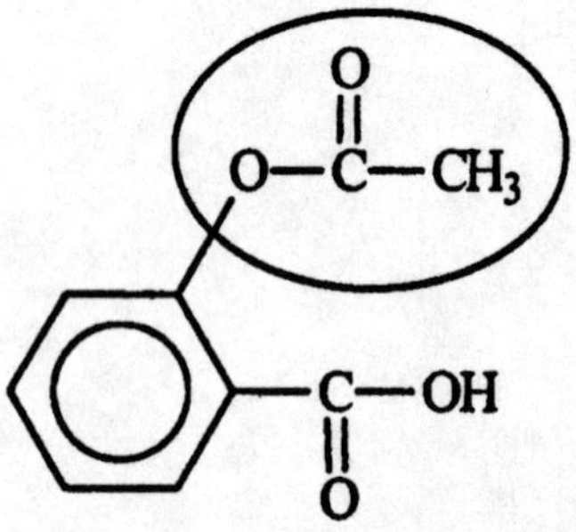

a. aldehyde
b. ketone
c. ester
d. carboxylic acid

43. Enflurane is an effective gaseous anesthetic with relatively low flammability. Identify the functional
E group circled.

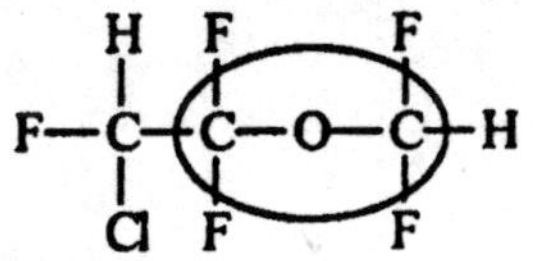

a. aldehyde
b. ketone
c. ester
d. ether

44. Anethole, a derivative of anise, is used in flavoring and as perfume in soap and toothpaste. Identify the
E functional group circled.

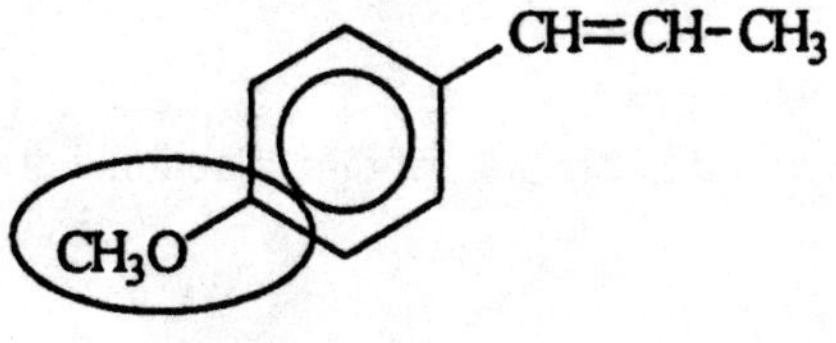

a. aldehyde
b. ketone
c. ester
d. ether

45. Glycerin is used in cosmetics as a moisturizer. Identify the functional group circled.
E

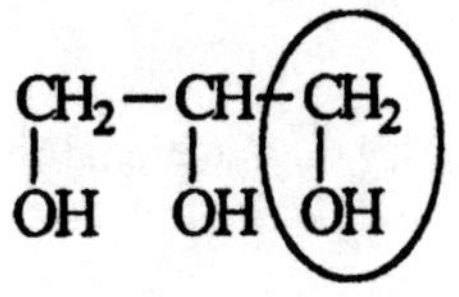

a. carboxylic acid
b. alcohol
c. ester
d. ether

46. Glucose is an important sugar in a person's metabolic cycle. Identify the functional group circled.
E

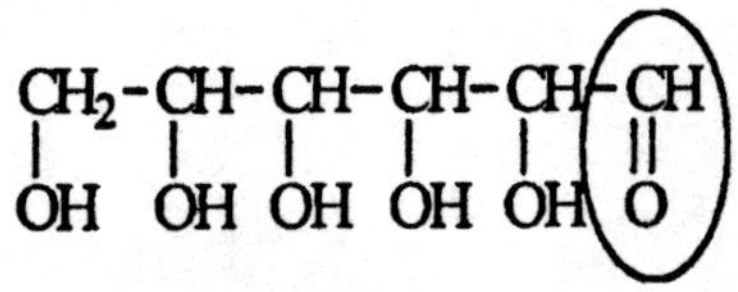

a. aldehyde
b. ketone
c. alcohol
d. ester

47. Testosterone is a male hormone. Identify the functional group circled.
E

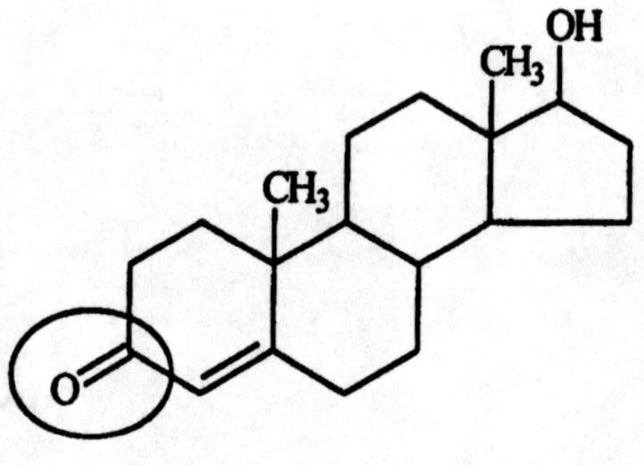

a. aldehyde
b. ketone
c. alcohol
d. ester

48. The compound shown below is responsible for the odor in rancid butter. Identify the functional group circled.
E

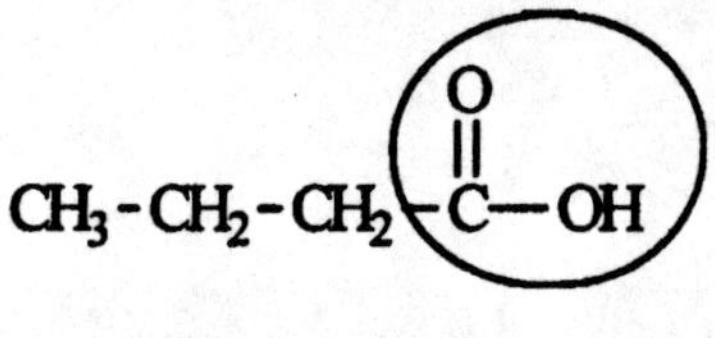

a. aldehyde
b. ketone
c. alcohol
d. carboxylic acid

49. Benzocaine is from a family of chemicals that are good local anesthetics. Identify the functional group circled.
M

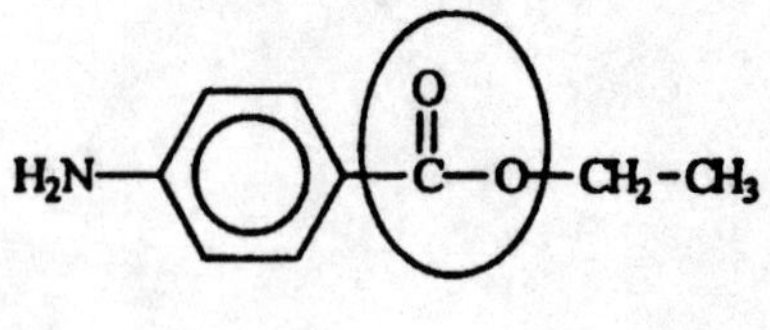

a. aldehyde
b. ketone
c. ester
d. carboxylic acid

50.
E

Putrescine is produced during the decay and protein breakdown of meats and is responsible for some of the odor found in them. Identify the functional group circled.

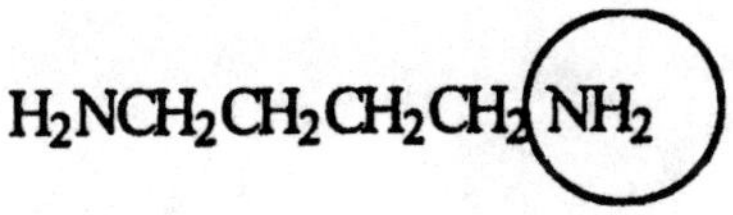

a. aldehyde
b. ketone
c. amide
d. amine

51.
E

Serotonin transmits nerve impulses through the body. Identify the functional group circled.

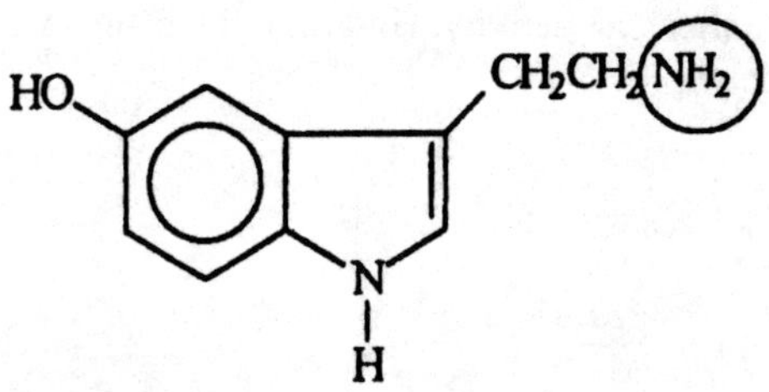

a. aldehyde
b. alcohol
c. amide
d. amine

52.
M

Acetominophen is a widely used and an effective pain reliever. Identify the functional group circled.

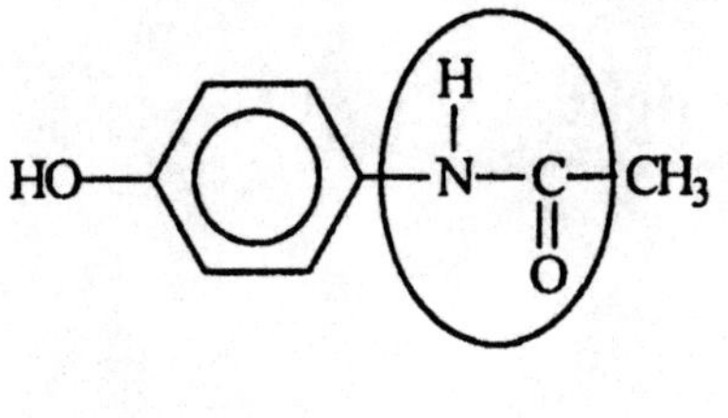

a. aldehyde
b. alcohol
c. amide
d. amine

53.
M

Urea carries waste nitrogen from the body in urine. Identify the functional group circled.

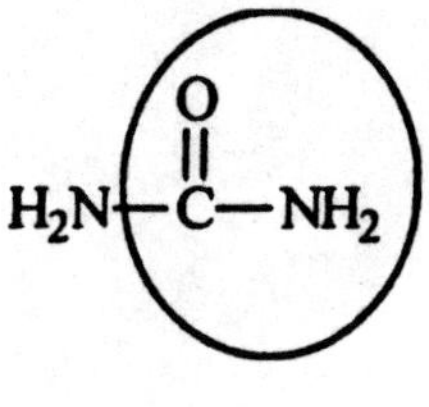

a. aldehyde
b. alcohol
c. amide
d. amine

54. The millipede ejects the compound shown below to protect itself from its enemies. Identify the functional group circled.
M

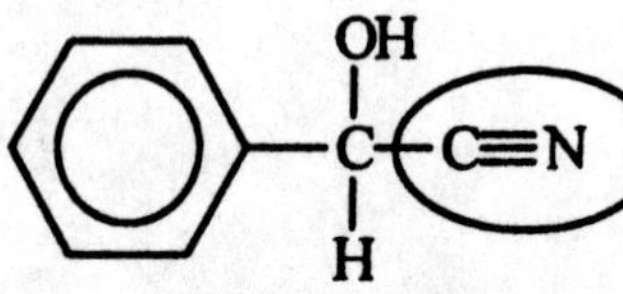

a. nitrile
b. alcohol
c. amide
d. amine

55. Amygdalin (Laetrile) was once touted for its potential anticancer activity. However, the claims were never scientifically verified. Identify the functional group circled.
M

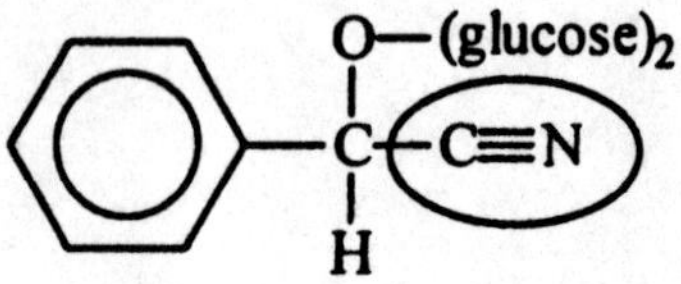

a. nitrile
b. ether
c. amide
d. amine

56. Identify the principal organic product of the reaction between butane and chlorine.
M

$$CH_3CH_2CH_2CH_3 + Cl_2 \xrightarrow{light}$$

a. CH_3Cl
b. CH_3CH_2Cl
c. $CH_3CHClCH_3$
d. $CH_3CH_2CHClCH_3$

57. Identify the principal organic product when benzyl chloride reacts with aqueous sodium hydroxide.
M

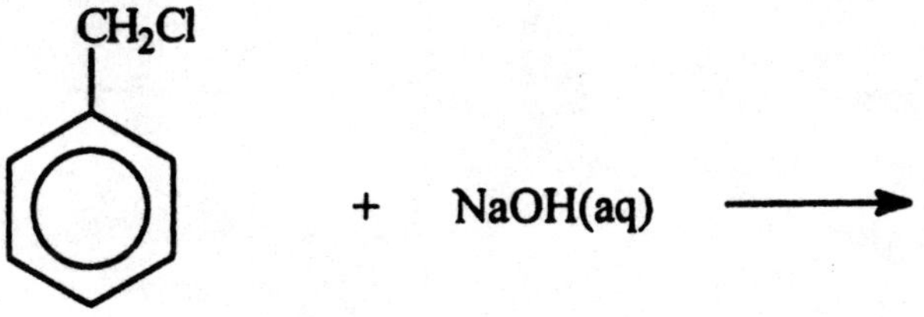

a. 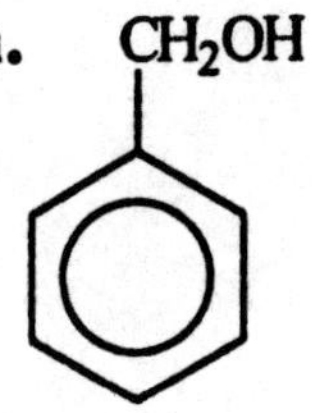

b. 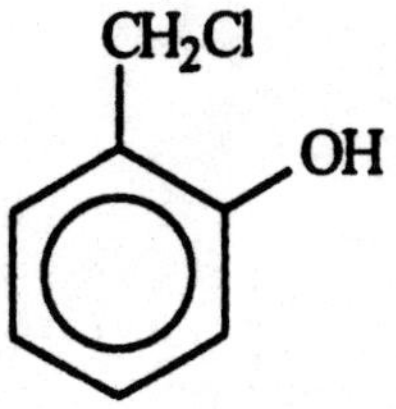

c. 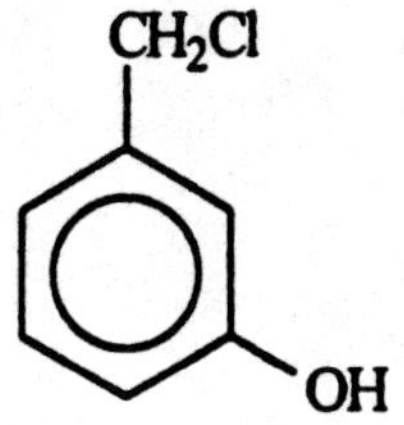

d. 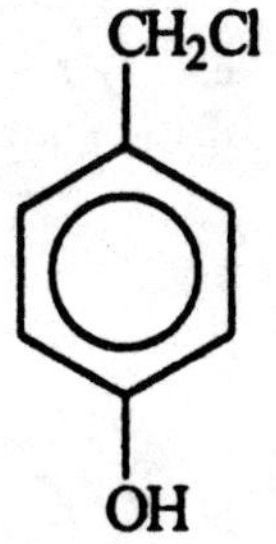

58. Identify the two principal products of the reaction between ammonia and ethyl propionate.
H

$$CH_3-CH_2-C(=O)-O-CH_2-CH_3 \quad + \quad NH_3 \longrightarrow$$

a. $CH_3-C(=O)-NH_2 \quad + \quad CH_3-CH_2-CH_2-OH$

b. $CH_3-CH_2-C(=O)-NH_2 \quad + \quad CH_3-CH_2-OH$

c. $CH_3-CH_2-NH_2 \quad + \quad CH_3-CH_2-C(=O)-OH$

d. $CH_3-CH_2-O-C(=O)-NH_2 \quad + \quad CH_3CH_3$

59. Identify the products of the following reaction.

H

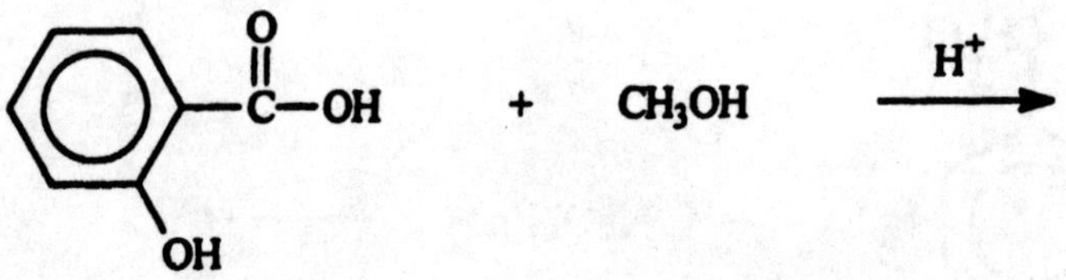

a. 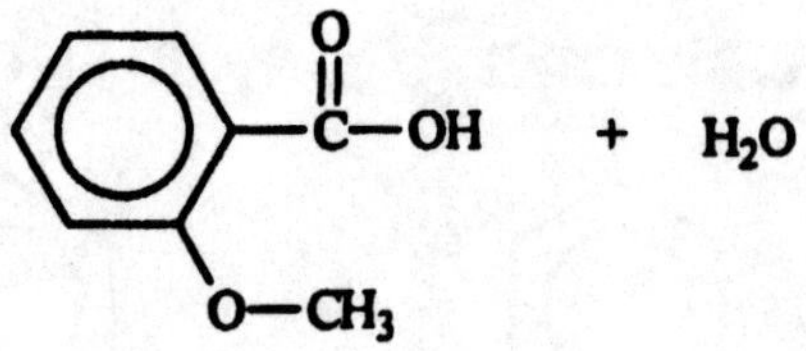

b. 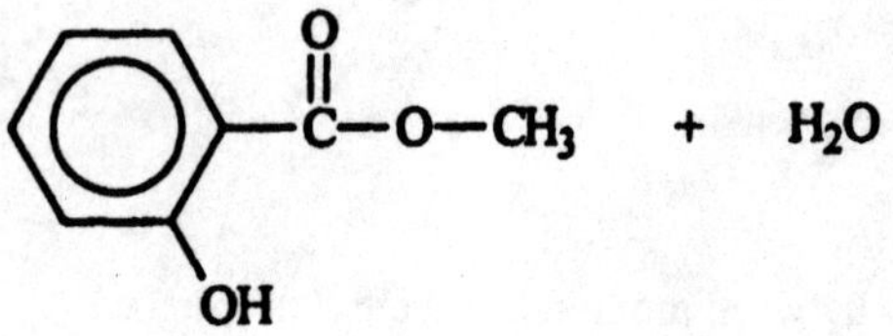

c. 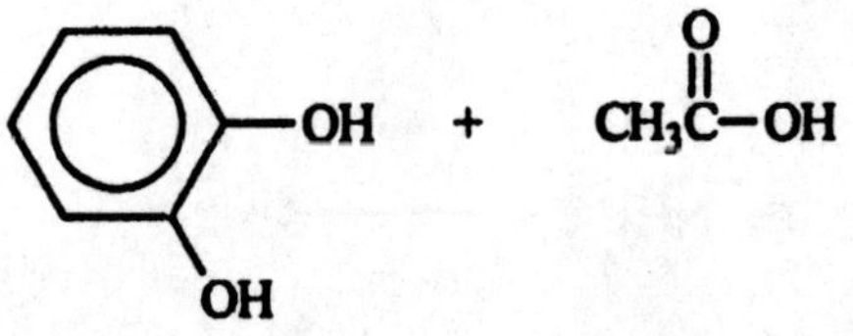

d. 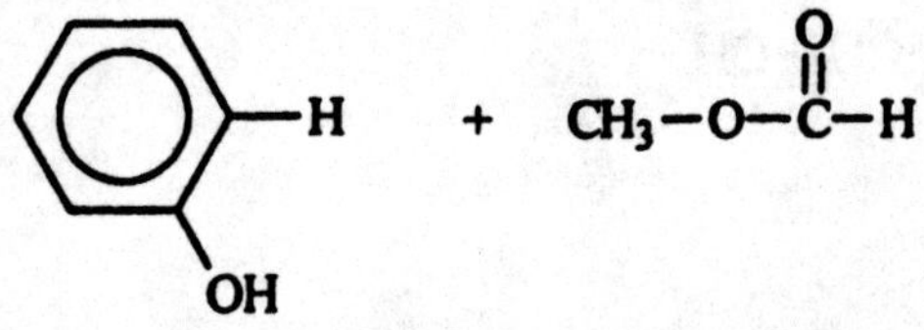

60. Identify the products for the reaction between cyclohexene and bromine.
M

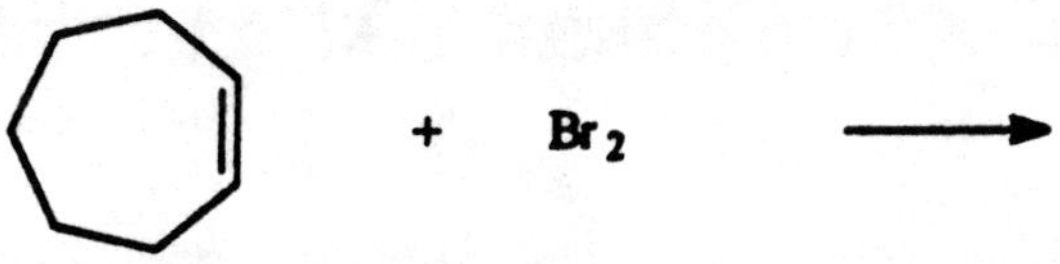

a. 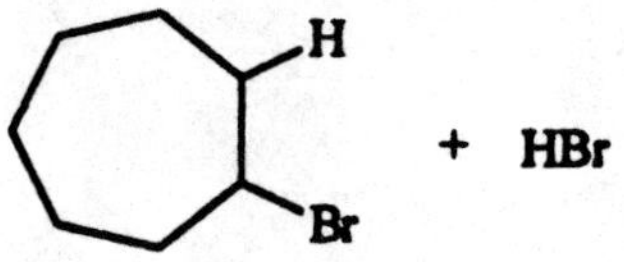

b. 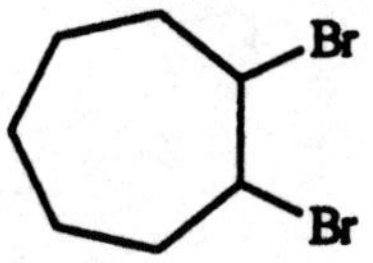

c. 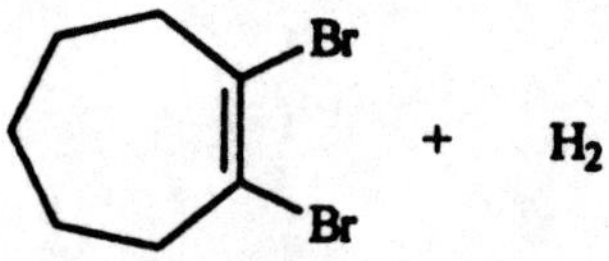

d. 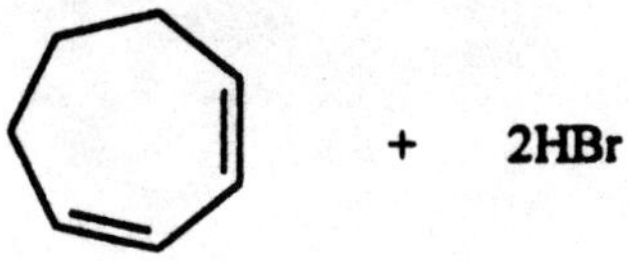

61. Identify the products of the reaction of 3-octene with chlorine.
M

$$CH_3CH_2CH_2CH_2CH{=}CHCH_2CH_3 \quad + \quad Cl_2 \longrightarrow$$

a. 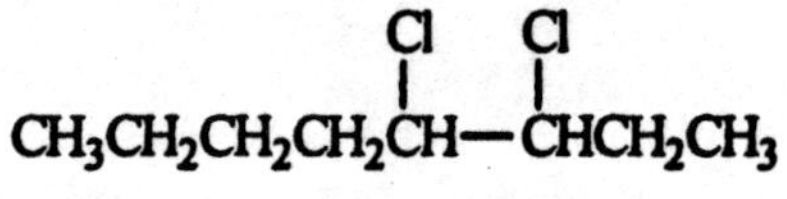

b. 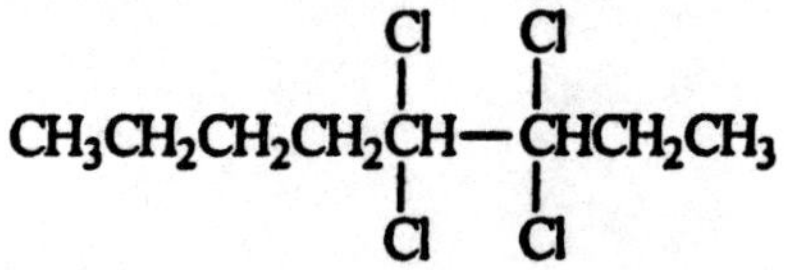

c. 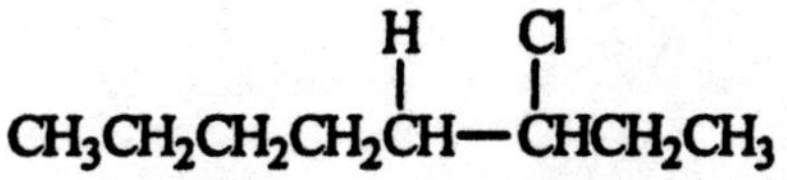

d. 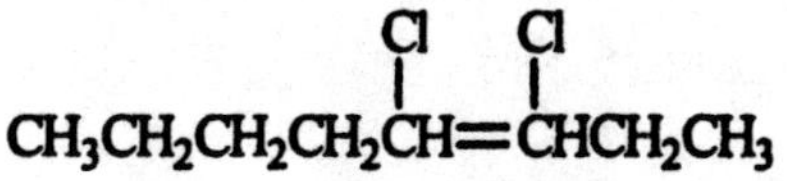

62. Identify the products of the reaction between 2-bromopentane and potassium ethoxide.
H

$$CH_3CH_2CH_2\overset{\overset{\large Br}{|}}{C}HCH_3 \quad + \quad CH_3CH_2{-}O^-K^+ \longrightarrow$$

a. $CH_3CH_2CH_2CH_2CH_3 \quad + \quad CH_3CH_2Br \quad + \quad KOH$

b. $CH_3CH_2CH_2{-}O{-}CH_2CH_3 \quad + \quad CH_3CH_3 \quad + \quad KBr$

c. $CH_3CH_2CH{=}CHCH_3 \quad + \quad CH_3CH_2OH \quad + \quad KBr$

d. $CH_3CH_2CH_2\underset{\underset{O-CH_2CH_3}{|}}{C}HCH_3 \quad + \quad KBr$

63. Identify the principal organic products for the following reaction.
H

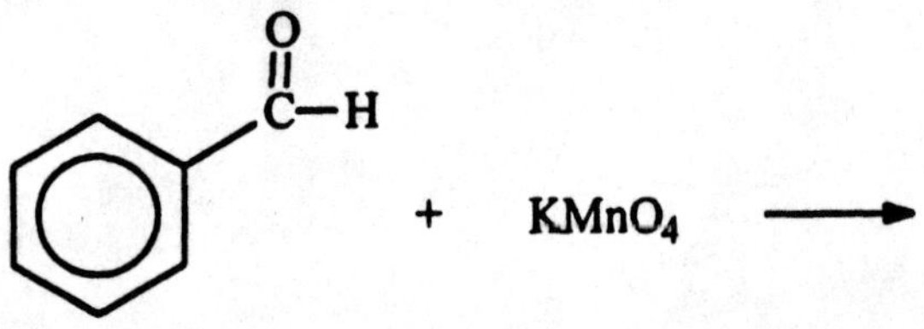

a. 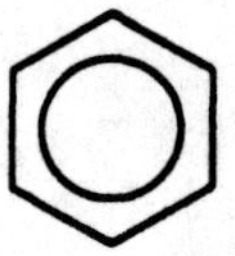 + CO_2

b. 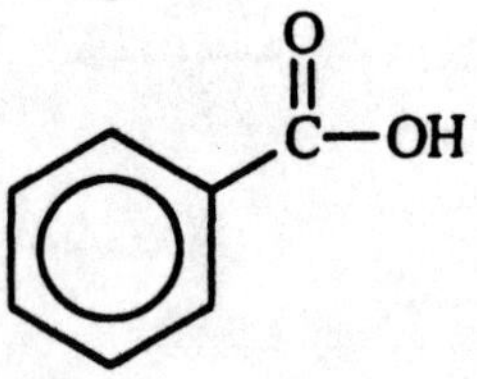

c. 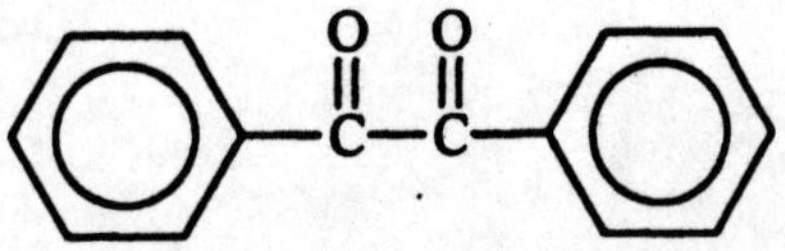

d. 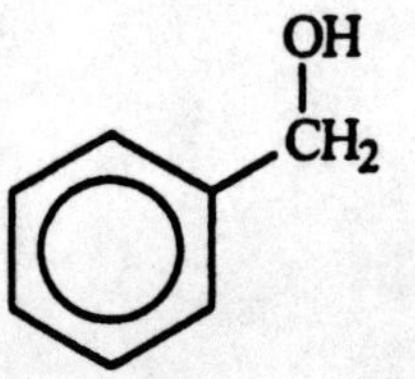

64. Identify the organic product when cyclohexanol reacts with excess potassium dichromate in the presence of
M sulfuric acid.

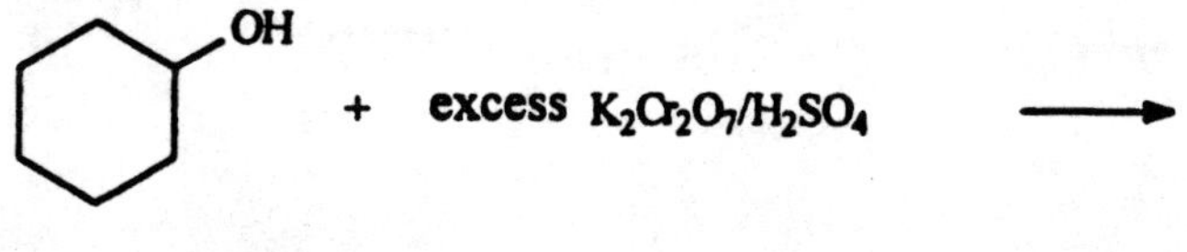

a. 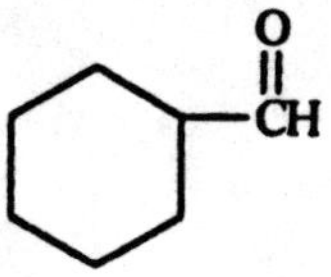

b. 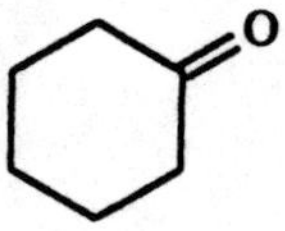

c. 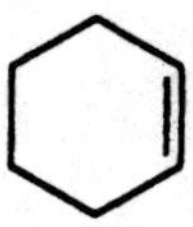

d. 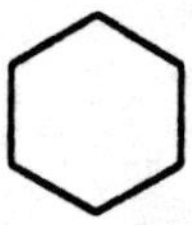

65. Identify the organic product when 3-cylcobutyl-1-propanol reacts with excess potassium dichromate in
H sulfuric acid.

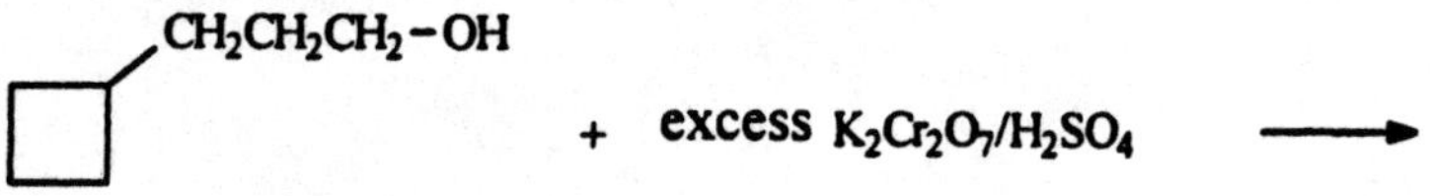

a. 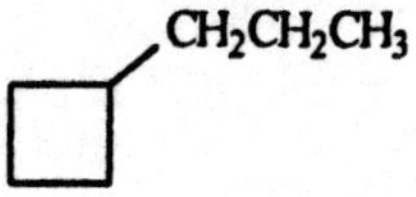

b. 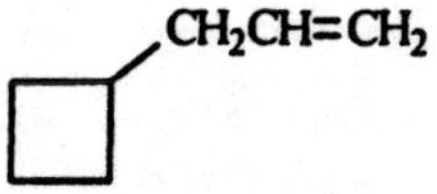

c. 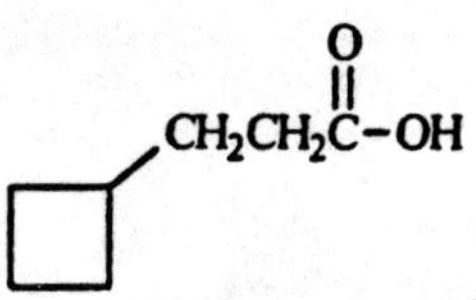

d. 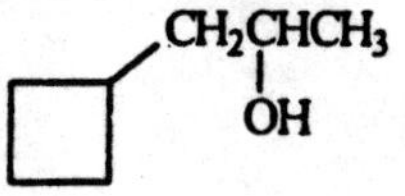

66. Identify the organic product for the reaction of 2-pentanol with sulfuric acid.
M

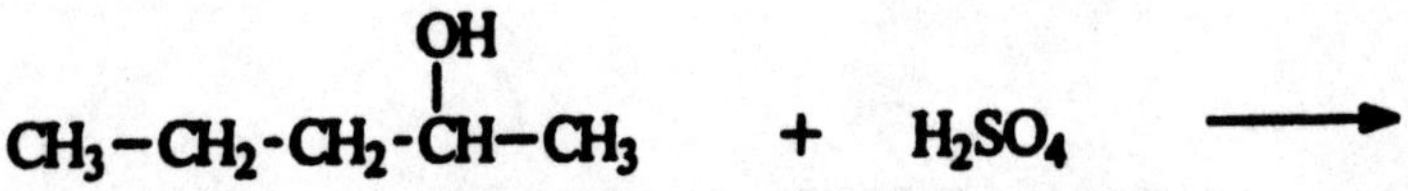

a. 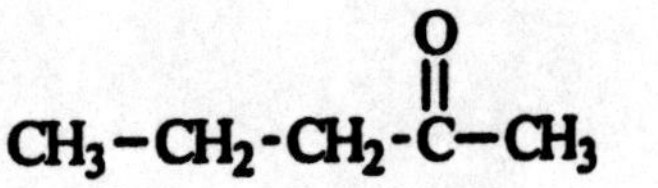

b. 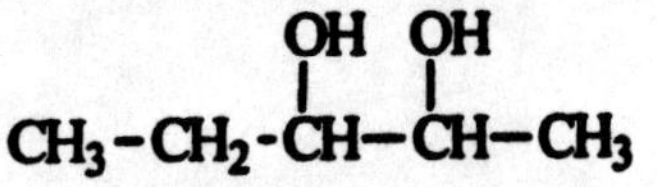

c. $CH_3-CH_2-CH_2-CH_2-CH_3$

d. $CH_3-CH_2-CH=CH-CH_3$

67. Identify the organic product when cyclopentanol reacts with sulfuric acid.
M

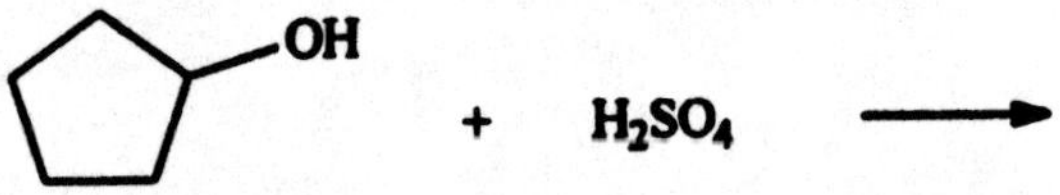

a. 

b. 

c. 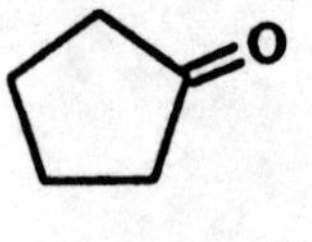

d. 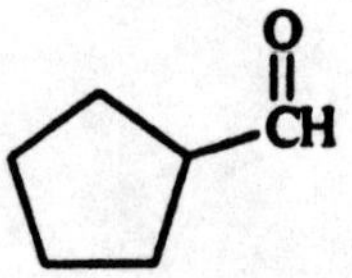

68. One characteristic of the monomers that form condensation polymers that is not common in monomers which form addition polymers is
M

a. pi bonds
b. the presence of two functional groups
c. alkyl side chains
d. ability to form free radicals

69. Which of the following polymers is a condensation polymer?
H

a. polystyrene
b. Teflon®
c. Dacron®
d. polypropylene

70. The most abundant organic chemical on earth is
M

a. glycogen
b. starch
c. cellulose
d. glucose

71. Each amino acid has two functional groups in common and one of 20 other groups attached to the α-carbon. The two functional groups are
E

a. carboxyl and amine
b. ester and amine
c. carboxyl and amide
d. alcohol and amine

72. Helical and sheet-like segments in proteins arise from
M

a. disulfide bridges
b. salt bridges
c. hydrogen bonding
d. dispersion forces within the protein's interior

73. The protein amino acid sequence, the RNA base sequence, and the DNA base sequence are interrelated. Which of the following arrangements is correct?
M

a. the RNA base sequence determines the DNA base sequence which, in turn, determines the protein amino acid sequence
b. the DNA base sequence determines the RNA base sequence which, in turn, determines the protein amino acid sequence
c. the DNA base sequence determines the protein amino acid sequence which, in turn, determines the RNA base sequence
d. the RNA base sequence determines the protein amino acid sequence which, in turn, determines the DNA base sequence

Short Answer Questions

74. Draw and name two non-alkenes with the formula C_4H_8.
E

75. Draw and name all stable molecules with the formula C_5H_{12}.
M

76. Draw and name four alkenes with the formula C_4H_8.
M

77. Name the compound with the molecular structure shown.
M

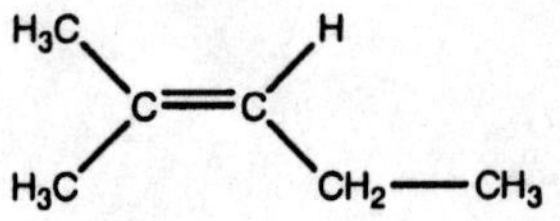

78. Draw the structure of *trans*-5-methyl-3-heptene.
M

79. Name the compounds:
E

(a) 

(b) 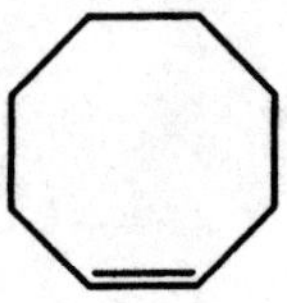

(c) 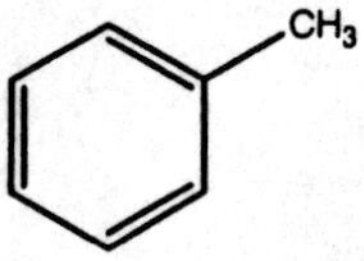

80. What is the simplest molecular formula for
H

a. an alkane capable of having structural isomers?
b. an alkane capable of having optical isomers?
c. an alkene capable of having geometrical isomers?

81. In one sentence, what is the general requirement for a molecule to be optically active?
M

82. a. Draw two different structures with the molecular formula C_2H_6O.
M b. Name the functional group in each structure.
c. Which one will have the higher boiling point, and why?

83. Name each of the following functional groups:
E

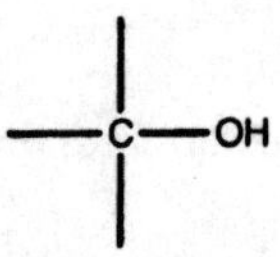

(a)

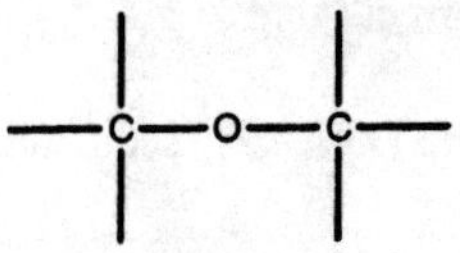

(b)

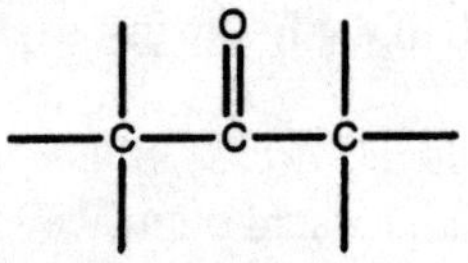

(c)

(d) 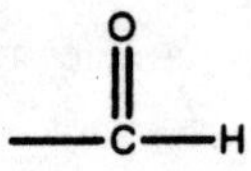

(e) 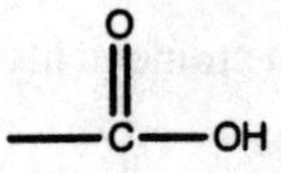

(f) 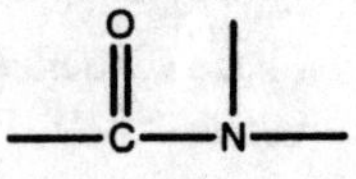

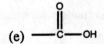

84. Write down the structure of the missing organic reactant or product(s) in the following reactions, and
M name the type of reaction involved (inorganic products may also be formed):

a. 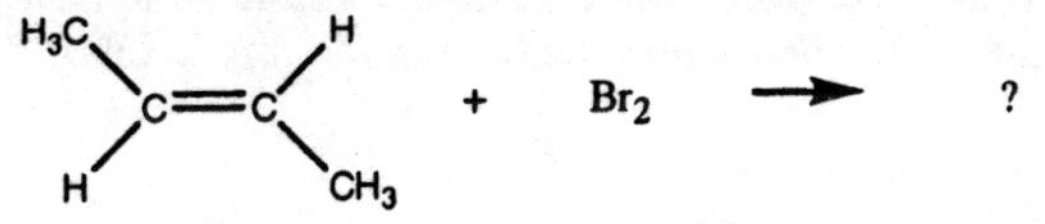

b. 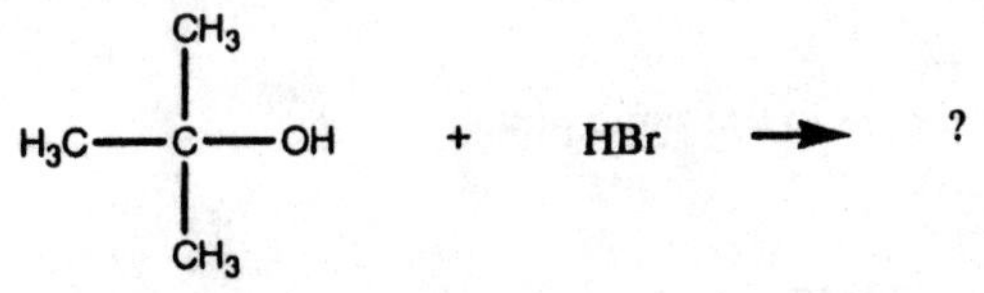

c. 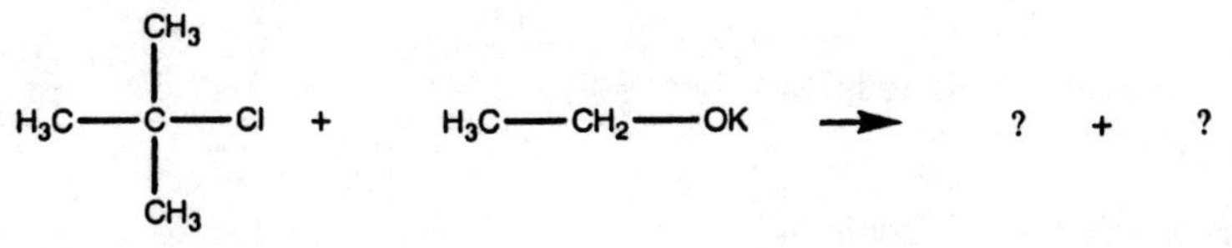

d. 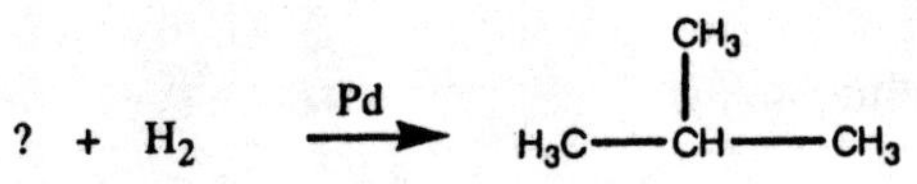

e. 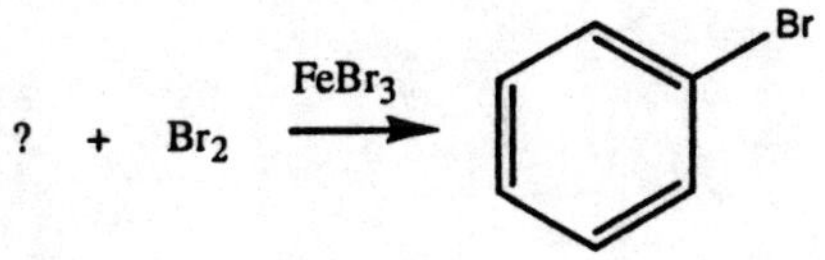

f. 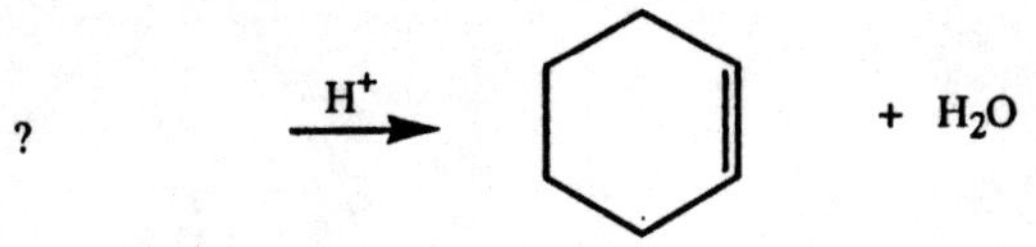

g. 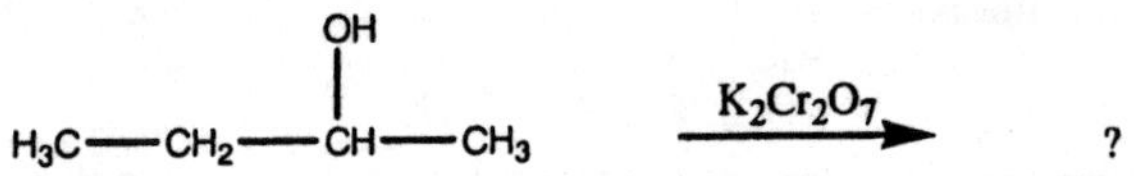

h. 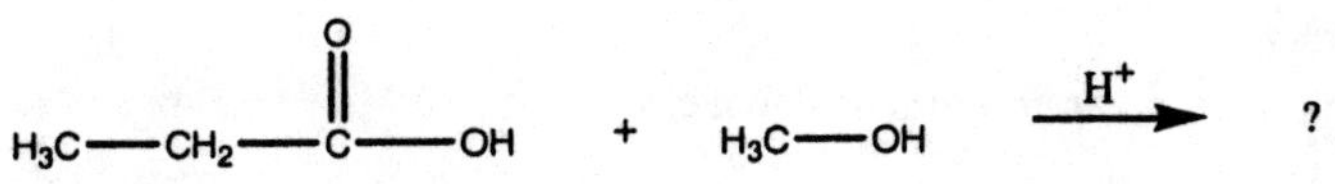

85. Describe and contrast fibrous and globular proteins in terms of their amino acid composition, their structure
M and their function.

86. Explain how a disulfide bridge can arise a protein molecule.
H

87. Explain how a salt bridge can arise a protein molecule.
H

88. Name the three component parts of a nucleotide.
E

89. Explain what is meant by "complementary" in the context of DNA strands.
M

90. Given that each 3-base sequence in DNA is a "code word" for a particular amino acid, how many different code words are possible using a 3-base sequence and the bases available in DNA?
M

True/False Questions

91. Ethane (C_2H_6) is much more reactive than disilane (Si_2H_6).
E

92. The carbon atoms in a molecule of cyclohexane lie in the same plane.
M

93. All the atoms in a molecule of benzene lie in the same plane.
M

94. A characteristic reaction of alkanes is addition.
E

95. A characteristic reaction of alkenes is addition.
E

96. A characteristic reaction of haloalkanes is substitution.
E

97. All alcohols are capable of hydrogen bonding.
M

98. All ketones are capable of hydrogen bonding.
M

99. Carboxylic acids are weak acids.
M

100. Amines are strong bases.
M

101. Secondary amines have the general formula RNH_2.
E

102. Esters can be formed by the dehydration-condensation of a carboxylic acid and an alcohol.
M

103. Amino acids in solution can undergo an internal acid-base reaction.
M

104. The backbone of protein molecules consists of repeating N-C-C-O units.
M

Organic Compounds and the Atomic Properties of Carbon
Chapter 15
Answer Key

1.	d	26.	b	51.	d
2.	c	27.	d	52.	c
3.	d	28.	c	53.	c
4.	b	29.	b	54.	a
5.	d	30.	d	55.	a
6.	c	31.	c	56.	d
7.	a	32.	c	57.	a
8.	c	33.	a	58.	b
9.	b	34.	a	59.	b
10.	d	35.	b	60.	b
11.	d	36.	b	61.	a
12.	b	37.	d	62.	c
13.	d	38.	d	63.	b
14.	c	39.	d	64.	b
15.	a	40.	a	65.	c
16.	c	41.	b	66.	d
17.	b	42.	c	67.	b
18.	c	43.	d	68.	b
19.	b	44.	d	69.	c
20.	d	45.	b	70.	c
21.	d	46.	a	71.	a
22.	c	47.	b	72.	c
23.	a	48.	d	73.	b
24.	a	49.	c		
25.	c	50.	d		

74.

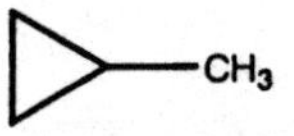

methylcyclopropane

cyclobutane

75.

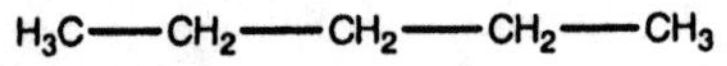

pentane

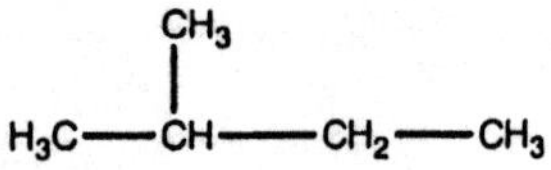

2-methylbutane

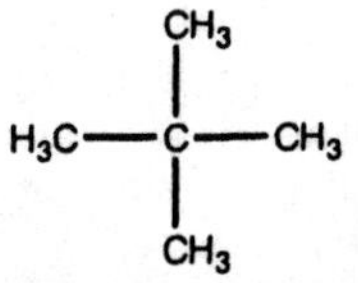

2,2-dimethylpropane

76.

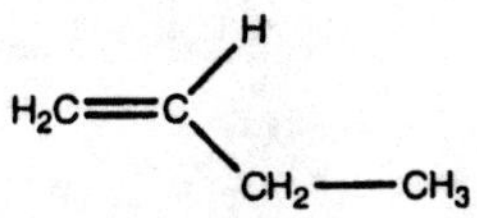

1-butene

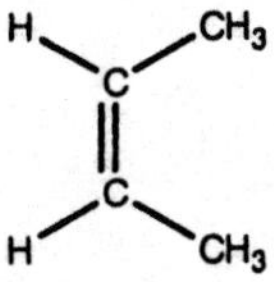

cis-2-butene

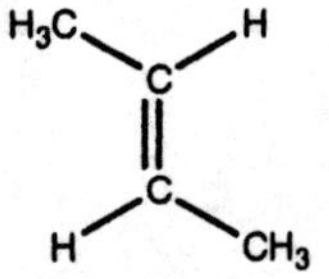

trans-2-butene

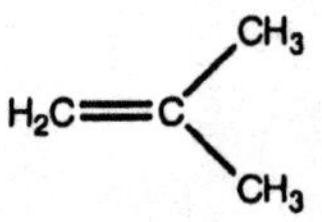

2-methyl-1-propene

77. 2-methyl-2-pentene

78.

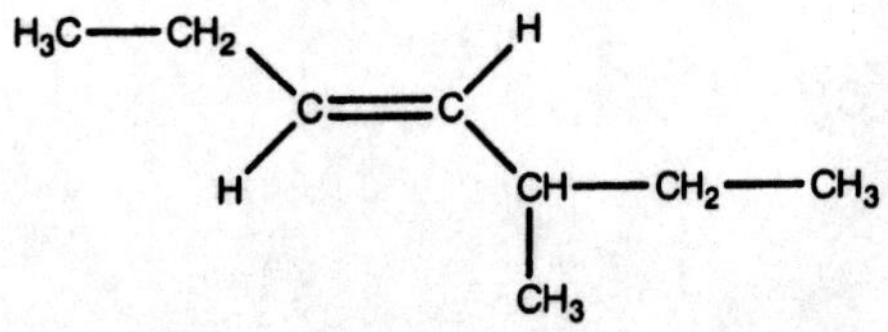

79. a. cyclobutane
    b. cyclooctene
    c. toluene (methylbenzene)

80. a. C_4H_{10}
    b. C_7H_{16}
    c. C_4H_8

81. In order to be optically active, the compound must have a non-superimposable mirror image.

82. a. $H_3C—CH_2—OH$ $H_3C—O—CH_3$
    b. alcohol and ether
    c. The alcohol will have the higher boiling point; it will have hydrogen bonding while the ether does not.

83. a. alcohol
    b. ether
    c. ketone
    d. aldehyde
    e. carboxylic acid
    f. amide

84.

a.  addition

b. 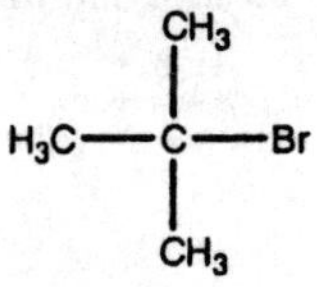 substitution

c. 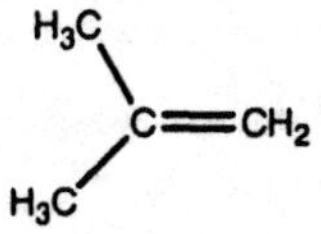 and 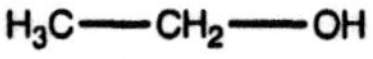 elimination

d. 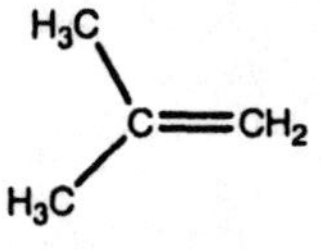 addition (reduction)

e.  substitution

f. 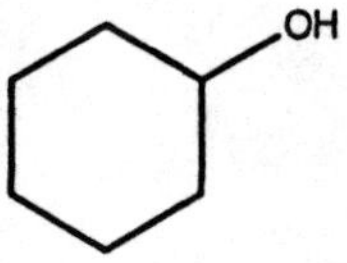 elimination

g. 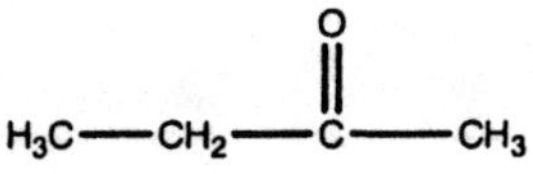 oxidation (elimination)

h. 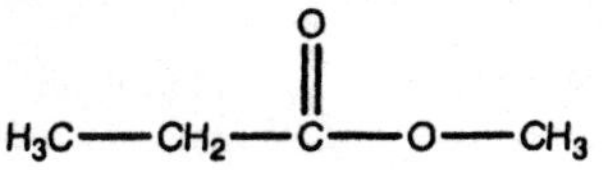 dehydration-condensation

85. Fibrous proteins are composed of relatively few amino acids in a repeating pattern. Their structures are helices or sheets, and they are found in hair, skin and muscle. Globular proteins use more amino acids in their structures, and they have various of irregular shapes. Their functions are various - messenger molecules, enzymes (catalysts), etc.

86. Disulfide bridges arise when two cysteine -S-H side chains react to form a covalent -S-S- link between the two regions of the protein backbone. The two regions of the protein backbone are thus held together.

87. A salt bridge is the result of ionic attraction between a side chain with a carboxylate (-COO^-) group and one with a protonated amine group (-NH_3^+). These regions of the molecule are then held together by this ionic attraction.

88. phosphate, base, sugar

89. Each nucleotide along a single strand of DNA has one of four possible bases - guanine (G), cytosine (C), adenine (A) and thymine (T). The double helix of DNA is held in place by hydrogen bonds between pairs of bases, one base being on each of the two strands. G always pairs with C and A always pairs with T. Thus, the base sequence on one strand determines the base sequence on the other strand, and the strands are "complementary" to each other.

90. $4 \times 4 \times 4 = 64$

91. F

92. F

93. T

94. F

95. T

96. T

97. T

98. F

99. T

100. F

101. F

102. T

103. T

104. F

Kinetics: Rates and Mechanisms of Chemical Reactions
Chapter 16

Multiple Choice Questions

1. The compound RX_3 decomposes according to the equation
E

$$RX_3 \rightarrow R + R_2X_3 + 3X_2$$

In an experiment the following data were collected for the decomposition at 100°C. What is the average rate of reaction over the entire experiment?

t(s)	$[RX_3]$(mol L^{-1})
0	0.85
2	0.67
6	0.41
8	0.33
12	0.20
14	0.16

a. 0.019 mol $L^{-1}s^{-1}$
b. 0.044 mol $L^{-1}s^{-1}$
c. 0.049 mol $L^{-1}s^{-1}$
d. 0.069 mol $L^{-1}s^{-1}$

2. Consider the following reaction
M

$$A(g) + \tfrac{5}{8}B(g) \rightarrow C(g) + \tfrac{3}{4}D(g)$$

If [C] is increasing at the rate of 4.0 mol $L^{-1}s^{-1}$, at what rate is [B] changing?

a. -0.40 mol $L^{-1}s^{-1}$
b. -2.5 mol $L^{-1}s^{-1}$
c. -4.0 mol $L^{-1}s^{-1}$
d. -6.4 mol $L^{-1}s^{-1}$

3. Consider the general reaction
M

$$5Br^-(aq) + BrO_3^-(aq) + 6H^+(aq) \rightarrow 3Br_2(aq) + 3H_2O(aq)$$

For this reaction, the rate when expressed as $\Delta[Br_2]/\Delta t$ is the same as

a. $-\Delta[H_2O]/\Delta t$
b. $3\Delta[BrO_3^-]/\Delta t$
c. $-5\Delta[Br^-]/\Delta t$
d. $-0.6\Delta[Br^-]/\Delta t$
e. none of the above

4.
M

Consider the reaction

$$2NH_3(g) \rightarrow N_2(g) + 3H_2(g)$$

If the rate $\Delta[H_2]/\Delta t$ is 0.030 mol L^{-1} s^{-1}, then $\Delta[NH_3]/\Delta t$ in the same units is

a. -0.045
b. -0.030
c. -0.020
d. -0.010
e. none of the above

5.
M

For the reaction

$$3A(g) + 2B(g) \rightarrow 2C(g) + 2D(g)$$

the following data were collected at constant temperature. Determine the correct rate law for this reaction.

Trial	Initial [A] (mol/L)	Initial [B] (mol/L)	Initial Rate (mol/(L·min))
1	0.200	0.100	6.00×10^{-2}
2	0.100	0.100	1.50×10^{-2}
3	0.200	0.200	1.20×10^{-1}
4	0.300	0.200	2.70×10^{-1}

a. Rate $= k[A][B]^2$
b. Rate $= k[A]^2[B]$
c. Rate $= k[A]^3[B]^2$
d. Rate $= k[A]^{1.5}[B]$

6.
M

For the reaction

$$A(g) + 2B(g) \rightarrow 2C(g) + 2D(g)$$

the following data were collected at constant temperature. Determine the correct rate law for this reaction.

Trial	Initial [A] (mol/L)	Initial [B] (mol/L)	Initial Rate (mol/(L·min))
1	0.125	0.200	7.25
2	0.375	0.200	21.75
3	0.250	0.400	14.50
4	0.375	0.400	21.75

a. Rate $= k[A]\,[B]$
b. Rate $= k[A]^2\,[B]$
c. Rate $= k[A]\,[B]^2$
d. Rate $= k[A]$

7. H For the reaction

$$2A + B + 2C \rightarrow D + E$$

the following initial rate data were collected at constant temperature. Determine the correct rate law for this reaction.

Trial	[A]	[B]	[C]	Rate
1	0.225	0.150	0.350	0.0217
2	0.320	0.150	0.350	0.0439
3	0.225	0.250	0.350	0.0362
4	0.225	0.150	0.600	0.01270

a. Rate = k[A][B][C]
b. Rate = $k\,[A]^2[B][C]$
c. Rate = $k\,[A]^2[B]^1[C]^{-1}$
d. Rate = $k\,[A]^1[B]^2[C]^{-1}$

8. M The rate constant for a reaction is 4.65 L $mol^{-1}s^{-1}$. The overall order of the reaction is:

a. zero
b. first
c. second
d. third

9. M Sulfuryl chloride, $SO_2Cl_2(g)$, decomposes at high temperature to form $SO_2(g)$ and $Cl_2(g)$. The rate constant at a certain temperature is $4.68 \times 10^{-5} s^{-1}$. The order of the reaction is:

a. zero
b. first
c. second
d. third

10. E When the reaction A $\rightarrow$ B + C is studied, a plot of 1n [A] vs. time gives a straight line with a negative slope. The order of the reaction is:

a. zero
b. first
c. second
d. third

11. E When the reaction A $\rightarrow$ B + C is studied, a plot 1/[A] vs. time gives a straight line with a positive slope. The reaction order is:

a. zero
b. first
c. second
d. third

12. M Which of the following sets of units could be appropriate for a zero-order rate constant?

a. s^{-1}
b. mol L^{-1} s^{-1}
c. L mol^{-1} s^{-1}
d. L^2 mol^{-2} s^{-1}
e. L^3 mol^{-3} s^{-1}

13. Which one of the following sets of units is appropriate for a second-order rate constant?
M

a. s^{-1}
b. mol L^{-1} s^{-1}
c. L mol^{-1} s^{-1}
d. mol^2 L^{-2} s^{-1}
e. L^2 mol^{-2} s^{-1}

14. For a third-order reaction, appropriate units for the rate constant could be
M

a. s^{-1}
b. mol L^{-1} s^{-1}
c. L mol^{-1} s^{-1}
d. L^2 mol^{-2} s^{-1}
e. L^3 mol^{-3} s^{-1}

15. A reaction has the following rate law
M

Rate = k[A][B]^2

In experiment 1, the concentrations of A and B are both 0.10 mol L^{-1}; in experiment 2, the concentrations are both 0.30 mol L^{-1}. If the temperature stays constant, what is the value of the ratio, Rate(2)/Rate(1)?

a. 3.0
b. 6.0
c. 9.0
d. 18
e. 27

16. Ammonium cyanate (NH_4CNO) reacts to form urea (NH_2CONH_2). At 65°C the rate constant, k, is 3.60 L $mol^{-1}s^{-1}$. The rate law for this reaction is
M

a. Rate = 3.60 L $mol^{-1}s^{-1}$ [NH_4CNO]
b. Rate = 3.60 L $mol^{-1}s^{-1}$ [NH_4CNO]^2
c. Rate = 0.28 mol L^{-1}s [NH_4CNO]
d. Rate = 0.28 mol L^{-1}s [NH_4CNO]^2

17. $2NOBr(g) \rightarrow 2NO(g) + Br_2(g)$
H

[NOBr](mol L^{-1})	Rate (mol $L^{-1}s^{-1}$)
0.0450	1.62×10^{-3}
0.0310	7.69×10^{-4}
0.0095	7.22×10^{-5}

Based on the initial rate data above, what is the value of the rate constant?

a. 0.0360 L $mol^{-1}s^{-1}$
b. 0.800 L $mol^{-1}s^{-1}$
c. 1.25 L $mol^{-1}s^{-1}$
d. 27.8 L $mol^{-1}s^{-1}$

18. H

A study of the decomposition reaction $3RS_2 \rightarrow 3R + 6S$ yields the following initial rate data

$[RS_2]$(mol L^{-1})	Rate (mol/(L·s))
0.150	0.0394
0.250	0.109
0.350	0.214
0.500	0.438

What is the rate constant for the reaction?

a. 0.0103 L $mol^{-1}s^{-1}$
b. 0.263 L $mol^{-1}s^{-1}$
c. 0.571 L $mol^{-1}s^{-1}$
d. 1.75 L $mol^{-1}s^{-1}$

19. M

Sucrose decomposes to fructose and glucose in acid solution. When ln [sucrose] is plotted vs. time, a straight line with slope of -0.208 hr^{-1} results. What is the rate law for the reaction?

a. Rate = 0.208 hr^{-1} $[sucrose]^2$
b. Rate = 0.208 hr^{-1} [sucrose]
c. Rate = 0.0433 hr $[sucrose]^2$
d. Rate = 0.0433 hr [sucrose]

20. M

Tetrafluoroethylene, C_2F_4, can be converted to octafluorocyclobutane which can be used as a refrigerant or an aerosol propellant. A plot of $1/[C_2F_4]$ vs. time gives a straight line with a slope of 0.0448 L $mol^{-1}s^{-1}$. What is the rate law for this reaction?

a. Rate = 0.0448 (L $mol^{-1}s^{-1}$)$[C_2F_4]$
b. Rate = 22.3 (mol L^{-1}s)$[C_2F_4]$
c. Rate = 0.0448 (L $mol^{-1}s^{-1}$)$[C_2F_4]^2$
d. Rate = 22.3 (mol L^{-1}s)$[C_2F_4]^2$

21. M

The reaction $A \rightarrow B$ is first-order overall and first-order with respect to the reactant A. The result of doubling the initial concentration of A will be to

a. shorten the half-life of the reaction
b. increase the rate constant of the reaction
c. decrease the rate constant of the reaction
d. shorten the time taken to reach equilibrium
e. double the initial rate

22. H

The decomposition of hydrogen peroxide is a first-order process with a rate constant of 1.06×10^{-3} min^{-1}. How long will it take for the concentration of H_2O_2 to drop from 0.0200 *M* to 0.0120 *M*?

a. 7.55 min
b. 481 min
c. 4550 min
d. 31,400min

23. H

Cyclopropane is converted to propene in a first-order process. The rate constant is 5.4×10^{-2} hr^{-1}. If the initial concentration of cyclopropane is 0.150 *M*, what will its concentration be after 22.0 hours?

a. 0.0457 *M*
b. 0.105 *M*
c. 0.127 *M*
d. 0.492 *M*

24. The rate law for the reaction $3A \rightarrow 2B$ is rate = k[A] with a rate constant of 0.0447 hr^{-1}. What is the half-life of the reaction?
M

a. 0.0224 hr
b. 0.0645 hr
c. 15.5 hr
d. 44.7hr

25. The rate law for the rearrangement of CH_3NC to CH_3CN at 800 K is rate = (1300 s^{-1})[CH_3NC]. What is the half-life for this reaction?
M

a. 5.3×10^{-4}s
b. 1.9×10^{-3}s
c. 520 s
d. 1920 s

26. The rate constant for the reaction $3A \rightarrow 4B$ is 6.00×10^{-3} L $mol^{-1}min^{-1}$. How long will it take the concentration of A to drop from 0.75 *M* to 0.25 *M?*
M

a. 2.2×10^{-3}min
b. 5.5×10^{-3}min
c. 180 min
d. 440 min

27. Butadiene, C_4H_6 (used to make synthetic rubber and latex paints) dimerizes to C_8H_{12} with a rate law of rate = 0.014 L/(mol·s) $[C_4H_6]^2$. What will be the concentration of C_4H_6 after 3.0 hours if the initial concentration is 0.025 *M*?
M

a. 0.0052 *M*
b. 0.024 *M*
c. 43 *M*
d. 190 *M*

28. The rate law for the reaction $3A \rightarrow C$ is
H

$$\text{Rate} = 4.36 \times 10^{-2}\ \text{L mol}^{-1}\ \text{hr}^{-1}[A]^2$$

What is the half-life for the reaction if the initial concentration of A is 0.250 *M*?

a. 0.0109 hr
b. 0.0629 hr
c. 15.9 hr
d. 91.7 hr

29. The decomposition of $SOCl_2$ is first-order in $SOCl_2$. If the half-life for the reaction is 4.1 hr, how long would it take for the concentration of $SOCl_2$ to drop from 0.36 *M* to 0.045 *M?*
H

a. 0.52 hr
b. 1.4 hr
c. 12 hr
d. 33 hr

30. H The reaction $CH_3NC(g) \rightarrow CH_3CN(g)$ is first-order with respect to methyl isocyanide, CH_3NC. If it takes 10.3 minutes for exactly one quarter of the initial amount of methyl isocyanide to react, what is the rate constant in units of min^{-1}?

a. -0.135
b. 0.0279
c. 0.089
d. 0.135
e. 35.8

31. M A reactant R is being consumed in a first-order reaction. What fraction of the initial R is consumed in 4.0 half-lives?

a. 0.94
b. 0.87
c. 0.80
d. 0.13
e. 0.063

32. M A first-order reaction has a half-life of 20.0 minutes. Starting with 1.00×10^{20} molecules of reactant at time $t = 0$, how many molecules remain unreacted after 100.0 minutes?

a. 3.13×10^{18}
b. 1.00×10^{4}
c. 2.00×10^{19}
d. 3.20×10^{16}
e. none of the above

33. M Carbon-14 is a radioactive isotope which decays with a half-life of 5730 years. What is the first-order rate constant for its decay, in units of $years^{-1}$?

a. 5.25×10^{-5}
b. 1.21×10^{-4}
c. 1.75×10^{-4}
d. 3.49×10^{-4}
e. 3.97×10^{3}

34. H The radioactive isotope tritium decays with a first-order rate constant k of 0.056 $year^{-1}$. What fraction of the tritium initially in a sample is still present 30 years later?

a. 0.19
b. 0.60
c. 0.15
d. 2.8×10^{-38}
e. none of the above

35. H Dinitrogen tetraoxide, N_2O_4, decomposes to nitrogen dioxide, NO_2, in a first-order process. If $k = 2.5 \times 10^{3}\ s^{-1}$ at -5°C and $k = 3.5 \times 10^{4}\ s^{-1}$ at 25°C, what is the activation energy for the decomposition?

a. 0.73 kJ/mol
b. 58 kJ/mol
c. 140 kJ/mol
d. 580 kJ/mol

36. Ammonia will react with oxygen in the presence of a copper catalyst to form nitrogen and water.
H From 164.5°Cto 179.0°C, the rate constant increases by a factor of 4.27. What is the activation energy of this oxidation reaction?

a. 24.5 kJ/mol
b. 165 kJ/mol
c. 242 kJ/mol
d. 1630 kJ/mol

37. What effect will a 10.0 kJ/mol decrease in the activation energy have on the rate of a reaction at 298 K?
M

a. the rate will increase by a factor of more than 50.
b. the rate will decrease by a factor of more than 50.
c. the rate will not change unless temperature increases.
d. more information is needed to determine the effect.

38. A rate constant obeys the Arrhenius equation, the factor A being 2.2×10^{13} s^{-1} and the activation
M energy being 150. kJ mol^{-1}. What is the value of the rate constant at 227°C, in s^{-1}?

a. 2.1×10^{13}
b. 6.7×10^{-22}
c. 1.5×10^{11}
d. 4.7×10^{-3}
e. none of the above

39. A reaction has an activation energy of 195.0 kJ/mol. When the temperature is increased from 200. °C
H to 220.°C, the rate constant will increase by a factor of

a. 1.1
b. 4.3×10^4
c. 3.2
d. 7.5
e. none of the above

40. The decomposition of dinitrogen pentaoxide to nitrogen dioxide and oxygen follows first-order kinetics and
H has an activation energy of 102 kJ/mol. By what factor will the fraction of collisions with energy greater than or equal to the activation energy increase if the reaction temperature goes from 30°C to 60°C?

a. 1.00
b. 1.10
c. 2.00
d. 38.4

41. The decomposition of dinitrogen pentaoxide has an activation energy of 102 kJ/mol and
E $\Delta H^\circ_{rxn} = +55$ kJ/mol. What is the activation energy for the reverse reaction?

a. 27 kJ/mol
b. 47 kJ/mol
c. 55 kJ/mol
d. 102 kJ/mol

42. The kinetics of the decomposition of dinitrogen pentaoxide is studied at 50°C and at75°C. Which of the
E following statements concerning the studies is correct?

a. The rate at 75°C will be greater than the rate at 50°C because the activation energy will be lower at 75°C than at 50°C.
b. The rate at 75°C will be greater than the rate at 50°C because the number of molecules with enough energy to react increases with increasing temperature.
c. The rate at 75°C will be less than the rate at 50°C because the molecules at higher speeds do not interact as well as those at lower speeds.
d. The rate at 75°C will be greater than at 50°C because the concentration of a gas increases with increasing temperature.

43. Reaction intermediates differ from activated complexes in that
M

a. they are stable molecules with normal bonds and are frequently isolated.
b. they are molecules with normal bonds rather than partial bonds and can occasionally be isolated.
c. they are intermediate structures which have characteristics of both reactants and products.
d. they are unstable and can never be isolated.

44. Consider the following mechanism for the oxidation of bromide ions by hydrogen peroxide in aqueous
H acid solution.

$H^+ + H_2O_2 \rightleftarrows H_2O^+\text{–}OH$ (rapid equilibrium)
$H_2O^+\text{–}OH + Br^- \rightarrow HOBr + H_2O$ (slow)
$HOBr + H^+ + Br^- \rightarrow Br_2 + H_2O$ (fast)

What is the overall reaction equation for this process?

a. $2H_2O^+\text{–}OH + 2Br^- \rightarrow H_2O_2 + Br_2 + 2H_2O$
b. $2H^+ + 2Br^- + H_2O_2 \rightarrow Br_2 + 2H_2O$
c. $2H^+ + H_2O_2 + Br^- + HOBr \rightarrow H_2O^+\text{–}OH + Br_2 + H_2O$
d. $H_2O^+\text{–}OH + Br^- + H^+ \rightarrow Br_2 + H_2O$

45. What is the molecularity of the following elementary reaction?
E

$NH_2Cl(aq) + OH^-(aq) \rightarrow NHCl^-(aq) + H_2O(l)$

a. unimolecular
b. bimolecular
c. termolecular
d. tetramolecular

46. Consider the following mechanism for the oxidation of bromide ions by hydrogen peroxide in aqueous
M acid solution.

$H^+ + H_2O_2 \rightleftarrows H_2O^+\text{–}OH$ (rapid equilibrium)
$H_2O^+\text{–}OH + Br^- \rightarrow HOBr + H_2O$ (slow)
$HOBr + H^+ + Br^- \rightarrow Br_2 + H_2O$ (fast)

Which of the following rate laws is consistent with the mechanism?

a. Rate = $k[H_2O_2][H^+]^2[Br^-]$
b. Rate = $k\,[H_2O^+\text{–}OH][Br^-]$
c. Rate = $k[H_2O_2][H^+][Br^-]$
d. Rate = $k[HOBr][H^+][Br^-][H_2O_2]$

47. Which of the following affects the activation energy of a reaction?
E

a. temperature of the reactants
b. concentrations of reactants
c. presence of a catalyst
d. surface area of reactants

48. A catalyst is effective because
E

a. it increases the number of molecules with energy equal to or greater than the activation energy.
b. it lowers the activation energy for the reaction.
c. it increases the number of collisions between molecules.
d. it increases the temperature of the molecules in the reaction.

49. Enzyme catalysis has many features in common with ordinary catalysis. Which of the following is not one of the common features?
H

a. Substrates bind to active sites through intermolecular forces such as hydrogen bonding
b. Rates of reaction are enhanced by more than 100 million times
c. Rates of reaction are affected by the concentration of reactant bound to the catalyst
d. Heterogeneous catalysis is possible

50. When a catalyst is added to a reaction mixture, it
E

a. increases the rate of collisions between reactant molecules
b. provides reactant molecules with more energy
c. slows down the rate of the back reaction
d. provides a new pathway (mechanism) for the reaction
e. does none of the above

51. The gas-phase reaction $CH_3NC \rightarrow CH_3CN$ has been studied in a closed vessel, and the rate equation was found to be: rate = $-\Delta[CH_3NC]/\Delta t = k[CH_3NC]$. Which one of the following actions is least likely to cause a change in the rate of the reaction?
M

a. lowering the temperature
b. adding a catalyst
c. using a larger initial amount of CH_3NC in the same vessel
d. using a bigger vessel, but the same initial amount of CH_3NC
e. continuously removing CH_3CN as it is formed

Short Answer Questions

52. In the gas phase at 500.°C, cyclopropane reacts to form propene in a first-order reaction. The figure alongside shows the concentration of cyclopropane plotted versus time. Use the graph to calculate approximate values of
E

a. the rate of the reaction, 600. seconds after the start
b. the half-life of the reaction, $t_{1/2}$.

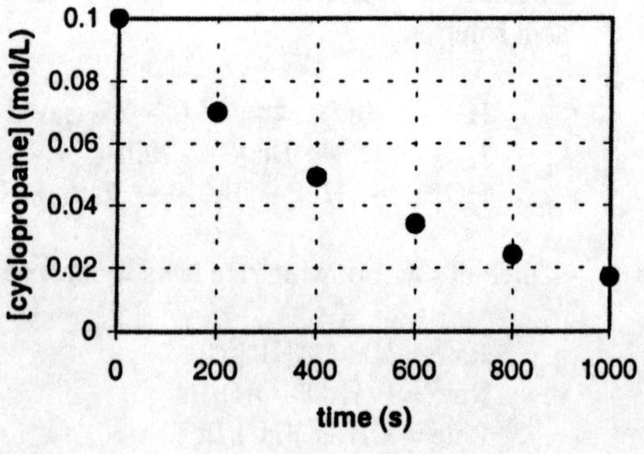

53. H You are studying the rate of the reaction $2A \rightarrow B$ and have obtained measurements of the concentration of A at times t = 100, 200, 300, ..., 1000 seconds from the start of the reaction. Carefully describe how you would plot a graph and use it to:

a. prove that the reaction is second-order with respect to A
b. determine the second-order rate constant k.

54. M In the gas phase at 500.°C, cyclopropane reacts to form propene in a first-order reaction. The figure shows the natural logarithm of the concentration of cyclopropane (in mol/L) plotted versus time.

a. Explain how this plot confirms that the reaction is first order.
b. Calculate the first-order rate constant, k.
c. Determine the initial concentration of cyclopropane in this experiment.

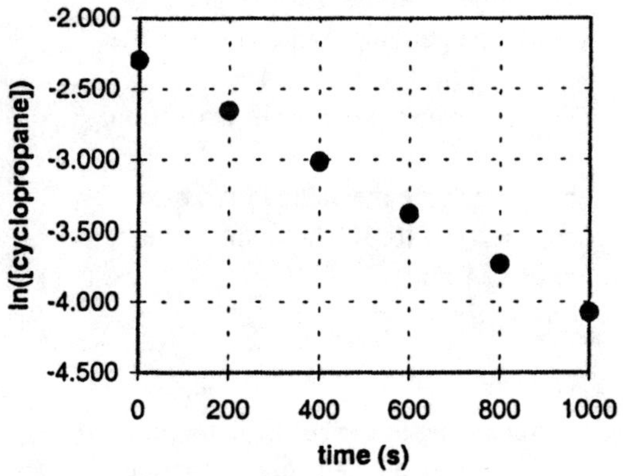

55. H A chemical reaction of the general type

$$A \rightarrow 2B$$

is first-order, with a rate constant of $1.52 \times 10^{-4}\ s^{-1}$.

a. Calculate the half-life of A.
b. Assuming the initial concentration of A is 0.067 mol L^{-1}, calculate the time needed for the concentration to fall to 0.010 mol L^{-1}.

56. M The gas-phase conversion of 1,3-butadiene to 1,5-cyclooctadiene, $2C_4H_6 \rightarrow C_8H_{12}$ was studied, providing data for the plot shown alongside, of 1/[butadiene] versus time.

a. Explain how this plot confirms that the reaction is second order.
b. Calculate the second-order rate constant, k.
c. Determine the initial concentration of 1,3-butadiene in this experiment.

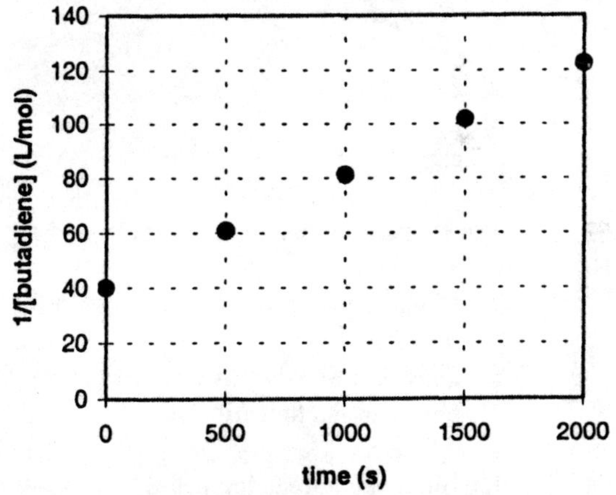

57. H You are required to determine the energy of activation (E_a) of a reaction. Briefly describe the experimental measurements you would make and how you would obtain the activation energy from a suitable linear plot of the experimental data.

58. M At 25.0°C, a rate constant has the value 5.21×10^{-8} L mol^{-1} s^{-1}. If the activation energy is 75.2 kJ/mol, calculate the rate constant when the temperature is 50.0°C.

59.
H

Cyclobutane decomposes to ethene in a first-order reaction. From measurements of the rate constant (k) at various absolute temperatures (T), the accompanying Arrhenius plot was obtained ($\ln k$ versus $1/T$).

a. Calculate the energy of activation, E_a.

b. Determine the value of the rate constant at 740. K. (In the plot, the units of k are s^{-1}.)

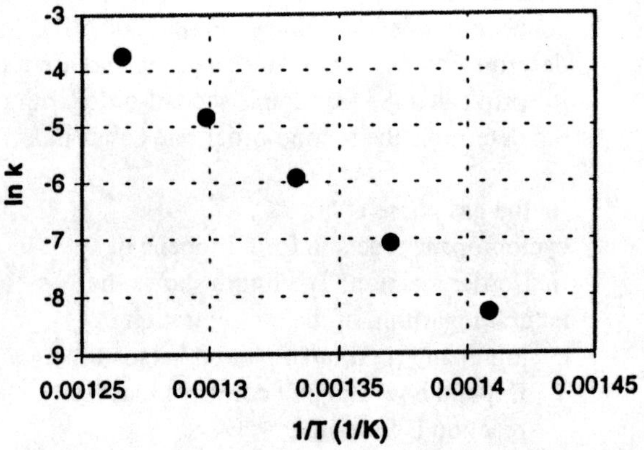

60.
E

According to the collision theory of reaction rates, what are the three requirements which must be met before an elementary reaction between two molecules can occur?

61.
H

Briefly outline the key arguments in the collision theory of reaction rates for the elementary reaction $C + D \rightarrow$ products. Show that this theory predicts a second-order rate law, and how it predicts the form of the rate constant k.

62.
M

In the collision theory of reaction rates, the rate constant for a bimolecular reaction can be written as

$$k = z \cdot p \cdot \exp(-E_a/RT)$$

In one sentence each, clearly explain the physical meaning (interpretation) of the three factors in the above expression

a. z
b. p
c. $\exp(-E_a/RT)$

63.
H

The elementary reaction $HBr(g) + Br(g) \rightarrow H(g) + Br_2(g)$ is endothermic.

a. Would you expect the rate constant for the back reaction to be smaller or larger than that for the forward reaction? Explain, briefly.
b. Draw a fully-labelled reaction energy diagram for this reaction, showing the locations of the reactants, products and transition state.

64.
H

Is a bimolecular reaction necessarily second-order? Is a second-order reaction necessarily bimolecular? Answer, with explanations and clarifications.

65.
M

For each of the following terms/concepts, give a brief explanation or definition. Where possible, use examples.

a. order of a reaction
b. elementary reaction
c. reaction intermediate

66.
M

Briefly list the features/properties common to all catalysts and how they work. Draw a labeled reaction energy diagram as part of your answer.

67. Consider the general gas-phase reaction of a molecular substance, A
H

1. $A \rightarrow B$

At very low pressures many such reactions occur by the following mechanism:

2. $A + A \rightarrow A^* + A$ (slow)
3. $A^* \rightarrow B$ (fast)

A* represents a molecule with sufficient energy to overcome the activation energy barrier.

a. Which of the three reactions above is/are elementary?
b. Where appropriate, identify the molecularity of the reactions.
c. Show that the proposed mechanism is consistent with reaction 1, the observed reaction.
d. Given the mechanism above, suggest a likely rate law for reaction (1).

True/False Questions

68. The rate law cannot be predicted from the stoichiometry of a reaction.
M

69. The units of the rate constant depend on the order of the reaction.
E

70. The units of the rate of reaction depend on the order of the reaction.
M

71. The half-life of a first-order reaction does not depend on the initial concentration of reactant.
E

72. The half-life of a second-order reaction does not depend on the initial concentration of reactant.
M

73. The greater the energy of activation, E_a, the faster will be the reaction.
E

74. An elementary reaction is a simple, one-step process.
E

75. All second-order reactions are bimolecular reactions.
H

76. All bimolecular reactions are second-order reactions.
M

77. The rate of a reaction is determined by the rate of the fastest step in the mechanism.
E

78. A transition state is a species (or state) corresponding to an energy maximum on a reaction energy diagram.
M

79. A reaction intermediate is a species corresponding to a local energy maximum on a reaction energy diagram.
M

Kinetics: Rates and Mechanisms of Chemical Reactions
Chapter 16
Answer Key

1.	c	18.	d	35.	b
2.	b	19.	b	36.	b
3.	d	20.	c	37	a
4.	c	21.	e	38.	d
5.	b	22.	b	39.	d
6.	d	23.	a	40.	d
7.	c	24.	c	41.	b
8.	c	25.	a	42.	b
9.	b	26.	d	43.	b
10.	b	27.	a	44.	b
11.	c	28.	d	45.	b
12.	b	29.	c	46.	c
13.	c	30.	b	47.	c
14.	d	31.	a	48.	b
15.	e	32.	a	49.	b
16.	b	33.	b	50.	d
17.	b	34.	a	51.	e

52. a. Rate = - slope = $6 \pm 1 \times 10^{-5}$ mol L^{-1} s^{-1}
    b. 390 ± 20 s

53. a. Plot 1/[A] versus time. If a straight line results, the reaction is second-order.
    b. The rate constant k is the slope of the plot in (a).

54. a. The fact that a plot of ln k versus t gives a straight line proves that it is first-order.
    b. k = - slope = 1.8×10^{-5} s^{-1}
    c. 0.10 mol L^{-1}

55. a. $t_{1/2\_} = 4.56 \times 10^3$ s
    b. 1.25×10^4 s

56. a. The fact that a plot of 1/[butadiene] versus t gives a straight line proves that it is second-order.
    b. k = slope = 8.2×10^{-2} L mol^{-1} s^{-1}
    c. 0.025 mol L^{-1}

57. Obtain rate constants k over a suitable range of absolute temperatures T. A plot of ln k versus $1/T$ should be linear, and its slope is - E_a/R.

58. 5.44×10^{-7} s^{-1}

59. a. 260 ± 20 kJ/mol
    b. 1.6×10^{-3} s^{-1}

60. Molecules must collide with each other; the molecules must have sufficient energy to overcome the activation energy barrier; the molecules must have the correct orientation.

61. For reaction to occur, the molecules must collide. Collision rate = $z \cdot [C][D]$
For a collision to lead to reaction, the molecules must have an energy at least equal to the energy of activation, E_a. The fraction of molecules with sufficient energy is $\exp(-E_a/RT)$. Even if a collision has sufficient energy, reaction will not occur unless the molecules are correctly oriented w.r.t. each other. The steric factor, p, gives the fraction of collisions with correct orientation. Thus, the total rate of successful collisions between C and D is given by $z \cdot [C][D] \cdot p \cdot \exp(-E_a/RT)$. This is a second-order rate law, being first-order in both C and D. According to the rate equation, the rate of reaction is $k \cdot [C][D]$. Comparison of the two expressions for the rate shows that k is equal to $z \cdot p \cdot \exp(-E_a/RT)$.

62. a. z is a collision rate constant, such that the rate of collision between C and D = $z \cdot [C][D]$.
b. p is the steric factor, which is the fraction of collisions between C and D in which the molecules are suitably oriented for reaction.
c. $\exp(-E_a/RT)$ is the fraction of collisions which have sufficient energy to overcome the activation energy barrier.

63. a. The back reaction is exothermic. Since the reactions share a common transition state, the activation energy for the back reaction must be smaller than that for the forward reaction. Other things being equal, the rate constant for the back reaction will be larger.

b.

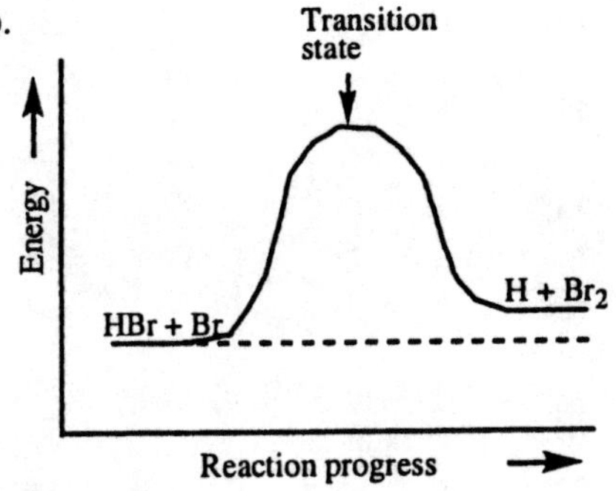

64. For elementary reactions, the order follows from the molecularity. Therefore a bimolecular reaction is necessarily second-order. A second-order reaction may result from a complex mechanism involving steps of different orders, and it is thus meaningless to attempt to deduce molecularity from reaction order.

65. a. Order refers to the power to which the concentration of a reactant is raised in the rate equation.
b. An elementary reaction is one occurring in a single step on the molecular level.
c. A reaction intermediate is an unstable species which appears in a reaction mechanism, but is neither a reactant nor a product.

66. Catalysts speed up a reaction by providing an alternative route of lower activation energy. They do not alter the equilibrium position of a reaction.

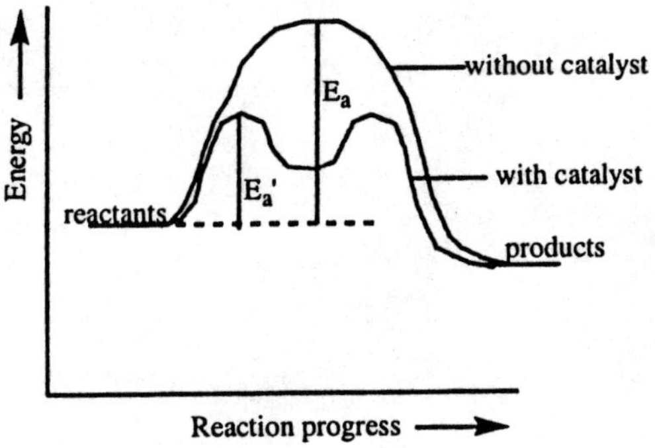

67. a. Reactions 2 and 3 are elementary.
    b. Reaction 2 is bimolecular; reaction 3 is unimolecular.
    c. When reactions 2 and 3 are added, the result is A $\rightarrow$ B, i.e. the same as reaction 1.
    d. Reaction 2, being slow, is the rate-determining step of the mechanism. The overall rate will be the rate of this reaction, i.e. rate = $k_2[A]^2$.

68. T

69. T

70. F

71. T

72. F

73. F

74. T

75. F

76. T

77. F

78. T

79. F

Equilibrium: The Extent of Chemical Reactions
Chapter 17

Multiple Choice Questions

1. When a chemical system is at equilibrium,
E
   a. the concentrations of the reactants are equal to the concentrations of the products.
   b. the concentrations of the reactants and products have reached constant values.
   c. the forward and reverse reactions have stopped.
   d. the reaction quotient, Q, has reached a maximum.

2. Which of the following has an effect on the magnitude of the equilibrium constant?
E
   a. activation energy of the forward reaction
   b. concentrations of the reactants and products
   c. presence of a catalyst
   d. change in temperature

3. In order to write the correct mass-action expression for a reaction one must
E
   a. know the kinetic rate law for the reaction.
   b. know the mechanism for the reaction.
   c. have a properly balanced chemical equation.
   d. have values for the concentrations of the reactants.

4. The equilibrium constants, K_c and K_p, will equal one another when
E
   a. all of the reactants and products are gases.
   b. the number of moles of gaseous products equals the number of moles of gaseous reactants.
   c. the number of moles of gaseous products is greater than the number of moles of gaseous reactants.
   d. the number of moles of gaseous products is smaller than the number of moles of gaseous reactants.

5. The reaction quotient for a reaction has a value of 2000. Which of the following statements is accurate?
E
   a. The reaction must proceed to the left to establish equilibrium.
   b. The reaction must proceed to the right to establish equilibrium.
   c. When the system is at equilibrium, the concentrations of the products will be much larger than the concentrations of the reactants
   d. The concentrations of the products are much larger than the concentrations of the reactants.

6. The reaction quotient, Q_c, for a reaction has a value of 75 while the equilibrium constant, K_c, has a value of
E 195. Which of the following statements is accurate?
   a. The reaction must proceed to the left to establish equilibrium.
   b. The reaction must proceed to the right to establish equilibrium.
   c. The concentrations of the products will be much smaller than the concentrations of the reactants when the system is at equilibrium.
   d. The concentrations of the products will be about the same as the concentrations of the reactants when the system is at equilibrium.

7. Write the mass-action expression, Q_c, for the following chemical reaction equation.
M

$$2C_6H_6(g) + 15O_2(g) \rightleftarrows 12CO_2(g) + 6H_2O(g)$$

a. $\dfrac{[CO_2]\,[H_2O]}{[C_6H_6]\,[O_2]}$

b. $\dfrac{[CO_2]^{12}\,[H_2O]^6}{[C_6H_6]^2\,[O_2]^{15}}$

c. $\dfrac{[C_6H_6]\,[O_2]}{[CO_2]\,[H_2O]}$

d. $\dfrac{[C_6H_6]^2\,[O_2]^{15}}{[CO]^{12}\,[H_2O]^6}$

8. Write the mass-action expression, Q_p, for the following reaction.
M

$$SbF_5(g) + 4Cl_2(g) \rightleftarrows SbCl_3(g) + 5ClF(g)$$

a. $\dfrac{(P_{Cl_2})\cdot(P_{SbF_5})}{(P_{SbCl_3})\cdot(P_{ClF})}$

b. $\dfrac{(P_{Cl_2})^4\cdot(P_{SbF_3})}{(P_{SbCl_3})\cdot(P_{ClF})^5}$

c. $\dfrac{(P_{SbCl_3})\cdot(P_{ClF})}{(P_{Cl_2})\cdot(P_{SbF_5})}$

d. $\dfrac{(P_{SbCl_3})\cdot(P_{ClF})^5}{(P_{Cl_2})^4\cdot(P_{SbF_5})}$

9. Write the mass-action expression, Q_c, for the following chemical reaction.
M

$$NO(g) + \tfrac{1}{2}Br_2(g) \rightleftarrows NOBr(g)$$

a. $\dfrac{[NOBr]^2}{[NO]^2[Br_2]}$

b. $\dfrac{[NOBr]}{[NO]\,[Br_2]^{1/2}}$

c. $\dfrac{[NO]\,[Br_2]^{1/2}}{[NOBr]}$

d. $\dfrac{[NO]^2[Br_2]}{[NOBr]^2}$

10. Write the mass-action expression, Q_c, for the following chemical reaction.
M

$$ClO_2^-(aq) \rightleftarrows 2ClO_3^-(aq) + Cl^-(aq)$$

a. $\dfrac{[ClO_3^-]^2[Cl^-]}{[ClO_2^-]^3}$

b. $\dfrac{2[ClO_3^-][Cl^-]}{3[ClO_2^-]}$

c. $\dfrac{[ClO_2^-]^3}{[ClO_3^-]^2[Cl^-]}$

d. $\dfrac{3[ClO_2^-]}{2[ClO_3^-]\,[Cl^-]}$

11. Write the mass-action expression, Q_c, for the following chemical reaction.
M

$$Cu^{2+}(aq) + 4NH_3(aq) \rightleftarrows Cu(NH_3)_4^{2+}(aq)$$

a. $\dfrac{[Cu(NH_3)_4^{2+}]}{[Cu^{2+}]\,[NH_3]}$

b. $\dfrac{[Cu(NH_3)_4^{2+}]}{[Cu^{2+}]\,[NH_3]^4}$

c. $\dfrac{[Cu^{2+}]\,[NH_3]}{[Cu(NH_3)_4^{2+}]}$

d. $\dfrac{[Cu^{2+}]\,[NH_3]^4}{[Cu(NH_3)_4^{2+}]}$

12. Write the mass-action expression, Q_c, for the following chemical reaction.
M

$$2Cu^{2+}(aq) + 4I^-(aq) \rightleftarrows 2CuI(s) + I_2(aq)$$

a. $\dfrac{[CuI]^2[I_2]}{[Cu^{2+}]^2[I^-]^4}$

b. $\dfrac{[Cu^{2+}]^2[I^-]^4}{[CuI]^2[I_2]}$

c. $\dfrac{[I_2]}{[Cu^{2+}]^2[I^-]^4}$

d. $\dfrac{[Cu^{2+}]^2[I^-]^4}{[I_2]}$

13. Write the mass-action expression, Q_c, for the following chemical reaction.
M

$$Zn(s) + 2Ag^+(aq) \rightleftarrows Zn^{2+}(aq) + 2Ag(s)$$

a. $\dfrac{[Zn^{2+}]\,[Ag(s)]^2}{[Zn(s)]\,[Ag^+]^2}$

b. $\dfrac{[Zn(s)]\,[Ag^+]^2}{[Zn^{2+}]\,[Ag(s)]^2}$

c. $\dfrac{[Zn^{2+}]}{[Ag^+]^2}$

d. $\dfrac{[Ag^+]^2}{[Zn^{2+}]}$

14. Write the mass-action expression, Q_c, for the following chemical reaction.
M

$$Fe^{3+}(aq) + 3OH^-(aq) \rightleftarrows Fe(OH)_3(s)$$

a. $\dfrac{[Fe(OH)_3]}{[Fe^{3+}]\,[OH^-]^3}$

b. $\dfrac{[Fe^{3+}]\,[OH^-]^3}{[Fe(OH)_3]}$

c. $\dfrac{1}{[Fe^{3+}]\,[OH^-]^3}$

d. $[Fe^{3+}]\,[OH^-]^3$

15. Write the mass-action expression, Q_c, for the following chemical reaction.
M

$$PbO(s) + CO(g) \rightleftarrows Pb(l) + CO_2(g)$$

a. $\dfrac{[CO_2]}{[CO]}$

b $\dfrac{[CO]}{[CO_2]}$

c. $\dfrac{[Pb]\,[CO_2]}{[PbO]\,[CO]}$

d. $\dfrac{[Pb]\,[CO_2]}{[CO]}$

16. Write the mass-action expression, Q_c, for the following chemical reaction.
M

$$4H_3O^+(aq) + 2Cl^-(aq) + MnO_2(s) \rightleftarrows Mn^{2+}(aq) + 6H_2O(l) + Cl_2(g)$$

a. $\dfrac{[H_3O^+]^4[Cl^-]^2[MnO_2]}{[Mn^{2+}][H_2O]^6[Cl_2]}$

b. $\dfrac{[Mn^{2+}][H_2O]^6[Cl_2]}{[H_3O^+]^4[Cl^-]^2[MnO_2]}$

c $\dfrac{[H_3O^+]^4[Cl\text{-}]^2}{[Mn^{2+}][Cl_2]}$

d $\dfrac{[Mn^{2+}][Cl_2]}{[H_3O^+]^4[Cl^-]^2}$

17. Write the mass-action expression, Q_c, for the following chemical reaction.
M

$$MgO(s) + SO_2(g) + \tfrac{1}{2}O_2(g) \rightleftarrows MgSO_4(s)$$

a. $\dfrac{[MgO][SO_2][O_2]^{1/2}}{[MgSO_4]}$

b. $\dfrac{[MgSO_4]}{[MgO][SO_2][O_2]^{1/2}}$

c. $\dfrac{1}{[SO_2][O_2]^{1/2}}$

d. $[SO_2][O_2]^{1/2}$

18. Write the mass-action expression, Q_c, for the following chemical reaction.
M

$$Sn^{2+}(aq) + \tfrac{1}{2}O_2(g) + 3H_2O(l) \rightleftarrows SnO_2(s) + 2H_3O^+(aq)$$

a. $\dfrac{[H_3O^+]^2}{[Sn^{2+}][O_2]^{1/2}}$

b. $\dfrac{[Sn^{2+}][O_2]^{1/2}}{[H_3O^+]^2}$

c. $\dfrac{[SnO_2][H_3O^+]^2}{[H_2O][Sn^{2+}][O_2]^{1/2}}$

d. $\dfrac{[H_3O^+]^2}{[H_2O][Sn^{2+}][O_2]^{1/2}}$

19. Consider the reactions of cadmium with the thiosulfate anion.
M

$Cd^{2+}(aq) + S_2O_3^{2-}(aq) \rightleftarrows Cd(S_2O_3)(aq)$ $K_1 = 8.3 \times 10^3$
$Cd(S_2O_3)(aq) + S_2O_3^{2-}(aq) \rightleftarrows Cd(S_2O_3)_2^{2-}(aq)$ $K_2 = 2.5 \times 10^2$

What is the value for the equilibrium constant for the following reaction?

$Cd^{2+}(aq) + 2S_2O_3^{2-}(aq) \rightleftarrows Cd(S_2O_3)_2^{2-}(aq)$

a. 33
b. 8.1×10^3
c. 8.6×10^3
d. 2.1×10^6

20. Hydrogen sulfide will react with water as shown in the following reactions.
M

$H_2S(g) + H_2O(l) \rightleftarrows H_3O^+(aq) + HS^-(aq)$ $K_1 = 1 \times 10^{-7}$
$HS^-(aq) + H_2O(l) \rightleftarrows H_3O^+(aq) + S^{2-}(aq)$ $K_2 = ?$
$H_2S(g) + 2H_2O(l) \rightleftarrows 2H_3O^+(aq) + S^{2-}(aq)$ $K_3 = 1.3 \times 10^{-20}$

What is the value of K_2?

a. 1.3×10^{-27}
b. 2.3×10^{-7}
c. 1.3×10^{-13}
d. 7.7×10^{12}

21. At 500°C the equilibrium constant, K_p, is 4.00×10^{-4} for the equilibrium:
M

$2HCN(g) \rightleftarrows H_2(g) + C_2N_2(g)$

What is K_p for the following reaction?

$H_2(g) + C_2N_2(g) \rightleftarrows 2HCN(g)$

a. 2.00×10^{-4}
b. -4.00×10^{-4}
c. 2.50×10^3
d. 4.00×10^4

22. M About half of the sodium carbonate produced is used in making glass products because it lowers the melting point of sand, the major component of glass. When sodium carbonate is added to water it hydrolyses according to the following reactions.

$$CO_3^{2-}(aq) + H_2O(l) \rightleftarrows HCO_3^-(aq) + OH^-(aq) \qquad K_1$$
$$HCO_3^-(aq) + H_2O(l) \rightleftarrows H_2CO_3(aq) + OH^-(aq) \qquad K_2$$

These can be combined to yield

$$CO_3^{2-}(aq) + 2H_2O(l) \rightleftarrows H_2CO_3(aq) + 2OH^-(aq) \qquad K_3$$

What is the value of K_3?

a. $K_1 \times K_2$
b. $K_1 \div K_2$
c. $K_1 + K_2$
d. $K_1 - K_2$

23. M The equilibrium constant, K_c, for the decomposition of $COBr_2$

$$COBr_2(g) \rightleftarrows CO(g) + Br_2(g)$$

is 0.190. What is K_c for the following reaction?

$$2CO(g) + 2Br_2(g) \rightleftarrows 2COBr_2(g)$$

a. 0.0361
b. 2.63
c. 10.5
d. 27.7

24. M The equilibrium constant for the reaction of bromine with chlorine to form bromine monochloride is 58.0 at a certain temperature.

$$Br_2(g) + Cl_2(g) \rightleftarrows 2BrCl(g)$$

What is the equilibrium constant for the following reaction?

$$BrCl(g) \rightleftarrows \tfrac{1}{2}Br_2(g) + \tfrac{1}{2}Cl_2(g)$$

a. 2.97×10^{-4}
b. 1.72×10^{-2}
c. 3.45×10^{-2}
d. 1.31×10^{-1}

25. Consider the following two equilibria and their respective equilibrium constants:
M

(1) $NO(g) + \frac{1}{2}O_2(g) \rightleftarrows NO_2(g)$
(2) $2NO_2(g) \rightleftarrows 2NO(g) + O_2(g)$

Which one of the following is the correct relationship between the equilibrium constants K_1 and K_2?

a. $K_2 = 2/K_1$
b. $K_2 = (1/K_1)^2$
c. $K_2 = -K_1/2$
d. $K_2 = 1/(2K_1)$
e. $K_2 = 1/(2K_1)^2$

26. The equilibrium constant for reaction (1) below is 276. Under the same conditions, what is the equilibrium constant of reaction (2)?
M

(1) $\frac{1}{2}X_2(g) + \frac{1}{2}Y_2(g) \rightleftarrows XY(g)$
(2) $2XY(g) \rightleftarrows X_2(g) + Y_2(g)$

a. 6.02×10^{-2}
b. 7.25×10^{-3}
c. 3.62×10^{-3}
d. 1.31×10^{-5}
e. none of the above

27. Consider the equilibrium reaction: $N_2O_4(g) \rightleftarrows 2NO_2(g)$
M
Which of the following correctly describes the relationship between K_c and K_p for the reaction?

a. $K_p = K_c$
b. $K_p = RT \times K_c$
c. $K_p = (RT \times K_c)^{-1}$
d. $K_p = K_c / RT$
e. $K_p = RT/K_c$

28. Consider the equilibrium reaction: $H_2(g) + Br_2(g) \rightleftarrows 2HBr(g)$
E
Which of the following correctly describes the relationship between K_c and K_p for the reaction?

a. $K_p = K_c$
b. $K_p = (RT)K_c$
c. $K_p = (RT)^2K_c$
d. $K_p = K_c/RT$
e. $K_p = K_c/(RT)^2$

29. The equilibrium constant, K_p, has a value of 6.5×10^{-4} at 308 K for the reaction of chlorine with nitric oxide.
H

$2NO(g) + Cl_2(g) \rightleftarrows 2NOCl(g)$

What is the value of K_c?

a. 2.5×10^{-7}
b. 6.5×10^{-4}
c. 1.6×10^{-2}
d. 1.7

30. The reaction of nitrogen with oxygen to form nitric oxide can be represented by the following equation.
H

$$N_2(g) + O_2(g) \rightleftarrows 2NO(g)$$

At 2000°C, the equilibrium constant, K_c, has a value of 4.10×10^{-4}. What is the value of K_p?

a. 2.17×10^{-8}
b. 4.10×10^{-4}
c. 7.65×10^{-2}
d. 7.75

31. Nitrogen dioxide decomposes according to the reaction
H

$$2NO_2(g) \rightleftarrows 2NO(g) + O_2(g)$$

where $K_p = 4.48 \times 10^{-13}$ at 25°C. What is the value for K_c?

a. 1.81×10^{-16}
b. 1.83×10^{-14}
c. 4.48×10^{-13}
d. 1.11×10^{-9}

32. The equilibrium constant, K_p, for the reaction
M

$$H_2(g) + I_2(g) \rightleftarrows 2HI(g)$$

is 55.2 at 425°C. A rigid cylinder at that temperature contains 0.127 atm of hydrogen, 0.134 atm of iodine, and 1.055 atm of hydrogen iodide. Is the system at equilibrium?

a. Yes.
b. No, the forward reaction must proceed to establish equilibrium.
c. No, the reverse reaction must proceed to establish equilibrium.
d. More data is needed to draw a conclusion.

33. The equilibrium constant, K_p, for the reaction
M

$$CO(g) + H_2O(g) \rightleftarrows CO_2(g) + H_2(g)$$

at 986°C is 0.63. A rigid cylinder at that temperature contains 1.2 atm of carbon monoxide, 0.20 atm of water vapor, 0.30 atm of carbon dioxide, and 0.27 atm of hydrogen. Is the system at equilibrium?

a. Yes.
b. No, the forward reaction must proceed to establish equilibrium.
c. No, the reverse reaction must proceed to establish equilibrium.
d. More data is needed to draw a conclusion.

34. Nitric oxide and bromine were allowed to react in a sealed container. When equilibrium was reached
M P_{NO} = 0.526 atm, P_{Br_2} = 1.59 atm, and P_{NOBr} = 7.68 atm. Calculate K_p for the reaction.

$$2NO(g) + Br_2(g) \rightleftarrows 2NOBr(g)$$

a. 7.45×10^{-3}
b. 0.109
c. 9.18
d. 134

35. Compounds A, B, and C react according to the following equation.
M

$$3A(g) + 2B(g) \rightleftarrows 2C(g)$$

At 100°C a mixture of these gases at equilibrium showed that [A] = 0.855 *M*, [B] = 1.23 *M*, and [C] = 1.75 *M*. What is the value of K_c for this reaction?

a. 0.309
b. 0.601
c. 1.66
d. 3.24

36. Consider the reversible reaction: $2NO_2(g) \rightleftarrows N_2O_4(g)$
E If the concentrations of both NO_2 and N_2O_4 are 0.016 mol L^{-1}, what is the value of Q_c in units of (mol $L^{-1})^n$?

a. 63
b. 2.0
c. 1.0
d. 0.50
e. 0.016

37. 10.0 mL of a 0.100 mol L^{-1} solution of a metal ion M^{2+} is mixed with 10.0 mL of a 0.100 mol l^{-1}
M solution of a substance L. The following equilibrium is established:

$$M^{2+}(aq) + 2L(aq) \rightleftarrows ML_2^{2+}(aq)$$

At equilibrium the concentration of L is found to be 0.0100 mol L^{-1}. What is the equilibrium concentration of ML_2^{2+}, in mol L^{-1}?

a. 0.100
b. 0.050
c. 0.025
d. 0.0200
e. 0.0100

38. A mixture 0.500 mole of carbon monoxide and 0.400 mole of bromine was placed into a rigid 1.00-L
M container and the system was allowed to come to equilibrium. The equilibrium concentration of $COBr_2$ was 0.233 *M*. What is the value of K_c for this reaction?

$$CO(g) + Br_2(g) \rightleftarrows COBr_2(g)$$

a. 5.23
b. 1.165
c. 0.858
d. 0.191

39. M A mixture of 0.600 mol of bromine and 1.600 mol of iodine is placed into a rigid 1.000-L container at 350°C.

$$Br_2(g) + I_2(g) \rightleftarrows 2IBr(g)$$

When the mixture has come to equilibrium, the concentration of iodine monobromide is 1.190 *M*. What is the equilibrium constant for this reaction at 350°C?

a. 3.55×10^{-3}
b. 1.47
c. 282
d. 325

40. H The equilibrium constant K_c for the reaction

$$PCl_3(g) + Cl_2(g) \rightleftarrows PCl_5(g)$$

is 49 at 230°C. If 0.700 mol of PCl_3 is added to 0.700 mol of Cl_2 in a 1.00-L reaction vessel at 230°C, what is the concentration of PCl_3 when equilibrium has been established?

a. 0.108 *M*
b. 0.296 *M*
c. 0.592 *M*
d. 0.828 *M*

41. H The equilibrium constant K_c for the reaction

$$A(g) + B(g) \rightleftarrows C(g)$$

is 0.75 at 150°C. If 0.800 mol of A is added to 0.600 mol of B in a 1.00-L container at 150°C, what will be the equilibrium concentration of C?

a. 0.19 *M*
b. 0.36 *M*
c. 0.41 *M*
d. 0.61 *M*

42. H Nitric oxide is formed in automobile exhaust when nitrogen and oxygen in air react at high temperatures.

$$N_2(g) + O_2(g) \rightleftarrows 2NO(g)$$

The equilibrium constant K_p for the reaction is 0.0025 at 2127°C. If a container is charged with 8.00 atm of nitrogen and 5.00 atm of oxygen and the mixture is allowed to reach equilibrium, what will be the equilibrium partial pressure of nitrogen?

a. 0.16 atm
b. 0.31 atm
c. 7.7 atm
d. 7.8 atm

43. At 25°C, the equilibrium constant K_c for the reaction in the solvent CCl_4
H

$$2BrCl \rightleftarrows Br_2 + Cl_2$$

is 0.145. If the initial concentration of chlorine is 0.030 *M* and of bromine monochloride is 0.020 *M*, what is the equilibrium concentration of bromine?

a. 1.35×10^{-3} *M*
b. 2.70×10^{-3} *M*
c. 8.82×10^{-3} *M*
d. 9.70×10^{-2} *M*

44. At 25°C, the equilibrium constant K_c for the reaction
H

$$2A(aq) \rightleftarrows B(aq) + C(aq)$$

is 65. If 2.50 mol of A is added to enough water to prepare 1.00 L of solution, what will the equilibrium concentration of A be?

a. 0.14 *M*
b. 0.28 *M*
c. 1.18 *M*
d. 2.4 *M*

45. At a certain temperature the reaction
H

$$CO_2(g) + H_2(g) \rightleftarrows CO(g) + H_2O(g)$$

has $K_c = 2.50$. If 2.00 mol of carbon dioxide and 1.5 mol of hydrogen are placed in a 5.00 L vessel and equilibrium is established, what will be the concentration of carbon monoxide?

a. 0.091 *M*
b. 0.191 *M*
c. 0.209 *M*
d. 1.05 *M*

46. At 25°C, the equilibrium constant K_c for the reaction
H

$$2A(g) \rightleftarrows B(g) + C(g)$$

is 0.035. A mixture of 8.00 moles of B and 12.00 moles of C in a 20.0 L container is allowed to come to equilibrium. What is the equilibrium concentration of A?

a. [A] = 0.339 *M*
b. [A] = 0.678 *M*
c. [A] = 6.78 *M*
d. [A] = 13.56 *M*

47. M At 850°C, the equilibrium constant K_p for the reaction

$$C(s) + CO_2(g) \rightleftarrows 2CO(g)$$

has a value of 10.7. If the total pressure in the system at equilibrium is 1.000 atm, what is the partial pressure of carbon monoxide?

a. 0.489 atm
b. 0.667 atm
c. 0.921 atm
d. 0.978 atm

48. M Ammonium iodide dissociates reversibly to ammonia and hydrogen iodide.

$$NH_4I(s) \rightleftarrows NH_3(g) + HI(g)$$

At 400°C, $K_p = 0.215$. Calculate the partial pressure of ammonia at equilibrium when a sufficient quantity of ammonium iodide is heated to 400°C.

a. 0.103 atm
b. 0.232 atm
c. 0.464 atm
d. 2.00 atm

49. E The reaction system

$$POCl_3(g) \rightleftarrows POCl(g) + Cl_2(g)$$

is at equilibrium. Which of the following statements describes the behavior of the system if POCl is added to the container?

a. The forward reaction will proceed to establish equilibrium.
b. The reverse reaction will proceed to establish equilibrium.
c. The partial pressures of $POCl_3$ and POCl will remain steady while the partial pressure of chlorine increases.
d. The partial pressure of chlorine remains steady while the partial pressures of $POCl_3$ and POCl increase.

50. E The reaction system

$$POCl_3(g) \rightleftarrows POCl(g) + Cl_2(g)$$

is at equilibrium. Which of the following statements describes the behavior of the system if the partial pressure of chlorine is reduced by 50%?

a. $POCl_3$ will be consumed as equilibrium is established.
b. POCl will be consumed as equilibrium is established.
c. Chlorine will be consumed as equilibrium is established.
d. The partial pressure of POCl will decrease while the partial pressure of Cl_2 increases as equilibrium is established.

51. The reaction system
E

$$CS_2(g) + 4H_2(g) \rightleftarrows CH_4(g) + 2H_2S(g)$$

is at equilibrium. Which of the following statements describes the behavior of the system if the partial pressure of hydrogen is doubled?

a. As equilibrium is reestablished, the partial pressure of carbon disulfide increases.
b. As equilibrium is reestablished, the partial pressure of methane, CH_4, decreases.
c. As equilibrium is reestablished, the partial pressure of hydrogen decreases.
d. As equilibrium is reestablished, the partial pressure of hydrogen sulfide decreases.

52. The reaction system
E

$$CS_2(g) + 4H_2(g) \rightleftarrows CH_4(g) + 2H_2S(g)$$

is at equilibrium. Which of the following statements describes the behavior of the system if the partial pressure of carbon disulfide is reduced?

a. As equilibrium is reestablished, the partial pressure of carbon disulfide increases.
b. As equilibrium is reestablished, the partial pressure of hydrogen decreases.
c. As equilibrium is reestablished, the partial pressure of methane, CH_4, increases.
d. As equilibrium is reestablished, the partial pressures of hydrogen and hydrogen sulfide decrease.

53. Magnesium hydroxide is used in several antacid formulations. When it is added to water it dissociates into
M magnesium and hydroxide ions.

$$Mg(OH)_2(s) \rightleftarrows Mg^{2+}(aq) + 2OH^-(aq)$$

The equilibrium constant at 25°C is 8.9×10^{-12}. One hundred grams of magnesium hydroxide is added to 1.00 L of water and equilibrium is established. What happens to the hydroxide ion concentration if 10 grams of $Mg(OH)_2$ are added to the mixture?

a. The hydroxide ion concentration will decrease.
b. The hydroxide ion concentration will increase.
c. The hydroxide ion concentration will be unchanged.
d. More information is needed to make a valid judgment.

54. Sodium hydrogen carbonate decomposes above 110°C to form sodium carbonate, water, and carbon dioxide.
M

$$2NaHCO_3(s) \rightleftarrows Na_2CO_3(s) + H_2O(g) + CO_2(g)$$

One thousand grams of sodium hydrogen carbonate are added to a reaction vessel, the temperature is increased to 200°C, and the system comes to equilibrium. What happens to the partial pressure of carbon dioxide in this system if 50 g of sodium carbonate are added?

a. The partial pressure of carbon dioxide will increase.
b. The partial pressure of carbon dioxide will decrease.
c. The partial pressure of carbon dioxide will be unchanged.
d. More information is needed to make an accurate prediction.

55. Methanol can be synthesized by combining carbon monoxide and hydrogen.
E

$$CO(g) + 2H_2(g) \rightleftarrows CH_3OH(g)$$

A reaction vessel contains the three gases at equilibrium with a total pressure of 1.00 atm. What will happen to the partial pressure of hydrogen if enough argon is added to raise the total pressure to 1.4 atm?

a. The partial pressure of hydrogen will decrease.
b. The partial pressure of hydrogen will increase.
c. The partial pressure of hydrogen will be unchanged.
d. More information is needed to make an accurate prediction.

56. At 450°C, *tert*-butyl alcohol decomposes into water and isobutene.
E

$$(CH_3)_3COH(g) \rightleftarrows (CH_3)_2CCH_2(g) + H_2O(g)$$

A reaction vessel contains these compounds at equilibrium. What will happen if the volume of the container is reduced by 50% at constant temperature?

a. The forward reaction will proceed to reestablish equilibrium.
b. The reverse reaction will proceed to reestablish equilibrium.
c. No change occurs.
d. The equilibrium constant will increase.

57. A container was charged with hydrogen, nitrogen, and ammonia gases at 120°C and the system was allowed to reach equilibrium. What will happen if the volume of the container is increased at constant temperature?
E

$$3H_2(g) + N_2(g) \rightleftarrows 2NH_3(g)$$

a. There will be no effect.
b. More ammonia will be produced at the expense of hydrogen and nitrogen.
c. Hydrogen and nitrogen will be produced at the expense of ammonia.
d. The equilibrium constant will increase.

58. Magnesium carbonate dissociates to magnesium oxide and carbon dioxide at elevated temperatures.
E

$$MgCO_3(s) \rightleftarrows MgO(s) + CO_2(g)$$

A reaction vessel contains these compounds in equilibrium at 300°C. What will happen if the volume of the container is reduced by 25% at 300°C?

a. The partial pressure of carbon dioxide present at equilibrium will increase.
b. The partial pressure of carbon dioxide present at equilibrium will decrease.
c. The partial pressure of carbon dioxide at equilibrium will be unchanged.
d. More information is needed in order to make a valid judgment.

59. The reaction of nitric oxide to form dinitrogen oxide and nitrogen dioxide is exothermic.
E

$$3NO(g) \rightleftarrows N_2O(g) + NO_2(g) + \text{heat}$$

What effect will be seen if the temperature of the system at equilibrium is raised by 25°C?

a. The partial pressure of NO will increase.
b. The partial pressure of NO will decrease.
c. The partial pressure of NO_2 will increase.
d. The partial pressures of NO and N_2O will increase.

60. Methanol can be synthesized by combining carbon monoxide and hydrogen.
E

$$CO(g) + 2H_2(g) \rightleftarrows CH_3OH(g) \qquad \Delta H^\circ_{rxn} = -90.7 \text{ kJ}$$

A reaction vessel contains these compounds at equilibrium. What effect will be seen when equilibrium is re-established after decreasing the temperature by 45°C?

a. The partial pressure of methanol will decrease.
b. The partial pressures of hydrogen and methanol will decrease.
c. The partial pressure of carbon monoxide will decrease.
d. The partial pressure of hydrogen will increase.

61. Hydrogen bromide will dissociate into hydrogen and bromine gases.
E

$$2HBr(g) \rightleftarrows H_2(g) + Br_2(g) \qquad \Delta H^\circ_{rxn} = 68 \text{ kJ}$$

What effect will a temperature increase of 50°C have on this system at equilibrium?

a. The partial pressure of hydrogen bromide will increase.
b. The partial pressure of hydrogen will increase.
c. The partial pressure of hydrogen bromide and bromine will increase.
d. There will be no effect on the partial pressure of any of the gases.

62. Ethane can be formed by reacting acetylene with hydrogen.
E

$$C_2H_2(g) + 2H_2(g) \rightleftarrows C_2H_6(g) \qquad \Delta H^\circ_{rxn} = -311 \text{ kJ}$$

Under which reaction conditions would you expect to have the greatest equilibrium yield of ethane?

a. high temperature, high pressure
b. low temperature, high pressure
c. high temperature, low pressure
d. low temperature, low pressure

63. Nitrogen dioxide can dissociate to nitric oxide and oxygen.
E

$$2NO_2(g) \rightleftarrows 2NO(g) + O_2(g) \qquad \Delta H^\circ_{rxn} = +114 \text{ kJ}$$

Under which reaction conditions would you expect to produce the largest amount of oxygen?

a. high temperature, high pressure
b. low temperature, high pressure
c. high temperature, low pressure
d. low temperature, low pressure

64. The following reaction is at equilibrium at one atmosphere, in a closed container.
M

$$NaOH(s) + CO_2(g) \rightleftarrows NaHCO_3(s)$$

Which, if any, of the following actions will decrease the total amount of CO_2 gas present at equilibrium?

a. adding N_2 gas to double the pressure
b. adding more solid NaOH
c. decreasing the volume of the container
d. removing half of the solid $NaHCO_3$
e. none of the above

65. The following reaction is at equilibrium at a pressure of 1 atm, in a closed container.
M

$$NaOH(s) + CO_2(g) \rightleftarrows NaHCO_3(s) \qquad (\Delta H^\circ < 0)$$

Which, if any, of the following actions will decrease the concentration of CO_2 gas present at equilibrium?

a. Adding N_2 gas to double the pressure
b. Adding more solid NaOH
c. Increasing the volume of the container
d. Lowering the temperature
e. None of the above

66. The following reaction is at equilibrium in a closed container.
M

$$CuSO_4.5H_2O(s) \rightleftarrows CuSO_4(s) + 5H_2O(g)$$

Which, if any, of the following actions will lead to an increase in the pressure of H_2O present at equilibrium?

a. increasing the volume of the container
b. decreasing the volume of the container
c. adding a catalyst
d. removing some solid $CuSO_4$
e. none of the above

67. The following reaction is at equilibrium in a sealed container.
M

$$N_2(g) + 3H_2(g) \rightleftarrows 2NH_3(g) \qquad (\Delta H^\circ < 0)$$

Which, if any, of the following actions will increase the value of the equilibrium constant, K_c?

a. adding a catalyst
b. adding more N_2
c. increasing the pressure
d. lowering the temperature
e. none of the above

68. Stearic acid, nature's most common fatty acid, dimerizes when dissolved in hexane:
H

$$2C_{17}H_{35}COOH \rightleftarrows (C_{17}H_{35}COOH)_2 \qquad \Delta H^\circ_{rxn} = -172 \text{ kJ}$$

The equilibrium constant for this reaction at 28°C is 2900. Estimate the equilibrium constant at 38°C.

a. 4.7×10^5
b. 2.6×10^4
c. 3.2×10^2
d. 18

69. Hydrogen sulfide can be formed in the following reaction:
H

$$H_2(g) + \tfrac{1}{2}S_2(g) \rightleftarrows H_2S(g) \qquad \Delta H^\circ_{rxn} = -92 \text{ kJ}$$

The equilibrium constant K_p = 106 at 1023 K. Estimate the value of K_p at 1218 K.

a. 5.05
b. 18.8
c. 34.7
d. 598

<u>Short Answer Questions</u>

70. Write the expression for K_c and K_p for the reaction
E

$$PH_3BCl_3(s) \rightleftarrows PH_3(g) + BCl_3(g)$$

71. Ammonia is synthesized in the Haber process:
H

$$N_2(g) + 3H_2(g) \rightleftarrows 2NH_3(g)$$

K_p for this reaction is 1.49×10^{-5} atm^{-2} at 500.°C. Calculate K_c at this temperature.

72. At a high temperature, the following reaction has an equilibrium constant of 1.0×10^2.
M

$$H_2(g) + F_2(g) \rightleftarrows 2HF(g)$$

If 1.00 mol of each of H_2 and F_2 are allowed to come to equilibrium in a 10.0 L vessel, calculate the equilibrium amounts of H_2 and HF.

73. When 0.152 mol of solid PH_3BCl_3 is introduced into a 3.0 L container at a certain temperature,
M 8.44×10^{-3} mol of PH_3 is present at equilibrium:

$$PH_3BCl_3(s) \rightleftarrows PH_3(g) + BCl_3(g)$$

Construct a reaction table for the process, and use it to calculate K_c at this temperature.

74. Consider the equilibrium
H

$$H_2(g) + Br_2(g) \rightleftarrows 2HBr(g)$$

To a 20.0 L flask are added 0.100 moles of H_2 and 0.200 moles of HBr. The equilibrium constant for this reaction is 989. Calculate the number of moles of Br_2 in the flask when equilibrium is established. Make any reasonable approximation, clearly stating what that approximation is.

75. M
Consider the following gas-phase equilibrium reaction:

$N_2(g) + O_2(g) \rightleftarrows 2NO(g)$ $K_c = 4.10 \times 10^{-4}$ at 2000°C.

If 1.0 mol of NO is introduced into a 1.0 L container at 2000°C, what is the concentration of NO when equilibrium is reached?

76. M
Consider the equilibrium:

$A(s) \rightleftarrows B(s) + C(g)$ $(\Delta H° > 0)$

Predict and explain how or whether the following actions would affect this equilibrium.

a. adding more solid A
b. lowering the temperature
c. increasing the pressure on the system by reducing its volume
d. adding helium gas to increase the total pressure

77. M
a. State Le Chatelier's principle
b. The following reaction is at equilibrium in a closed container:

$2Fe(OH)_3(s) \rightleftarrows Fe_2O_3(s) + H_2O(g)$ $(\Delta H° > 0)$

What effects, if any, will the following actions have on the position of equilibrium? In each case, state the direction of any shift in equilibrium, and give your reasons in one sentence.

(i) adding more $Fe(OH)_3$
(ii) raising the temperature
(iii) adding a catalyst

78. H
The Haber process for ammonia synthesis is exothermic:

$N_2(g) + 3H_2(g) \rightleftarrows 2NH_3(g)$ $\Delta H° = -92$ kJ

If the equilibrium constant K_c for this process at 500.°C is 6.0×10^{-2}, what is its value at 300.°C?

True/False Questions

79. M
Although a system may be at equilibrium, the rate constants of the forward and reverse reactions will in general be different.

80. E
When a reaction system reaches equilibrium, the forward and reverse reactions stop.

81. E
Once a reaction system reaches equilibrium, the concentrations of reactions and products no longer change.

82. M
There is a direct correlation between the speed of a reaction and its equilibrium constant.

83. M
For a gas-phase equilibrium, a change in the pressure of any single reactant or product will change K_p.

84. E
For a gas-phase equilibrium, a change in the pressure of any single reactant or product will affect the amounts of other substances involved in the equilibrium.

85. For a solution equilibrium, a change in concentration of a reactant or product does not change K_c.
E

86. For some gas-phase reactions, $K_p = K_c$.
E

87. If $Q > K$, more products need to be formed as the reaction proceeds to equilibrium.
E

88. Changing the amount of reactant or product in an equilibrium reaction will always change the equilibrium position, regardless of the physical state of the substance involved.
M

89. Unless $\Delta H° = 0$, a change in temperature will affect the value of the equilibrium constant K_c.
M

Equilibrium: The Extent of Chemical Reactions
Chapter 17
Answer Key

1.	b	24.	d	47.	c
2.	d	25.	b	48.	c
3.	c	26.	d	49.	b
4.	b	27.	b	50.	a
5.	d	28.	a	51.	c
6.	b	29.	c	52.	a
7.	b	30.	b	53.	c
8.	d	31.	b	54.	c
9.	b	32.	c	55.	c
10.	a	33.	b	56.	b
11.	b	34.	d	57.	c
12.	c	35.	d	58.	c
13.	c	36.	a	59.	a
14.	c	37.	d	60.	c
15.	a	38.	a	61.	b
16.	d	39.	c	62.	b
17.	c	40.	a	63.	c
18.	a	41.	a	64.	c
19.	d	42.	d	65.	d
20.	c	43.	a	66.	e
21.	c	44.	a	67.	d
22.	a	45.	c	68.	c
23.	d	46.	b	69.	b

70. $K_c = [PH_3][BCl_3]$ $\quad$ $K_p = p(PH_3)\cdot p(BCl_3)$

71. $K_c = 6.00 \times 10^{-2}\ L^2/mol^2$

72. 0.17 mol of H_2 and 1.7 mol of HF

73. $PH_3BCl_3(s) \rightleftarrows PH_3(g) + BCl_3(g)$

	$PH_3BCl_3(s)$	$PH_3(g)$	$BCl_3(g)$
Initial amount (mol)	0.152	0	0
Change (mol)	-0.144	+0.144	+0.144
Final amount (mol)	0.00844	0.144	0.144

$$K_c = (0.144\ mol/3.0\ L)^2 = 2.3 \times 10^{-3}$$

74. ~ 4.0×10^{-4} mol of Br_2 present at equilibrium. The approximation is to neglect the small changes in amounts of H_2 and HBr that occur in reaching equilibrium.

75. $[NO] = 1.0 \times 10^{-2}\ mol\ L^{-1}$

76. a. No effect. Pure solids and liquids have constant concentration, regardless of total amount.
    b. By a shift in the exothermic direction, the "disturbance" is reduced. More A will form.
    c. More A will form, since this will decrease the amount of gas present and thus reduce the effect of the disturbance.
    d. No effect. Helium does not participate in the reaction, and the added helium will not change the partial pressure of C.

77. a. If a stress is applied to a system at equilibrium, the equilibrium position shifts so as to reduce that stress.
    b. (i) No effect. Pure solids and liquids have constant concentration, regardless of total amount.
    (ii) The equilibrium will move in the endothermic direction, i.e. more products will form.
    (iii) No effect, since a catalyst will speed up the forward and back reactions equally.

78. 8.9

79. T

80. F

81. T

82. F

83. F

84. T

85. T

86. F

87. F

88. T

89. T

Acid-Base Equilibria
Chapter 18

Multiple Choice Questions

1. The substance H_2SO_3 is considered
M
   a. a weak Arrhenius acid
   b. a weak Arrhenius base
   c. a strong Arrhenius acid
   d. a strong Arrhenius base

2. The substance HOBr is considered
M
   a. a weak Arrhenius acid
   b. a weak Arrhenius base
   c. a strong Arrhenius acid
   d. a strong Arrhenius base

3. The substance $Mg(OH)_2$ is considered
M
   a. a weak Arrhenius acid
   b. a weak Arrhenius base
   c. a strong Arrhenius acid
   d. a strong Arrhenius base

4. The substance $Ba(OH)_2$ is considered
M
   a. a weak Arrhenius acid
   b. a weak Arrhenius base
   c. a strong Arrhenius acid
   d. a strong Arrhenius base

5. The substance $Ca(OH)_2$ is considered
M
   a. a weak Arrhenius acid
   b. a weak Arrhenius base
   c. a strong Arrhenius acid
   d. a strong Arrhenius base

6. The substance $HBrO_2$ is considered
M
   a. a weak Arrhenius acid
   b. a weak Arrhenius base
   c. a strong Arrhenius acid
   d. a strong Arrhenius base

7. Which, if any, of the following acids is strong?
H

a. phosphoric
b. carbonic
c. acetic
d. water
e. none of the above

8. Which one of the following is a strong acid?
M

a. H_2CO_3
b. H_2SO_3
c. H_2SO_4
d. H_3PO_4
e. CH_3COOH

9. Which one of the following will give a solution with a pH > 7, but is not an Arrhenius base in the strict sense?
M

a. CH_3NH_2
b. NaOH
c. CO_2
d. $Ca(OH)_2$
e. CH_4

10. The substance NH_3 is considered
M

a. a weak acid
b. a weak base
c. a strong acid
d. a strong base

11. The substance $(CH_3CH_2)_2NH$ is considered
M

a. a weak acid
b. a weak base
c. a strong acid
d. a strong base

12. The substance $HClO_4$ is considered
E

a. a weak acid
b. a weak base
c. a strong acid
d. a strong base

13. Which of the following is the strongest acid?
M

a. CH_3COOH
b. HF
c. HI
d. H_2SO_3

14. Which of the following is the strongest base?
E

a. CH_3NH_2
b. LiOH
c. $B(OH)_3$
d. $Al(OH)_3$

15. Which of the following pairs has the stronger acid listed first?
M

a. HBr, HI
b. $HClO_2$, $HClO_3$
c. H_2SeO_4, H_2SeO_3
d. HF, HCl

16. Which of the following pairs has the stronger acid listed first?
M

a. H_2AsO_3, H_2AsO_4
b. HI, HBr
c. HClO, $HClO_3$
d. H_2S, HCl

17. Which of the following acids has the lowest pH?
E

0.1 *M* HBO, $pK_a = 2.43$
0.1 *M* HA, $pK_a = 4.55$
0.1 *M* HMO, $pK_a = 8.23$
0.1 *M* HST, $pK_a = 11.89$

a. HA
b. HST
c. HMO
d. HBO

18. Which of the following solutions contains the strongest acid?
E

0.1 *M* HA, pH = 6.85
0.1 *M* HD, pH = 7.22
0.1 *M* HE, pH = 8.34
0.1 *M* HJ, pH = 11.88

a. HE
b. HA
c. HJ
d. HD

19. What is the pH of a 0.20 *M* HCl solution?
E

a. 0.70
b. 1.61
c. 12.39
d. 13.30

20. What is the pH of a 0.75 *M* HNO_3 solution?
E

a. 0.12
b. 0.29
c. 0.63
d. 0.82

21. What is the pH of a 0.00200 *M* $HClO_4$ solution?
E

a. 0.995
b. 1.378
c. 2.699
d. 6.215

22. What is the pH of a 0.050 *M* HBr solution?
E

a. 0.89
b. 1.12
c. 1.30
d. 3.00

23. What is the pH of a 0.050 *M* LiOH solution?
M

a. 1.30
b. 3.00
c. 11.00
d. 12.70

24. What is the pH of a 0.0035 *M* KOH solution?
M

a. 2.46
b. 5.65
c. 8.35
d. 11.54

25. What is the pH of a 0.0125 *M* NaOH solution?
M

a. 0.972
b. 1.903
c. 12.097
d. 13.028

26. What is the pOH of a 0.0250 *M* HI solution?
M

a. 0.944
b. 1.602
c. 12.398
d. 13.056

27. What is the pOH of a 0.0085 *M* KOH solution?
E

a. 2.07
b. 4.77
c. 9.23
d. 11.93

28. What is the $[OH^-]$ for a solution at 25°C that has $[H_3O^+] = 2.35 \times 10^{-3}$ *M*?
E

a. 4.26×10^{-5} *M*
b. 2.35×10^{-11} *M*
c. 4.26×10^{-12} *M*
d. 2.35×10^{-17} *M*

29. What is the $[OH^-]$ for a solution at 25°C that has $[H_3O^+] = 8.23 \times 10^{-2}$ *M*?
E

a. 1.22×10^{-6}*M*
b. 8.23×10^{-12} *M*
c. 1.22×10^{-13} *M*
d. 8.23×10^{-16} *M*

30. What is the $[OH^-]$ for a solution at 25°C that has pH = 4.29?
M

a. 1.4×10^{-2}*M*
b. 5.1×10^{-5}*M*
c. 1.9×10^{-10}*M*
d. 7.3×10^{-13}*M*

31. What is the $[H_3O^+]$ for a solution at 25°C that has pOH = 5.640?
M

a. 2.34×10^{-4}*M*
b. 2.29×10^{-6}*M*
c. 4.37×10^{-9}*M*
d. 4.27×10^{-11}*M*

32. Select the pair of substances in which an acid is listed followed by its conjugate base.
E

a. NH_3, NH_4^+
b. HPO_4^{2-}, $H_2PO_4^-$
c. HCO_3^-, CO_3^{2-}
d. CH_3COOH, $CH_3COOH_2^+$

33. Select the pair of substances which is not a conjugate acid-base pair.
E

a. H_3O^+, H_2O
b. HNO_2, NO_2^-
c. H_2SO_4, HSO_4^-
d. H_2S, S^{2-}

34. Which one of the following pairs is not a conjugate acid-base pair?
M

a. H_2O/OH^-
b. H_2O_2/HO_2^-
c. OH^-/O^{2-}
d. $H_2PO_4^-/HPO_4^{2-}$
e. HCl/H^+

35. According to Brønsted and Lowry, which one of the following is not a conjugate acid-base pair?
M

a. H_3O^+/OH^-
b. $CH_3OH_2^+/CH_3OH$
c. HI/I^-
d. HSO_4^-/SO_4^{2-}
e. H_2/H^-

36. The acid dissociation constant K_a equals 1.26×10^{-2} for HSO_4^- and is 5.6×10^{-10} for NH_4^+. Which statement
M about the following equilibrium is correct?

$$HSO_4^-(aq) + NH_3(aq) \rightleftarrows SO_4^{2-}(aq) + NH_4^+(aq)$$

a. The reactants will be favored because ammonia is a stronger base than the sulfate anion.
b. The products will be favored because the hydrogen sulfate ion is a stronger acid than the ammonium ion
c. Neither reactants or products will be favored because all of the species are weak acids or bases.
d. The initial concentrations of the hydrogen sulfate ion and ammonia must be known before any prediction can be made.

37. A student adds 0.1 mol of oxalic acid and 0.1 mol of sodium dihydrogen phosphate to enough water to make
M 1.0 L of solution. The following equilibrium is established with the concentrations of the products greater than the concentrations of the reactants. Which of the statements about the equilibrium system is correct?

$$H_2C_2O_4(aq) + H_2PO_4^-(aq) \rightleftarrows HC_2O_4^-(aq) + H_3PO_4(aq)$$

a. Oxalic acid is a weaker acid than phosphoric acid.
b. The hydrogen oxalate anion, $HC_2O_4^-$, is a stronger base than the dihydrogen phosphate anion, $H_2PO_4^-$.
c. Phosphoric acid is a weaker acid than oxalic acid.
d. The dihydrogen phosphate anion, $H_2PO_4^-$, is a stronger acid than oxalic acid.

38. An aqueous solution of phosphoric acid and sodium nitrite is prepared and the following equilibrium
M is established.

$$H_3PO_4(aq) + NO_2^-(aq) \rightleftarrows H_2PO_4^-(aq) + HNO_2(aq)$$

The equilibrium constant K_c for this reaction is greater than one. Which of the following statements is correct?

a. Phosphoric acid is a weaker acid than nitrous acid.
b. Phosphoric acid is a stronger acid than nitrous acid.
c. The nitrite anion is a weaker base than the dihydrogen phosphate anion.
d. The dihydrogen phosphate anion is a stronger acid than nitrous acid.

39. Butyric acid is responsible for the odor in rancid butter. A solution of 0.25 *M* butyric acid has a pH of 2.71.
M What is the K_a for the acid?

a. 0.36
b. 2.4×10^{-2}
c. 7.8×10^{-3}
d. 1.5×10^{-5}

40. A 0.15 M solution of chloroacetic acid has a pH of 1.86. What is the value of K_a for this acid?
M

a. 0.16
b. 0.099
c. 0.0014
d. 0.00027

41. A 0.050 M solution of the weak acid HA has $[H_3O^+] = 3.77 \times 10^{-4}$ M. What is the K_a for the acid?
M

a. 7.5×10^{-3} M
b. 2.8×10^{-6} M
c. 7.0×10^{-8} M
d. 2.6×10^{-11} M

42. Formic acid, which is a component of insect venom, has a $K_a = 1.8 \times 10^{-4}$. What is the $[H_3O^+]$ in a solution
H that is initially 0.10 M formic acid, HCOOH?

a. 1.3×10^{-2} M
b. 4.2×10^{-3} M
c. 8.4×10^{-3} M
d. 1.8×10^{-5} M

43. Picric acid has been used in the leather industry and in etching copper. However, its laboratory use has been
H restricted because it dehydrates on standing and can become shock sensitive. It has an acid dissociation constant of 0.42. What is the $[H_3O^+]$ for a 0.20 M solution of picric acid?

a. 0.022 M
b. 0.052 M
c. 0.15 M
d. 0.29 M

44. Lactic acid has a pK_a of 3.08. What is the approximate degree of dissociation of a 0.35 M solution of
H lactic acid?

a. 1.1%
b. 2.2%
c. 4.8%
d. 14%

45. A 1.25 M solution of the weak acid HA is 9.2% dissociated. What is the pH of the solution?
M

a. 0.64
b. 0.94
c. 1.13
d. 2.16

46. Farmers who raise cotton once used arsenic acid, H_3AsO_4, as a defoliant at harvest time. Arsenic acid is a
M polyprotic acid with $K_1 = 2.5 \times 10^{-4}$, $K_2 = 5.6 \times 10^{-8}$, and $K_3 = 3 \times 10^{-13}$. What is the pH of a 0.500 M solution of arsenic acid?

a. 0.85
b. 1.96
c. 3.90
d. 4.51

47. H Arsenic acid, H_3AsO_4, is used industrially to manufacture insecticides. Arsenic acid is a polyprotic acid with $K_1 = 2.5 \times 10^{-4}$, $K_2 = 5.6 \times 10^{-8}$, and $K_3 = 3 \times 10^{-13}$. What is the concentration of the $HAsO_4^{2-}$ in a solution whose initial arsenic acid concentration was 0.35 *M*?

a. 9.4×10^{-3} *M*
b. 2.5×10^{-4} *M*
c. 8.8×10^{-5} *M*
d. 5.6×10^{-8} *M*

48. H What is the pH of a 0.050 *M* triethylamine, $(C_2H_5)_3N$, solution? $K_b = 5.3 \times 10^{-4}$

a. 2.31
b. 5.32
c. 8.68
d. 11.69

49. H Hydroxylamine, $HONH_2$, readily forms salts such as hydroxylamine hydrochloride which are used as antioxidants in soaps. Hydroxylamine has K_b of 9.1×10^{-9}. What is the pH of a 0.025 *M* $HONH_2$ solution?

a. 2.90
b. 4.82
c. 9.18
d. 11.10

50. E What is the value of K_b for the formate anion, $HCOO^-$? $K_a(HCOOH) = 2.1 \times 10^{-4}$

a. 2.1×10^{-4}
b. 6.9×10^{-6}
c. 4.8×10^{-11}
d. 2.1×10^{-18}

51. E What is the value of K_b for the cyanide anion, CN^-? $K_a(HCN) = 4 \times 10^{-10}$

a. 5×10^{-3}
b. 3×10^{-5}
c. 4×10^{-10}
d. 4×10^{-24}

52. E What is the value of K_a for the methylammonium ion, $CH_3NH_3^+$? $K_b(CH_3NH_2) = 4.4 \times 10^{-4}$

a. 4.4×10^{-4}
b. 4.8×10^{-6}
c. 2.3×10^{-11}
d. 4.4×10^{-18}

53. H What is the pH of a 0.0100 *M* sodium benzoate solution? $K_b(C_7H_5O_2^-) = 1.5 \times 10^{-10}$

a. 0.38
b. 5.91
c. 8.09
d. 13.62

54. What is the pH of a 0.010 *M* triethanolammonium chloride, $(HOC_2H_2)_3NHCl$, solution?
H K_b, $((HOC_2H_2)_3N) = 5.9 \times 10^{-7}$

a. 2.75
b. 4.89
c. 9.11
d. 11.25

55. Which of the following aqueous solutions will have the highest pH?
E

a. 0.1 *M* CH_3COOH, $pK_a = 4.7$
b. 0.1 *M* $CuCl_2$, $pK_a = 7.5$
c. 0.1 *M* $H_3C_6H_5O_7$, $pK_a = 3.1$
d. 0.1 *M* $ZnCl_2$, $pK_a = 9.0$

56. Which of the following solutions is the most acidic?
E

a. 0.1 *M* $Al(NO_3)_3$, $K_a = 1 \times 10^{-5}$
b. 0.1 *M* $Be(NO_3)_2$, $K_a = 4 \times 10^{-6}$
c. 0.1 *M* $Pb(NO_3)_2$, $K_a = 3 \times 10^{-8}$
d. 0.1 *M* $Ni(NO_3)_2$, $K_a = 1 \times 10^{-10}$

57. A solution is prepared by adding 0.10 mol of sodium fluoride, NaF, to 1.00 L of water. Which statement
M about the solution is correct?

a. The solution is basic.
b. The solution is neutral.
c. The solution is acidic.
d. The values for K_a and K_b for the species in solution must be known before a prediction can be made.

58. A solution is prepared by adding 0.10 mol of potassium acetate, KCH_3COO, to 1.00 L of water. Which
M statement about the solution is correct?

a. The solution is basic.
b. The solution is neutral.
c. The solution is acidic.
d. The values for K_a and K_b for the species in solution must be known before a prediction can be made.

59. A solution is prepared by adding 0.10 mol of sodium sulfide, Na_2S, to 1.00 L of water. Which statement
M about the solution is correct?

a. The solution is basic.
b. The solution is neutral.
c. The solution is acidic.
d. The values for K_a and K_b for the species in solution must be known before a prediction can be made.

60. A solution is prepared by adding 0.10 mol of iron(III) nitrate, $Fe(NO_3)_3$, to 1.00 L of water. Which statement
M about the solution is correct?

a. The solution is basic.
b. The solution is neutral.
c. The solution is acidic.
d. The values for K_a and K_b for the species in solution must be known before a prediction can be made.

61. Ammonium chloride is used as an electrolyte in dry cells. Which of the following statements about a 0.10 M
M solution of NH_4Cl, is correct?

a. The solution is basic.
b. The solution is neutral.
c. The solution is acidic.
d. The values for K_a and K_b for the species in solution must be known before a prediction can be made.

62. A solution is prepared by adding 0.10 mol of lithium nitrate, $LiNO_3$, to 1.00 L of water. Which statement
M about the solution is correct?

a. The solution is basic.
b. The solution is neutral.
c. The solution is acidic.
d. The values for K_a and K_b for the species in solution must be known before a prediction can be made.

63. A solution is prepared by adding 0.10 mol of potassium chloride, KCl, to 1.00 L of water. Which statement
M about the solution is correct?

a. The solution is basic.
b. The solution is neutral.
C. The solution is acidic.
d. The values for K_a and K_b for the species in solution must be known before a prediction can be made.

64. An aqueous solution is prepared by dissolving the salt formed by the neutralization of a weak acid by a weak
M base. Which statement about the solution is correct?

a. The solution is basic.
b. The solution is neutral.
c. The solution is acidic.
d. The values for K_a and K_b for the species in solution must be known before a prediction can be made.

65. Which one of the following substances will give an aqueous solution of pH < 7?
M

a. KI
b. NH_4Br
c. Na_2CO_3
d. CH_3COONa
e. CH_3OH

66. Which one of the following substances will give an aqueous solution of pH closest to 7?
M

a. KNO_3
b. CO_2
c. NH_4I
d. NH_3
e. CH_3NH_2

67. Which of the following is considered a Lewis acid?
M

a. CH_3NH_2
b. BCl_3
c. F^-
d. BF_4^-

68. Which of the following would be considered a Lewis base?
E

a. BCl_3
b. Cu^{2+}
c. Cl^-
d. Mn^{2+}

69. Iodine trichloride, ICl_3, will react with a chloride ion to form ICl_4^-. Which species acts as a Lewis acid in this reaction?
M

a. ICl_4^-
b. ICl_3
c. Cl^-
d. None of the species is a Lewis acid.

70. Calcium oxide, CaO, also known as quick lime, will react with carbon dioxide to form calcium carbonate, $CaCO_3$. Which species acts as a Lewis acid in the reaction?
M

a. Ca^{2+}
b. O^{2-}
c. CO_2
d. $CaCO_3$

71. Which one of the following is a Lewis acid but not a Brønsted-Lowry acid?
M

a. Fe^{3+}
b. H_3O^+
c. HSO_4^-
d. NH_3
e. H_2O

Short Answer Questions

72. Define an acid according to the Arrhenius theory, and write a balanced equation to support this definition.
E

73. Define a base according to the Brønsted-Lowry theory. Write a balanced equation in support of this definition, in which the base is acting as a Brønsted-Lowry base but not as an Arrhenius base.
M

74. Formic acid, HCOOH, is a weak acid with a pK_a of 3.74. Draw up a reaction table for the reaction of 0.300 *M* formic acid with water, and calculate the pH of this solution.
H

75. The pH of a 0.200 *M* solution of the weak base pyridine, C_5H_5N, is 8.59. Draw up a reaction table for the reaction of 0.200 *M* pyridine with water, and calculate K_b.
H

76. (a) Write a balanced equation representing the reaction of the acid, $H_2PO_4^-$ with the base, water.
E (b) Write the expression for K_a of $H_2PO_4^-$ in terms of concentrations of relevant species.

77. The English are fond of soggy French fries ("chips") wrapped in old newspaper and generously drenched in vinegar, which is a 0.83 *M* solution of acetic acid. If the acetic acid in vinegar is 0.47 % dissociated, calculate K_a for this acid.
H

78. Describe what is meant by the "leveling effect". Use a real acid as an example, and write an appropriate equation.
M

79. Define an acid according to the Lewis theory. Write a balanced equation in support of this definition,
M in which the acid is neither an Arrhenius acid nor a Brønsted-Lowry acid.

80. Consider the reaction: $BF_3 + F^- \rightarrow BF_4^-$
H Can this ever be considered to be an acid-base reaction? Support your answer with appropriate arguments.

True/False Questions

81. Arrhenius bases raise the hydroxide ion concentration when dissolved in water.
E

82. It is not possible to have a pH lying outside the range 0 to 14.
M

83. $K_w = 1.0 \times 10^{-14}$, regardless of temperature.
M

84. If a strong acid such as HCl is diluted sufficiently with water, the pH will be higher than 7.
M

85. All Brønsted-Lowry bases contain the hydroxide ion, OH^-.
M

86. All Brønsted-Lowry bases have at least one lone pair of electrons.
M

87. All weak acids have strong conjugate bases.
H

88. All strong acids have weak conjugate bases
E

89. The ammonium ion, NH_4^+, is a weak acid.
M

90. A solution of sodium acetate (CH_3COONa) in water is weakly basic.
H

91. The strongest base which can exist in water is the hydroxide ion.
E

92. The chloride ion, Cl^-, is a typical Lewis acid.
E

Acid-Base Equilibriua
Chapter 18
Answer Key

1.	a	26.	c	51.	b
2.	a	27.	a	52.	c
3.	d	28.	c	53.	c
4.	d	29.	c	54.	b
5.	d	30.	c	55.	d
6.	a	31.	c	56.	a
7.	e	32.	c	57.	a
8.	c	33.	d	58.	a
9.	a	34.	e	59.	a
10.	b	35.	a	60.	c
11.	b	36.	b	61.	c
12.	c	37.	c	62.	b
13.	c	38.	b	63.	b
14.	b	39.	d	64.	d
15.	c	40.	c	65.	b
16.	b	41.	b	66.	a
17.	d	42.	b	67.	b
18.	b	43.	c	68.	c
19.	a	44.	c	69.	b
20.	a	45.	b	70.	c
21.	c	46.	b	71.	a
22.	c	47.	d		
23.	d	48.	d		
24.	d	49.	c		
25.	c	50.	c		

72. An acid is a substance which contains hydrogen and raises the H_3O^+ concentration when dissolved in water. Representing the acid by HA,

$$HA(s, l \text{ or } g) + H_2O(l) \rightleftarrows H_3O^+(aq) + A^-(aq)$$

73. According to Brønsted and Lowry, a base is a substance which accepts a proton in a reaction.

$$CH_3NH_2(aq) + H_2O(l) \rightleftarrows CH_3NH_3^+(aq) + OH^-(aq)$$

or

$$NH_3(g) + HCl(g) \rightleftarrows NH_4Cl(s)$$

74.

	$HCOOH(aq)$	$+ H_2O(l)$	$\rightleftarrows HCOO^-(aq)$	$+ H_3O^+(aq)$
Initial conc, M	0.300	-	0	10^{-7}
Change in conc, M	$-x$	-	x	x
Equilibrium conc, M	$0.300 - x$	-	x	$10^{-7} + x$

Assuming $0.300 - x \approx 0.300$ and $10^{-7} + x \approx x$, the pH is 2.13

75.

	$C_5H_5N(l)$ +	$H_2O(l)$ ⇄	$C_5H_5NH^+(aq)$ +	$OH^-(aq)$
Initial conc, *M*	0.300	-	0	10^{-7}
Change in conc, *M*	$-x$	-	x	x
Equilibrium conc, *M*	$0.200 - x$	-	x	$10^{-7} + x$

pH = 10.95

76. $H_2PO_4^-(aq) + H_2O(l) \rightleftarrows HPO_4^{2-}(aq) + H_3O^+(aq)$

$$K_a = \frac{[HPO_4^{2-}][H_3O^+]}{[H_2PO_4^-]}$$

77. $K_a = 1.8 \times 10^{-5}$

78. In aqueous solution, the strongest acid which can exist is the hydronium ion, and the strongest base is the hydroxide ion. In water, all acids which are stronger than hydronium will react to produce hydronium ion; stronger bases than hydroxide will react to produce hydroxide. Thus, all strong acids are reduced in strength (leveled) to that of hydronium; all strong bases are leveled to the strength of hydroxide.

$$HCl(g) + H_2O(l) \rightarrow H_3O^+(aq) + Cl^-(aq)$$

79. According to Lewis, an acid is an electron pair acceptor in a chemical reaction.

$$BF_3 + F^- \rightarrow BF_4^-$$

Here, BF_3 is the Lewis acid, accepting a pair of electrons from F^-.

80. Yes, according to Lewis, this is an acid-base reaction. The Lewis acid, BF_3, accepts a pair of electrons from the base, F^-.

81. T

82. F

83. F

84. F

85. F

86. T

87. F

88. T

89. T

90. T

91. T

92. F

Ionic Equilibria in Aqueous Systems
Chapter 19

Multiple Choice Questions

1. Which of the following aqueous mixtures would be a buffer system?
E

a. HCl, NaCl
b. HNO_3, $NaNO_3$
c. H_3PO_4, $H_2PO_4^-$
d. H_2SO_4, CH_3COOH

2. Which of the following aqueous mixtures would be a buffer system?
E

a. CH_3COOH, NaH_2PO_4
b. H_2CO_3, HCO_3^-
c. $H_2PO_4^-$, HCO_3^-
d. HSO_4^-, HSO_3^-

3. Equal volumes of the following pairs of solutions are mixed. Which pair will produce a buffer solution?
M

a. 0.10 mol L^{-1} HCl and 0.05 mol L^{-1} NaOH
b. 0.10 mol L^{-1} HCl and 0.15 mol L^{-1} NH_3
c. 0.10 mol L^{-1} HCl and 0.05 mol L^{-1} NH_3
d. 0.10 mol L^{-1} HCl and 0.20 mol L^{-1} CH_3COOH
e. 0.10 mol L^{-1} HCl and 0.20 mol L^{-1} NaCl

4. Which one of the following aqueous solutions, when mixed with an equal volume of 0.10 mol L^{-1}
E aqueous NH_3, will produce a buffer solution?

a. 0.10 mol L^{-1} HCl
b. 0.20 mol L^{-1} HCl
c. 0.10 mol L^{-1} CH_3COOH
d. 0.050 mol L^{-1} NaOH
e. 0.20 mol L^{-1} NH_4Cl

5. Which one of the following pairs of 0.100 mol L^{-1} solutions, when mixed, will produce a buffer solution?
E

a. 50. mL of aqueous CH_3COOH and 25. mL of aqueous HCl
b. 50. mL of aqueous CH_3COOH and 100. mL of aqueous NaOH
c. 50. mL of aqueous NaOH and 25. mL of aqueous HCl
d. 50. mL of aqueous CH_3COONa and 25. mL of aqueous NaOH
e. 50. mL of aqueous CH_3COOH and 25. mL of aqueous CH_3COONa

6. Which of the following has the highest buffer capacity?
E

a. 0.10 M $H_2PO_4^-$/0.10 M HPO_4^{2-}
b. 0.50 M $H_2PO_4^-$/0.10 M HPO_4^{2-}
c. 0.10 M $H_2PO_4^-$/0.50 M HPO_4^{2-}
d. 0.50 M $H_2PO_4^-$/0.50 M HPO_4^{2-}

7. Which of the following acids should be used to prepare a buffer with a pH of 4.5?
M

a. HOC_6H_4OCOOH, $K_a = 1.0 \times 10^{-3}$
b. $C_6H_4(COOH)_2$, $K_a = 2.9 \times 10^{-4}$
c. CH_3COOH, $K_a = 1.8 \times 10^{-5}$
d. $C_5H_5O_5COOH^{-2}$, $K_a = 4.0 \times 10^{-6}$

8. Citric acid has an acid dissociation constant of 8.4×10^{-4}. It would be most effective for preparation of a buffer with a pH of
M

a. 2
b. 3
c. 4
d. 5

9. A buffer is to be prepared by adding solid sodium acetate to 0.10 M CH_3COOH. Which of the following concentrations of sodium acetate will produce the most effective buffer?
M

a. 2.5 M $NaCH_3COO$
b. 2.0 M $NaCH_3COO$
c. 1.5 M $NaCH_3COO$
d. 0.30 M $NaCH_3COO$

10. An acetate buffer has a pH of 4.40. Which of the following changes will cause the pH to decrease?
E

a. dissolving a small amount of solid sodium acetate
b. adding a small amount of dilute hydrochloric acid
c. adding a small amount of dilute sodium hydroxide
d. dissolving a small amount of solid sodium chloride

11. A phosphate buffer ($H_2PO_4^-/HPO_4^{2-}$) has a pH of 8.3. Which of the following changes will cause the pH to increase?
E

a. dissolving a small amount of Na_2HPO_4
b. dissolving a small amount of NaH_2PO_4
c. adding a small amount of dilute hydrochloric acid
d. adding a small amount of dilute phosphoric acid

12. What will be the effect of adding 0.5 mL of 0.1 M NaOH to 100 mL of an acetate buffer in which $[CH_3COOH] = [CH_3COO^-] = 0.5$ M?
E

a. the pH will increase slightly
b. the pH will increase significantly
c. the pH will decrease slightly
d. the pH will decrease significantly

13. What will be the effect of adding 0.5 mL of 0.1 M HCl to 100 mL of a phosphate buffer in which $[H_2PO_4^-] = [HPO_4^{2-}] = 0.35$ M?
E

a. the pH will increase slightly
b. the pH will increase significantly
c. the pH will decrease slightly
d. the pH will decrease significantly

14. M

Buffer solutions with the component concentrations shown below were prepared. Which of them should have the lowest pH?

a. $[CH_3COOH] = 0.25\ M$, $[CH_3COO^-] = 0.25\ M$
b. $[CH_3COOH] = 0.75\ M$, $[CH_3COO^-] = 0.75\ M$
c. $[CH_3COOH] = 0.75\ M$, $[CH_3COO^-] = 0.25\ M$
d. $[CH_3COOH] = 0.25\ M$, $[CH_3COO^-] = 0.75\ M$

15. M

Buffer solutions with the component concentrations shown below were prepared. Which of them should have the highest pH?

a. $[H_2PO_4^-] = 0.50\ M$, $[HPO_4^{2-}] = 0.50\ M$
b. $[H_2PO_4^-] = 1.0\ M$, $[HPO_4^{2-}] = 1.0\ M$
c. $[H_2PO_4^-] = 1.0\ M$, $[HPO_4^{2-}] = 0.50\ M$
d. $[H_2PO_4^-] = 0.50\ M$, $[HPO_4^{2-}] = 1.0\ M$

16. M

A buffer is prepared by adding 0.5 mol of solid sodium hydroxide to 1.0 L of 1.0 *M* acetic acid (CH_3COOH). What is the pH of the buffer? (Assume that the sodium hydroxide does not affect the volume of the solution.)

a. The pH will be greater than the pK_a for acetic acid.
b. The pH will be less than the pK_a for acetic acid.
c. The pH will be equal to the pK_a for acetic acid.
d. More information is needed to solve the problem.

17. M

A buffer is prepared by adding 100 mL of 0.2 *M* hydrochloric acid to 100 mL of 0.4 *M* sodium formate. What is the pH of the buffer?

a. The pH will be greater than the pK_a of formic acid.
b. The pH will be less than the pK_a of formic acid.
c. The pH will be equal to the pK_a of formic acid.
d. The pH will equal the pK_b of sodium formate.

18. M

A buffer is prepared by adding 100 mL of 0.50 *M* sodium hydroxide to 100 mL of 0.75 *M* propanoic acid. What is the pH of the buffer?

a. The pH will be greater than the pK_a of propanoic acid.
b. The pH will be less than the pK_a of propanoic acid.
c. The pH will be equal to the pK_a of propanoic acid.
d. The pH will equal the pK_b of sodium propanoate.

19. M

A solution is prepared by adding 500 mL of 0.3 *M* NaClO to 500 mL of 0.4 *M* HClO. What is the pH of this solution?

a. The pH will be greater than the pK_a of hypochlorous acid
b. The pH will be less than the pK_a of hypochlorous acid
c. The pH will be equal to the pK_a of hypochlorous acid
d. The pH will equal the pK_b of sodium hypochlorite.

20. M

What is the pH for a buffer that consists of 0.45 *M* CH_3COOH and 0.35 *M* CH_3COONa? $K_a = 1.8 \times 10^{-5}$

a. 4.49
b. 4.64
c. 4.85
d. 5.00

21. What is the pH for a solution that consists of 0.50 *M* $H_2C_6H_6O_6$ (ascorbic acid) and
M 0.75 *M* $NaHC_6H_6O_6$ (sodium ascorbate)? $K_a = 6.8 \times 10^{-5}$

a. 3.76
b. 3.99
c. 4.34
d. 4.57

22. What is the pH for a buffer that consists of 0.20 *M* NaH_2PO_4 and 0.40 *M* Na_2HPO_4? $K_a = 6.2 \times 10^{-8}$
M

a. 6.51
b. 6.91
c. 7.51
d. 7.90

23. What is the $[H_3O^+]$ in a buffer that consists of 0.30 *M* HCOOH and 0.20 *M* HCOONa? $K_a = 1.7 \times 10^{-4}$
M

a. 1.1×10^{-4} *M*
b. 2.6×10^{-4} *M*
c. 4.3×10^{-4} *M*
d. 6.7×10^{-5} *M*

24. What is the $[H_3O^+]$ in a solution that consists of 1.2 *M* HClO and 2.3 *M* NaClO? $K_a = 3.5 \times 10^{-8}$
M

a. 7.8×10^{-9} *M*
b. 6.7×10^{-8} *M*
c. 1.8×10^{-8} *M*
d. 1.6×10^{-7} *M*

25. What is the $[H_3O^+]$ in a solution that consists of 0.15 *M* $C_2N_2H_8$ (ethylene diamine) and 0.35 $C_2N_2H_9Cl$?
M $K_b = 4.7 \times 10^{-4}$

a. 1.1×10^{-3} *M*
b. 2.0×10^{-3} *M*
c. 5.0×10^{-11} *M*
d. 9.2×10^{-12} *M*

26. What is the $[H_3O^+]$ in a solution that consists of 1.5 *M* NH_3 and 2.5 NH_4Cl? $K_b = 1.8 \times 10^{-5}$
M

a. 1.1×10^{-5} *M*
b. 3.0×10^{-5} *M*
c. 3.3×10^{-10} *M*
d. 9.3×10^{-10} *M*

27. What is the pK_a for the acid HA if a solution of 0.65 *M* HA and 0.85 *M* NaA has a pH of 4.75?
M

a. 4.48
b. 4.63
c. 4.87
d. 5.02

28. A formic acid buffer containing 0.50 *M* HCOOH and 0.50 *M* HCOONa has a pH of 3.77.
H What will the pH be after 0.010 mol of NaOH has been added to 100.0 mL of the buffer?

a. 3.36
b. 3.59
c. 3.95
d. 4.18

29. An acetic acid buffer containing 0.50 *M* CH_3COOH and 0.50 *M* CH_3COONa has a pH of 4.74. What will the
H pH be after 0.0020 mol of HCl has been added to 100.0 mL of the buffer?

a. 4.66
b. 4.71
c. 4.78
d. 4.82

30. A buffer is prepared by adding 300.0 mL of 2.0 *M* NaOH to 500.0 mL of 2.0 *M* CH_3COOH. What is the pH
M of this buffer? $K_a = 1.8 \times 10^{-5}$

a. 4.57
b. 4.52
c. 4.97
d. 4.92

31. A buffer is prepared by adding 1.00 L of 1.0 *M* HCl to 750 mL of 1.5 *M* NaHCOO. What is the pH of this
M buffer? $K_a = 1.7 \times 10^{-4}$

a. 2.88
b. 3.72
c. 3.82
d. 4.66

32. A buffer is prepared by adding 150 mL of 1.0 *M* NaOH to 250 mL of 1.0 *M* NaH_2PO_4. How many moles of
H HCl must be added to this buffer solution to change the pH by 0.18 units?

a. 0.025
b. 0.063
c. 0.50
d. 1.0

33. At the equivalence point
E

a. the $[H_3O^+]$ equals the K_a of the acid.
b. the $[H_3O^+]$ equals the K_a of the indicator.
c. The amounts of acid and base which have been combined are in their stoichiometric ratio.
d. the pH is 7.0.

34. When a strong acid is titrated with a strong base, the pH at the equivalence point
E

a. is greater than 7.0.
b. is equal to 7.0.
c. is less than 7.0.
d. is equal to the pK_a of the acid.

35. When a weak acid is titrated with a strong base, the pH at the equivalence point
E

a. is greater than 7.0.
b. is equal to 7.0.
c. is less than 7.0.
d. is equal to the pK_a of the acid.

36. When a strong acid is titrated with a weak base, the pH at the equivalence point
E

a. is greater than 7.0.
b. is equal to 7.0.
c. is less than 7.0.
d. is equal to the pK_a of the acid.

37. When a weak acid is titrated with a weak base, the pH at the equivalence point
E

a. is greater than 7.0.
b. is equal to 7.0.
c. is less than 7.0.
d. is determined by the sizes of K_a and K_b.

38. Which one of the following is the best representation of the titration curve which will be obtained in

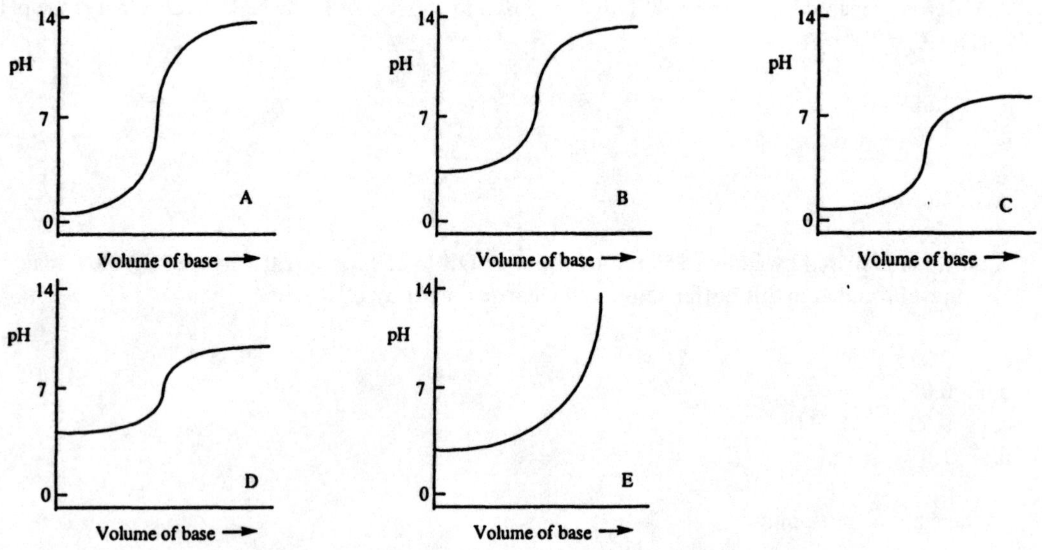

M the titration of a weak acid (0.10 mol L^{-1}) with a strong base of the same concentration?

39. Which one of the following is the best representation of the titration curve which will be obtained in

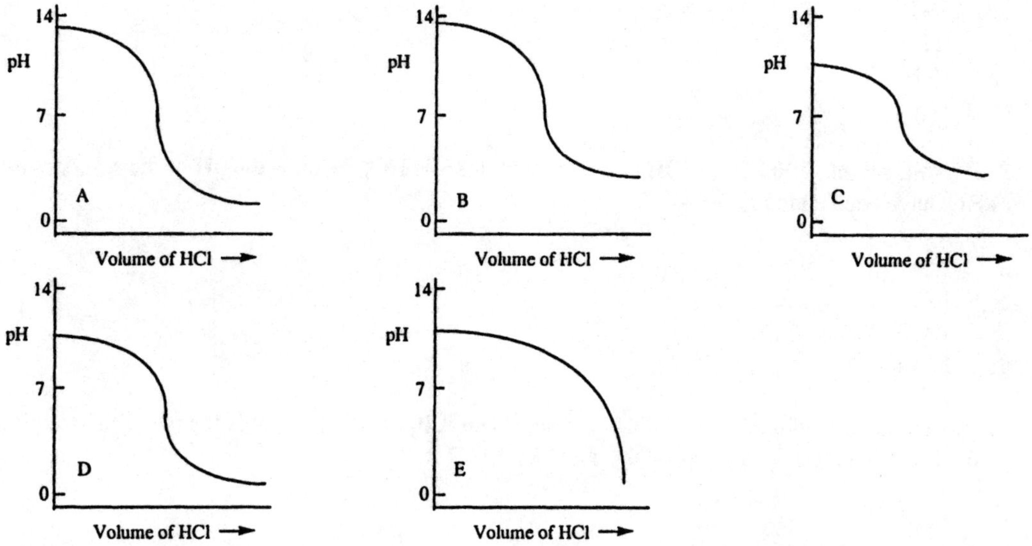

M the titration of a weak base (0.10 mol L^{-1}) with HCl of the same concentration?

40. The indicator propyl red has $K_a = 3.3 \times 10^{-6}$. What would be the approximate pH range over which it
M would change color?

a. 3.5-5.5
b. 4.5-6.5
c. 5.5-7.5
d. 6.5-8.5

41. Which of the following indicators would be the best to use when 0.050 *M* benzoic acid ($K_a = 6.6 \times 10^{-5}$) is
E titrated with 0.05 *M* NaOH?

a. bromphenol blue, pH range: 3.0-4.5
b. bromcresol green, pH range: 3.8-5.4
c. alizarin, pH range: 5.7-7.2
d. phenol red, pH range: 6.9-8.2

42. A 50.0-mL sample of 0.50 *M* HCl is titrated with 0.50 *M* NaOH. What is the pH of the solution after 28.0 mL
M of NaOH have been added to the acid?

a. 0.85
b. 0.75
c. 0.66
d. 0.49

43. A 20.0-mL sample of 0.25 *M* HNO_3 is titrated with 0.15 *M* Na OH. What is the pH of the solution after 30.0
M mL of NaOH have been added to the acid?

a. 2.00
b. 1.60
c. 1.05
d. 1.00

44. A 20.0-mL sample of 0.30 M HBr is titrated with 0.15 M NaOH. What is the pH of the solution after 40.3 mL
M of NaOH have been added to the acid?

a. 2.95
b. 3.13
c. 10.87
d. 11.05

45. A 35.0-mL sample of 0.20 M LiOH is titrated with 0.25 M HCl. What is the pH of the solution after 23.0 mL
M of HCl have been added to the base?

a. 1.26
b. 1.67
c. 12.33
d. 12.74

46. A 25.0-mL sample of 0.35 M HCOOH is titrated with 0.20 M KOH. What is the pH of the solution after
H 25.0 mL of KOH has been added to the acid? $K_a = 1.77 \times 10^{-4}$

a. 4.00
b. 3.88
c. 3.63
d. 3.51

47. A 10.0-mL sample of 0.75 M CH_3CH_2COOH is titrated with 0.30 M NaOH. What is the pH of the solution
H after 22.0 mL of NaOH have been added to the acid? $K_a = 1.3 \times 10^{-5}$

a. 5.75
b. 4.94
c. 4.83
d. 4.02

48. A 25.0-mL sample of 0.10 M $C_2H_3NH_2$ (ethylamine) is titrated with 0.15 M HCl. What is the pH of the
H solution after 9.00 mL of acid have been added to the amine? $K_b = 6.5 \times 10^{-4}$

a. 11.08
b. 10.88
c. 10.74
d. 10.55

49. A 25.0-mL sample of 1.00 M NH_3 is titrated with 0.15 M HCl. What is the pH of the solution after 15.00 mL
H of acid have been added to the ammonia solution? $K_b = 1.8 \times 10^{-5}$

a. 10.26
b. 9.30
c. 9.21
d. 8.30

50.
M

A 20.0-mL sample of 0.30 *M* HClO was titrated with 0.30 *M* NaOH. The following data was collected during the titration.

mL NaOH added	5.00	10.00	15.00	20.00
pH	6.98	7.46	7.93	10.31

What is the K_a for HClO?

a. 1.1×10^{-7}
b. 3.5×10^{-8}
c. 1.2×10^{-8}
d. 4.9×10^{-11}

51.
H

Which one of the following is the best representation of the titration curve which will be obtained in the titration of a weak diprotic acid H_2A (0.10 mol L^{-1}) with a strong base of the same concentration?

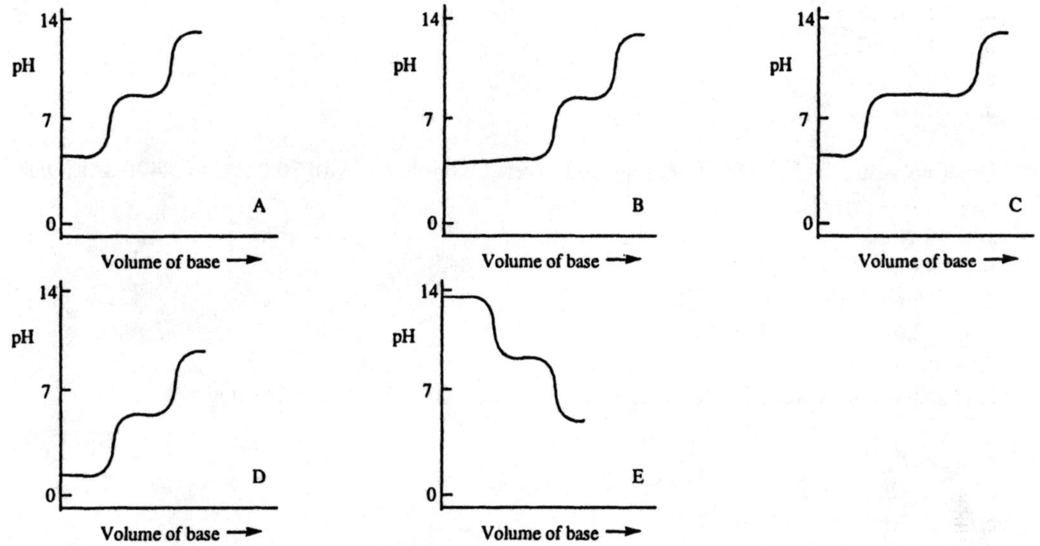

52.
H

A diprotic acid H_2A has $K_{a1} = 1 \times 10^{-4}$ and $K_{a2} = 1 \times 10^{-8}$. The corresponding base A^{2-} is titrated with aqueous HCl, both solutions being 0.1 mol L^{-1}. Which one of the following diagrams best represents the titration curve which will be seen?

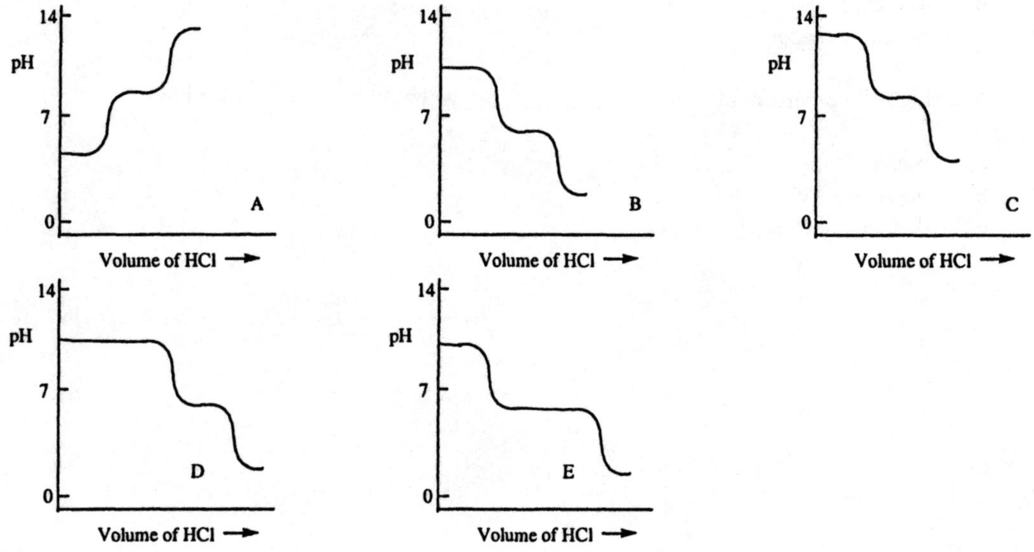

53. M A 20.0-mL sample of 0.50 *M* $H_2C_6H_6O_6$ (ascorbic acid, a diprotic acid) was titrated with 0.50 *M* NaOH. The following data were gathered during the titration.

mL NaOH added	10.00	20.00	30.00	40.00
pH	4.17	5.21	11.55	12.89

What is K_{a2} for ascorbic acid?

a. 6.8×10^{-5}
b. 6.2×19^{-6}
c. 2.8×10^{-12}
d. 1.3×10^{-13}

54. E What volume of 0.200 *M* KOH must be added to 17.5 mL of 0.135 *M* H_3PO_4 to reach the third equivalence point?

a. 11.8 mL
b. 17.5 mL
c. 23.6 mL
d. 35.4 mL

55. E What volume of 0.500 *M* H_2SO_4 is needed to react completely with 20.0 mL of 0.400 *M* LiOH?

a. 4.00
b. 8.00
c. 16.0
d. 32.0

56. H A change in pH will affect the solubility of which of the following compounds?

a. BaF_2
b. CuCl
c. CuBr
d. AgI

57. M The solubility of aluminum hydroxide _____________ when dilute nitric acid is added to it.

a. increases
b. decreases
c. does not change

58. E The solubility of silver chloride _____________ when dilute nitric is added to it.

a. increases
b. decreases
c. does not change

59. Write the ion product expression for magnesium fluoride, MgF_2.
E

a. $[Mg^{2+}][F^-]$
b. $[Mg^{2+}][F^-]^2$
c. $\dfrac{1}{[Mg^{2+}][F^-]}$
d. $\dfrac{1}{[Mg^{2+}][F^-]^2}$

60. Write the ion product expression for silver sulfide, Ag_2S.
E

a. $[Ag^+][S^{2-}]$
b. $[Ag^+][S^{2-}]^2$
c. $[Ag^+]^2[S^{2-}]$
d. $\dfrac{[Ag^+]^2[S^{2-}]}{[Ag_2S]}$

61. Write the ion product expression for calcium phosphate, $Ca_3(PO_4)_2$.
E

a. $[Ca^{2+}][PO_4^{3-}]$
b. $[Ca^{2+}]^2[PO_4^{3-}]^3$
c. $[Ca^{2+}]^3[PO_4^{3-}]^2$
d. $\dfrac{[Ca^{2+}]^2[PO_4^{3-}]^3}{[Ca_3(PO_4)_2]}$

62. The solubility of lead(II) chloride is 0.45 g/100 mL of solution. What is the K_{sp} of $PbCl_2$?
M

a. 4.9×10^{-2}
b. 1.7×10^{-5}
c. 4.2×10^{-6}
d. 8.5×10^{-6}

63. The solubility of calcium chromate is 1.56×10^{-3} g/100 mL of solution. What is the K_{sp} for $CaCrO_4$?
E

a. 2.4×10^{-4}
b. 1.5×10^{-5}
c. 7.6×10^{-6}
d. 1.0×10^{-8}

64. The solubility of silver chromate is 0.0287 g/1.0 L of solution. What is the K_{sp} for Ag_2CrO_4?
M

a. 2.4×10^{-5}
b. 9.5×10^{-5}
c. 2.6×10^{-12}
d. 6.5×10^{-13}

65. The solubility of magnesium phosphate is 2.27×10^{-3} g/1.0 L of solution. What is the K_{sp} for $Mg_3(PO_4)_2$?
H

a. 6.5×10^{-12}
b. 6.0×10^{-14}
c. 5.2×10^{-24}
d. 4.8×10^{-26}

66. Calculate the solubility of barium carbonate, $BaCO_3$, in pure water. $K_{sp} = 2.0 \times 10^{-9}$
E

a. 1.3×10^{-3} *M*
b. 2.2×10^{-5} *M*
c. 3.2×10^{-5} *M*
d. 4.5×10^{-5} *M*

67. Calculate the solubility of silver oxalate, $Ag_2C_2O_4$, in pure water. $K_{sp} = 1.0 \times 10^{-11}$
M

a. 1.4×10^{-4} *M*
b. 2.2×10^{-4} *M*
c. 5.4×10^{-5} *M*
d. 3.2×10^{-6} *M*

68. Calculate the solubility of strontium fluoride, SrF_2, in pure water. $K_{sp} = 2.6 \times 10^{-9}$
M

a. 1.4×10^{-3} *M*
b. 3.4×10^{-4} *M*
c. 8.7×10^{-4} *M*
d. 5.1×10^{-5} *M*

69. Calculate the solubility of silver phosphate, Ag_3PO_4, in pure water. $K_{sp} = 2.6 \times 10^{-18}$
M

a. 1.0×10^{-5} *M*
b. 1.8×10^{-5} *M*
c. 4.0×10^{-5} *M*
d. 1.5×10^{-6} *M*

70. Which of the following substances has the greatest solubility in water?
E

a. $MgCO_3$, $K_{sp} = 3.5 \times 10^{-8}$
b. $NiCO_3$, $K_{sp} = 1.3 \times 10^{-7}$
c. $AgIO_3$, $K_{sp} = 3.1 \times 10^{-8}$
d. $CuBr$, $K_{sp} = 5.0 \times 10^{-9}$

71. Which of the following substances has the greatest solubility in water?
E

a. BaF_2, $K_{sp} = 1.5 \times 10^{-6}$
b. $Ca(OH)_2$, $K_{sp} = 6.5 \times 10^{-6}$
c. $Zn(IO_3)_2$, $K_{sp} = 3.9 \times 10^{-6}$
d. Ag_2SO_4, $K_{sp} = 1.5 \times 10^{-5}$

72. Which of the following substances has the greatest solubility in water?
H

a. $Ba(IO_3)_2$, $K_{sp} = 1.5 \times 10^{-9}$
b. PbF_2, $K_{sp} = 3.6 \times 10^{-8}$
c. $SrSO_4$, $K_{sp} = 3.2 \times 10^{-7}$
d. $CuCl$, $K_{sp} = 1.9 \times 10^{-7}$

73. Calculate the solubility of magnesium sulfate, $MgSO_4$, when placed into a 0.10 M $MgCl_2$ solution.
H $K_{sp} = 5.9 \times 10^{-3}$

a. 4.2×10^{-2} M
b. 5.9×10^{-2} M
c. 7.7×10^{-2} M
d. 3.5×10^{-5} M

74. Calculate the solubility of silver chromate, Ag_2CrO_4, in 0.005 M Na_2CrO_4. $K_{sp} = 2.6 \times 10^{-12}$
H

a. 1.4×10^{-4} M
b. 1.1×10^{-5} M
c. 3.4×10^{-5} M
d. 1.6×10^{-6} M

75. Calculate the solubility of lead(II) iodide, PbI_2, in 0.025 M KI. $K_{sp} = 7.9 \times 10^{-9}$
M

a. 2.8×10^{-2} M
b. 4.5×10^{-2} M
c. 1.3×10^{-5} M
d. 8.9×10^{-5} M

76. A lab technician adds 0.015 mol of KOH to 1.00 L of 0.0010 M $Ca(NO_3)_2$.
M $K_{sp} = 6.5 \times 10^{-6}$ for $Ca(OH)_2$) Which of the following statements is correct?

a. Calcium hydroxide precipitates until the solution is saturated.
b. The solution is unsaturated and no precipitate forms.
c. The concentration of calcium ions is reduced by the addition of the hydroxide ions.
d. One must know K_{sp} for calcium nitrate to make meaningful predictions on this system.

77. A lab technician adds 0.20 mol of NaF to 1.00 L of 0.35 M cadmium nitrate, $Cd(NO_3)_2$. Which of the
M following statements is correct? $K_{sp} = 6.44 \times 10^{-3}$ for CdF_2

a. Cadmium fluoride precipitates until the solution is saturated.
b. The solution is unsaturated and no precipitate forms.
c. The solubility of cadmium fluoride is increased by the presence of additional fluoride ions.
d. One must know K_{sp} for cadmium nitrate to make meaningful predictions on this system.

78. What is the maximum amount of sodium sulfate that can be added to 1.00 L of 0.0020 M $Ca(NO_3)_2$ before
E precipitation of calcium sulfate begins? $K_{sp} = 2.4 \times 10^{-5}$ for calcium sulfate

a. 1.2×10^{-2} mol
b. 3.5×10^{-3} mol
c. 4.9×10^{-3} mol
d. 1.2×10^{-5} mol

79. Consider the dissolution of MnS in water ($K_{sp} = 3.0 \times 10^{-14}$).
M

$$MnS(s) + H_2O(l) \rightleftarrows Mn^{2+}(aq) + HS^-(aq) + OH^-(aq)$$

How is the solubility of manganese(II) sulfide affected by the addition of aqueous potassium hydroxide to the system?

a. the solubility will be unchanged
b. the solubility will decrease
c. the solubility will increase
d. the amount of KOH added must be known before its effect can be predicted

80. The lab technician Anna Lytic adds 2.20 mol KOH to 1.00 L of 0.5 *M* $Al(NO_3)_3$. What is the concentration of
H aluminum ions after the aluminum nitrate has reacted with the potassium hydroxide? $K_f = 3.0 \times 10^{33}$ for $Al(OH)_4^-$

a. 1.8×10^{-7} *M*
b. 9.1×10^{-18} *M*
c. 1.0×10^{-31} *M*
d. 7.1×10^{-36} *M*

81. A solution is prepared by adding 4.50 mol of sodium hydroxide to 1.00 L of 1.00 *M* $Co(NO_3)_2$. What is the
H equilibrium concentration of cobalt ions? $K_f = 5.0 \times 10^9$ for $Co(OH)_4^{2-}$

a. 1.1×10^{-2} *M*
b. 1.4×10^{-5} *M*
c. 3.2×10^{-9} *M*
d. 4.9×10^{-13} *M*

82. The concentration of the complex ion in each of following solutions is 1.00 *M*. In which of the solutions will
E the concentration of the uncomplexed metal ion be the greatest?

CdI_4^{2-}	$K_f = 1.0 \times 10^6$
$Cu(NH_3)_4^{2+}$	$K_f = 5.6 \times 10^{11}$
$Zn(OH)_4^{2-}$	$K_f = 3.0 \times 10^{15}$
$Be(OH)_4^{2-}$	$K_f = 4.0 \times 10^{18}$

a. Cd^{2+}
b. Cu^{2+}
c. Zn^{2+}
d. Be^{2+}

83. Calculate the solubility of zinc hydroxide, $Zn(OH)_2$, in 1.00 *M* NaOH. $K_{sp} = 3.0 \times 10^{-16}$ for $Zn(OH)_2$,
H! $K_f = 3.0 \times 10^{15}$ for $Zn(OH)_4^{2-}$

a. 0.24 *M*
b. 0.32 *M*
c. 0.37 *M*
d. 0.52 *M*

84. M A solution is prepared by mixing 50.0 mL of 0.50 *M* $Cu(NO_3)_2$ with 50.0 mL of 0.50 *M* $Co(NO_3)_2$. Sodium hydroxide is added to the mixture. Which hydroxide precipitates first and what concentration of hydroxide ions is needed to accomplish the separation? $K_{sp} = 2.2 \times 10^{-20}$ for $Cu(OH)_2$, $K_{sp} = 1.3 \times 10^{-15}$ for $Co(OH)_2$

a. $Cu(OH)_2$, $[OH^-] = 1.8 \times 10^{-7}\ M$
b. $Cu(OH)_2$, $[OH^-] = 1.1 \times 10^{-9}\ M$
c. $Co(OH)_2$, $[OH^-] = 6.9 \times 10^{-6}\ M$
d. $Co(OH)_2$, $[OH^-] = 2.6 \times 10^{-7}\ M$

85. H Which reagent would be used to separate $Ag^+(aq)$ (ion group 1) from $Cu^{2+}(aq)$ (ion group 2) in a standard qualitative analysis scheme?

a. 6 *M* HCl
b. H_2S at pH = 0.5
c. 6 *M* NH_3
d. $(NH_4)_2HPO_4$

86. H Ion group 2 sulfides can be separated from those from ion group 3 because

a. ion group 2 sulfides are insoluble in base while ion group 3 sulfides are insoluble in acid
b. ion group 3 sulfides are insoluble in base while ion group 2 sulfides are insoluble in acid
c. ion group 2 sulfides are soluble in a phosphate buffer while ion group 3 sulfides are not
d. ion group 3 sulfides are soluble in a phosphate buffer while ion group 3 sulfides are not

87. H Which reagent would be used to separate $Na^+(aq)$ (ion group 5) from $Mg^{-+}(aq)$(ion group 4) in a standard qualitative analysis scheme?

a. 6 *M* HCl
b. H_2S at pH = 0.5
c. NH_3/NH_4^+ buffer
d. $(NH_4)_2HPO_4$

Short Answer Questions

88. E What is the pH of 375 mL of solution containing 0.150 mol of propenoic acid (HA) and 0.250 mol of sodium propenoate (NaA)? (K_a for propenoic acid is 5.52×10^{-5}.)

89. M Formic acid is a monoprotic acid with a K_a value of 1.8×10^{-4} at 25°C.
a. Calculate the pH of a 0.200 mol L^{-1} solution of the acid, making any reasonable approximations.
b. If 0.0050 mol of NaOH is added to 100. mL of the solution in (a), calculate the pH of the resultant buffer.

90. H Hydrofluoric acid (HF) has a K_a value of 7.2×10^{-4}.
a. 0.250 mol of F^- ions (in the form of NaF) are added to 1.00 L of 0.100 mol L^{-1} aqueous HF. Calculate the resulting pH.
b. To the solution produced in (a) is added 10.0 mL of 5.00 mol L^{-1} NaOH. Calculate the resulting pH.

91. E Make a clear distinction between buffer range and buffer capacity.

92. M Hydrochloric acid (0.100 mol L^{-1}, 25.00 mL aliquot) is being titrated with sodium hydroxide of the same molarity. Calculate the solution pH after addition of 24.80 mL of the sodium hydroxide. Make any reasonable approximations.

93. Propanoic acid (CH_3CH_2COOH) has a K_a of 1.34×10^{-5}. A 25.00 mL sample of 0.1000 mol L^{-1}
H propanoic acid (in flask) is titrated with 0.1000 mol L^{-1} NaOH solution, added from a buret. Carry out the calculations of the quantities indicated below.

a. The pH after 0.00 mL of NaOH are added
b. The pH after 15.00 mL of NaOH are added
c. The hydroxide ion concentration after 26.00 mL of NaOH are added

94. Use a carefully drawn and labeled diagram of the titration curve to illustrate the titration of a weak
M diprotic acid, in which K_{a1} and K_{a2} are substantially different, with a strong base (base is the titrant). Label as many features of the diagram as possible.

95. Lead (II) iodide, PbI_2, is an ionic compound with a solubility product constant K_{sp} of 7.9×10^{-9}.
M Calculate the solubility of this compound in

a. pure water
b. 0.50 mol L^{-1} KI solution

96. Silver phosphate, Ag_3PO_4, is an ionic compound with a solubility product constant K_{sp} of 2.6×10^{-18}.
H Calculate the solubility of this compound in

a. pure water
b. 0.20 mol L^{-1} Na_3PO_4 solution

97. $Fe(NO_3)_3$ (0.00100 mol) and KSCN (0.200 mol) are added to water to make exactly 1 liter of
H solution. The red complex ion $FeSCN^{2+}$ is produced. Calculate the concentrations of $Fe^{3+}(aq)$ and $FeSCN^{2+}(aq)$ at equilibrium, if K_f of the $FeSCN^{2+}$ is 8.9×10^2.

98. Calculate the solubility of copper(II) carbonate, $CuCO_3$, in 1.00 mol L^{-1} NH_3. $K_{sp} = 3.0 \times 10^{-12}$ for
H $CuCO_3$, $K_f = 5.6 \times 10^{11}$ for $Cu(NH_3)_4^{2+}$

True/False Questions

99. A CH_3COOH/CH_3COO^- buffer can be produced by adding a strong acid to a solution of CH_3COO^- ions.
M

100. Increasing the concentrations of the components of a buffer solution will increase the buffer range.
E

101. Increasing the concentrations of the components of a buffer solution will increase the buffer capacity.
E

102. If the pH of a buffer solution is greater than the pK_a value of the buffer acid, the buffer will have
M more capacity to neutralize added base than added acid.

103. The end point in a titration is defined as the point when the indicator changes color.
E

104. The equivalence point in a titration is defined as the point when the indicator changes color.
E

105. For a diprotic acid H_2A, the relationship $K_{a1} > K_{a2}$ is always true.
M

106. The solubility of salt MX (solubility product constant K_{sp}) in water will always be less than that of
H salt MX_3 (solubility product constant K'_{sp}) provided that $K_{sp} < K'_{sp}$.

Ionic Equilibria in Aqueous Systems
Chapter 19
Answer Key

1.	c	26.	d	51	a	76.	b
2.	b	27.	b	52.	b	77.	a
3.	b	28.	c	53.	c	78.	a
4.	e	29.	b	54.	d	79.	b
5.	e	30.	d	55.	b	80.	c
6.	d	31.	a	56.	a	81.	c
7.	c	32.	a	57.	a	82.	a
8.	d	33.	c	58.	c	83.	a
9.	d	34.	b	59.	b	84.	b
10.	b	35.	a	60.	c	85.	a
11.	a	36.	c	61.	c	86.	b
12.	a	37.	d	62.	b	87.	d
13.	c	38.	b	63.	d		
14.	c	39.	d	64.	c		
15.	d	40.	b	65.	c		
16.	c	41.	d	66.	d		
17.	c	42.	a	67.	a		
18.	a	43.	a	68.	c		
19.	b	44.	c	69.	b		
20.	b	45.	c	70.	b		
21.	c	46.	b	71.	d		
22.	c	47.	a	72.	b		
23.	b	48.	b	73.	a		
24.	c	49.	a	74.	b		
25.	c	50.	b	75.	c		

88. 4.48

89. a. 2.22
    b. 3.27

90. a. 3.54
    b. 3.92

91. Buffer range is the range of pH over which the buffer acts effectively; in practice this is between p*K*a - 1 and p*K*a + 1. Buffer capacity relates to the amount of added acid or base which a buffer solution is able to neutralize. The greater the concentration of the buffer components, the greater the buffer capacity.

92. 3.40

93. a. 2.94
    b. 5.05
    c. 1.96×10^{-3} mol L^{-1}

94.

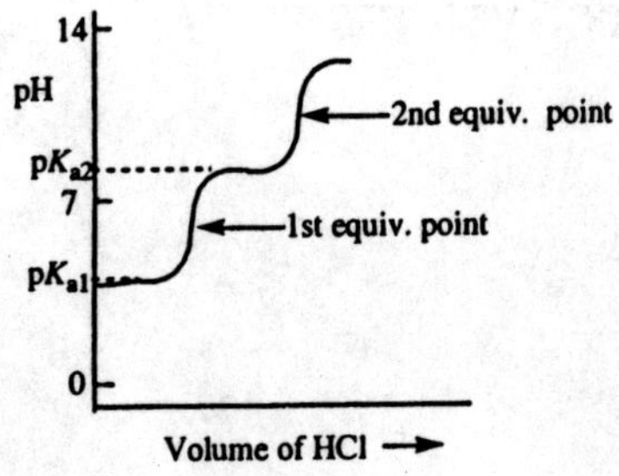

95. a. 1.3×10^{-3} mol L^{-1}
 b. 3.2×10^{-8} mol L^{-1}

96. a. 1.8×10^{-5} mol L^{-1}
 b. 7.8×10^{-7} mol L^{-1}

97. $[Fe^{3+}] = 5.6 \times 10^{-6}$ mol L^{-1} $[FeSCN^{2+}] = 0.100$ mol L^{-1}

98. 0.16 mol L^{-1}

99. T

100. F

101. T

102. F

103. T

104. F

105. T

106. F

Thermodynamics: Entropy, Free Energy, and the Direction of Chemical Reactions
Chapter 20

Multiple Choice Questions

1. Which of the following is an incorrect value for pure oxygen gas, $O_2(g)$ at 25°C?
E

a. $\Delta H_f^\circ = 0$
b. $\Delta G_f^\circ = 0$
c. $S^\circ = 0$

2. A certain process has $\Delta S_{univ} > 0$ at 25°C. What does one know about the process?
E

a. It is exothermic.
b. It is endothermic.
c. It is spontaneous at 25°C.
d. It will move rapidly toward equilibrium.

3. Which of the following is necessary for a process to be spontaneous?
E

a. $\Delta H_{sys} < 0$
b. $\Delta S_{sys} > 0$
c. $\Delta S_{surr} < 0$
d. $\Delta S_{univ} > 0$

4. Which of the following results in a decrease in entropy?
M

a. $O_2(g)$, 300 K $\rightarrow$ $O_2(g)$, 400 K
b. $H_2O(s)$, 0°C $\rightarrow$ $H_2O(l)$, 0°C
c. $N_2(g)$, 25°C $\rightarrow$ $N_2(aq)$, 25°C
d. $NH_3(l)$, -34.5°C $\rightarrow$ $NH_3(g)$, -34.5°C,

5. Which of the following should have the greatest molar entropy at 298 K?
M

a. $CH_4(g)$
b. $H_2O(l)$
c. $NaCl(s)$
d. $N_2O_4(g)$

6. Which of the following is true for a system at equilibrium?
M

a. $\Delta S^\circ_{sys} = \Delta S^\circ_{surr}$
b. $\Delta S^\circ_{sys} = -\Delta S^\circ_{surr}$
c. $\Delta S^\circ_{sys} = \Delta S^\circ_{surr} = 0$
d. $\Delta S^\circ_{univ} > 0$

7. Which of the following is true for an exothermic process?
M

a. $q_{sys} > 0, \Delta S_{surr} < 0$
b. $q_{sys} < 0, \Delta S_{surr} > 0$
c. $q_{sys} < 0, \Delta S_{surr} < 0$
d. $q_{sys} > 0, \Delta S_{surr} > 0$

8. Which of the following is true for an endothermic process?
M

a. $q_{sys} > 0, \Delta S_{surr} < 0$
b. $q_{sys} < 0, \Delta S_{surr} > 0$
c. $q_{sys} < 0, \Delta S_{surr} < 0$
d. $q_{sys} > 0, \Delta S_{surr} > 0$

9. Which of the following values is based on the Third Law of Thermodynamics?
M

a. $\Delta H_f^\circ = 0$ for Al(s) @ 298 K
b. $\Delta G_f^\circ = 0$ for $H_2(g)$ @ 298 K
c. $S^\circ = 51.446$ J/(mol·K) for Na(s) @ 298 K
d. $q_{sys} < 0$ for $H_2O\ (l) \rightarrow H_2O(s)$ @ 0°C

10. When a sky diver free-falls through the air, the process is
E

a. non-spontaneous because he is accelerating due to the force applied by gravity.
b. non-spontaneous because he is losing potential energy.
c. non-spontaneous because he had planned the jump for two weeks.
d. spontaneous.

11. Which of the following processes is spontaneous under the specified conditions?
M

a. $H_2O(l) \rightarrow H_2O(s)$ @ 25°C
b. $CO_2(s) \rightarrow CO_2(g)$ @ 0°C
c. $2H_2O(g) \rightarrow 2H_2(g) + O_2(g)$
d. Al(s) @ 25°C → Al(s) @ 75°C

12. Predict the sign of ΔS° for the following reaction.
E

$Pb(s) + Cl_2(g) \rightarrow PbCl_2(s)$

a. $\Delta S^\circ \approx 0$
b. $\Delta S^\circ < 0$
c. $\Delta S^\circ > 0$
d. More information is needed to make a reasonable prediction.

13. Predict the sign of ΔS° for the following reaction.
M

$HgS(s) + O_2(g) \rightarrow Hg(l) + SO_2(g)$

a. $\Delta S^\circ \approx 0$
b. $\Delta S^\circ < 0$
c. $\Delta S^\circ > 0$
d. More information is needed to make a reasonable prediction.

14. Predict the sign of ΔS° for the following reaction.
M

$$2H_2S(g) + 3O_2(g) \rightarrow 2H_2O(g) + 2SO_2(g)$$

a. $\Delta S^\circ \approx 0$
b. $\Delta S^\circ < 0$
c. $\Delta S^\circ > 0$
d. More information is needed to make a reasonable prediction.

15. Predict the sign of ΔS° for the following reaction.
M

$$CO(g) + H_2O(g) \rightarrow CO_2(g) + H_2(g)$$

a. $\Delta S^\circ \approx 0$
b. $\Delta S^\circ < 0$
c. $\Delta S^\circ > 0$
d. More information is needed to make a reasonable prediction.

16. Predict the sign of ΔS° for the following reaction.
E

$$2NH_3(g) + 2ClF_3(g) \rightarrow 6HF(g) + N_2(g) + Cl_2(g)$$

a. $\Delta S^\circ \approx 0$
b. $\Delta S^\circ < 0$
c. $\Delta S^\circ > 0$
d. More information is needed to make a reasonable prediction.

17. Predict the sign of ΔS° for the following reaction.
H

$$8H_2(g) + S_8(s) \rightarrow 8H_2S(g)$$

a. $\Delta S^\circ \approx 0$
b. $\Delta S^\circ < 0$
c. $\Delta S^\circ > 0$
d. More information is needed to make a reasonable prediction.

18. Predict the sign of ΔS° for the following reaction.
M

$$O_3(g) + NO(g) \rightarrow O_2(g) + NO_2(g)$$

a. $\Delta S^\circ \approx 0$
b. $\Delta S^\circ < 0$
c. $\Delta S^\circ > 0$
d. More information is needed to make a reasonable prediction.

19. Predict the sign of ΔS° for the following reaction.
M

$$C_2H_5OH(l) + 3O_2(g) \rightarrow 2CO_2(g) + 3H_2O(l)$$

a. $\Delta S^\circ \approx 0$
b. $\Delta S^\circ < 0$
c. $\Delta S^\circ > 0$
d. More information is needed to make a reasonable prediction.

20. Predict the sign of $\Delta S°$ for the following reaction.
E

$$K_2SO_4(s) \rightarrow 2K^+(aq) + SO_4^{2-}(aq)$$

a. $\Delta S° \approx 0$
b. $\Delta S° < 0$
c. $\Delta S° > 0$
d. More information is needed to make a reasonable prediction.

21. Predict the sign of $\Delta S°$ for the following reaction.
E

$$BaCl_2(aq) + Na_2SO_4(aq) \rightarrow BaSO_4(s) + 2NaCl(aq)$$

a. $\Delta S° \approx 0$
b. $\Delta S° < 0$
c. $\Delta S° > 0$
d. More information is needed to make a reasonable prediction.

22. Predict the sign of $\Delta S°$ for the following reaction.
E

$$CaO(s) + CO_2(g) \rightarrow CaCO_3(s)$$

a. $\Delta S° \approx 0$
b. $\Delta S° < 0$
c. $\Delta S° > 0$
d. More information is needed to make a reasonable prediction.

23. Which of the following pairs has the member with the greater molar entropy listed first? All systems are at 25°C
E

a. $CO(g)$, $CO_2(g)$
b. $NaCl(s)$, $NaCl(aq)$
c. $H_2S(g)$, $H_2S(aq)$
d. $Li(s)$, $Pb(s)$

24. You are given pure samples of ethane, $C_2H_6(g)$, and toluene, $C_7H_8(l)$. What prediction would you make concerning their standard molar entropies at 298 K?
H

a. $S°_{ethane} > S°_{toluene}$
b. $S°_{ethane} < S°_{toluene}$
c. $S°_{ethane} \approx S°_{toluene}$
d. Prediction would be difficult because they are in different phases.

25. You are given pure samples of pentane, $CH_3CH_2CH_2CH_2CH_3(l)$, and 1 ,3-pentadiene, $CH_2{=}\,CHCH{=}\,CHCH_3(l)$. What prediction would you make concerning their standard molar entropies at 298 K?
M

a. $S°_{pentane} > S°_{1,3\text{-pentadiene}}$
b. $S°_{pentane} < S°_{1,3\text{-pentadiene}}$
c. $S°_{pentane} \approx S°_{1,3\text{-pentadiene}}$
d. More information is needed to make reasonable predictions.

26. You are given pure samples of ammonia, $NH_3(g)$, and nitrogen trifluoride, $NF_3(g)$. What prediction would
E you make concerning their standard molar entropies at 298 K?

a. $S^\circ_{\text{ammonia}} > S^\circ_{\text{nitrogen trifluoride}}$
b. $S^\circ_{\text{ammonia}} < S^\circ_{\text{nitrogen trifluoride}}$
c. $S^\circ_{\text{ammonia}} \approx S^\circ_{\text{nitrogen trifluoride}}$
d. More information is needed to make reasonable predictions.

27. In which one of the following pairs will the first system have a higher entropy than the second?
E Assume *P* and *T* are the same for each pair, unless stated otherwise.

a. 1 mole $He(g)$; 1 mole $Kr(g)$
b. 1 mole $O_2(g)$; 2 mole $O(g)$
c. 1 mole $CH_4(g)$; 1 mole $C_2H_6(g)$
d. 1 mole $Xe(g)$ at 1 atmosphere; 1 mole $Xe(g)$ at 0.5 atmosphere
e. 20 one-dollar bills distributed randomly among 20 people; 20 one-dollar bills distributed randomly among 10 people

28. In which one of these pairs will the entropy of the first substance be greater than that of the second?
E Assume *P* and *T* are the same for each pair, unless stated otherwise.

a. 1 mole of $F_2(g)$; 1 mole of $Cl_2(g)$
b. 1 mole of $I_2(s)$; 1 mole of $I_2(g)$
c. 1 mole of $CaCO_3(s)$; 1 mole of $CaO(s)$ plus 1 mole of $CO_2(g)$
d. 1 mole of $H_2(g)$ at 25°C; 1 mole of $H_2(g)$ at 50°C
e. 1 mole of $O_3(g)$; 1 mole of $O_2(g)$

29. Calculate $\Delta S°$ for the reaction
E

$$SiCl_4(g) + 2Mg(s) \rightarrow 2MgCl_2(s) + Si(s)$$

Substance	$SiCl_4(g)$	$Mg(s)$	$MgCl_2(s)$	$Si(s)$
$S°$(J/K·mol)	330.73	32.68	89.62	18.83

a. -254.96 J/K
b. -198.02 J/K
c. 198.02 J/K
d. 254.96 J/K

30. Calculate $\Delta S°$ for the reaction
E

$$4Cr(s) + 3O_2(g) \rightarrow 2Cr_2O_3(s)$$

Substance	$Cr(s)$	$O_2(g)$	$Cr_2O_3(s)$
$S°$(J/K·mol)	23.77	205.138	81.2

a. -548.1 J/K
b. -147.7 J/K
c. 147.7 J/K
d. 548.1 J/K

31. Calculate $\Delta S°$ for the reaction
E

$$2Cl_2(g) + SO_2(g) \rightarrow SOCl_2(g) + Cl_2O(g)$$

Substance	$Cl_2(g)$	$SO_2(g)$	$SOCl_2(g)$	$Cl_2O(g)$
$S°$(J/K·mol)	223.0	248.1	309.77	266.1

a. -118.2 J/K
b. -104.8 J/K
c. 104.8 J/K
d. 118.2 J/K

32. Calculate $\Delta S°$ for the combustion of propane.
E

$$C_3H_8(g) + 5O_2(g) \rightarrow 3CO_2(g) + 4H_2O(g)$$

Substance	$C_3H_8(g)$	$O_2(g)$	$CO_2(g)$	$H_2O(g)$
$S°$(J/K·mol)	269.9	205.138	213.74	188.825

a. -100.9 J/K
b. -72.5 J/K
c. 72.5 J/K
d. 100.9 J/K

33. Elemental boron can be formed by reaction of boron trichloride with hydrogen.
M

$$BCl_3(g) + 1.5H_2(g) \rightarrow B(s) + 3HCl(g)$$

Substance	$BCl_3(g)$	$H_2(g)$	$B(s)$	$HCl(g)$
$S°$(J/K·mol)	?	130.6	5.87	186.8

If $\Delta S°$= 80.3 J/K, what is $S°$ for $BCl_3(g)$?

a. -18.2 J/K·mol
b. 18.2 J/K·mol
c. 290.1 J/K·mol
d. 450.6 J/K·mol

34. For a chemical reaction to be spontaneous only at high temperatures, which of the following conditions must
M be met?

a. $\Delta S° > 0, \Delta H° > 0$
b. $\Delta S° > 0, \Delta H° < 0$
c. $\Delta S° < 0, \Delta H° < 0$
d. $\Delta S° < 0, \Delta H° > 0$

35. For a chemical reaction to be spontaneous only at low temperatures, which of the following conditions must
M be met?

a. $\Delta S° > 0, \Delta H° > 0$
b. $\Delta S° > 0, \Delta H° < 0$
c. $\Delta S° < 0, \Delta H° < 0$
d. $\Delta S° < 0, \Delta H° > 0$

36. For a chemical reaction to be spontaneous at all temperatures, which of the following conditions must
M be met?

a. $\Delta S° > 0, \Delta H° > 0$
b. $\Delta S° > 0, \Delta H° < 0$
c. $\Delta S° < 0, \Delta H° < 0$
d. $\Delta S° < 0, \Delta H° > 0$

37. For a chemical reaction to be spontaneous at no temperature, which of the following conditions must be met?
M

a. $\Delta S° > 0, \Delta H° > 0$
b. $\Delta S° > 0, \Delta H° < 0$
c. $\Delta S° < 0, \Delta H° < 0$
d. $\Delta S° < 0, \Delta H° > 0$

38. For a process with $\Delta S < 0$, which one of the following statements is correct?
M

a. The process will definitely be spontaneous if $\Delta H < 0$
b. The process will be definitely be spontaneous if $\Delta H < T\Delta S$
c. The process can never be spontaneous
d. The process will definitely be spontaneous, regardless of ΔH
e. The process will definitely be spontaneous if $\Delta S_{surr} > 0$

39. Consider the following quantities used in thermodynamics: *E*, *H*, *q*, *w*, *S*, *G*. How many of them are
M state functions?

a. 1
b. 2
c. 3
d. 4
e. 6

40. In order for a process to be spontaneous,
E

a. ΔH must be less than 0
b. ΔS must be greater than 0
c. ΔG must be greater than 0
d. it should be rapid
e. $\Delta S_{sys} + \Delta S_{surr}$ must be greater than 0

41. Calculate $\Delta G°$ for the reaction
E

$SiCl_4(g) + 2Mg(s) \rightarrow 2MgCl_2(s) + Si(s)$

Substance	$SiCl_4(g)$	$Mg(s)$	$MgCl_2(s)$	$Si(s)$
ΔG°_f (kJ/mol)	-616.98	0	-591.79	0

a. -566.60 kJ
b. -25.19 kJ
c. 25.19 kJ
d. 566.60 kJ

42. Calculate $\Delta G°$ for the combustion of propane.
E

$C_3H_8(g) + 5O_2(g) \rightarrow 3CO_2(g) + 4H_2O(g)$

Substance	$C_3H_8(g)$	$O_2(g)$	$CO_2(g)$	$H_2O(g)$
ΔG_f° (kJ/mol)	-24.5	0	-394.4	-228.6

a. -598.5 kJ
b. 598.5 kJ
c. -2073.1 kJ
d. 2073.1 kJ

43. Elemental boron can be formed by reaction of boron trichloride with hydrogen.
E

$BCl_3(g) + 1.5H_2(g) \rightarrow B(s) + 3HCl(g)$

Calculate $\Delta G°$ for the reaction.

Substance	$BCl_3(g)$	$H_2(g)$	$B(s)$	$HCl(g)$
ΔG_f° (kJ/mol)	-388.7	0	0	-95.3

a. -293.4 kJ
b. 293.4 kJ
c. -102.8 kJ
d. 102.8 kJ

44. Calculate $\Delta G°$ for the reaction of ammonia with fluorine.
E

$2NH_3(g) + 5F_2(g) \rightarrow N_2F_4(g) + 6HF(g)$

Substance	$NH_3(g)$	$F_2(g)$	$N_2F_4(g)$	$HF(g)$
ΔG_f° (kJ/mol)	-16.4	0	79.9	-275.4

a. 179.1 kJ
b. -179.1 kJ
c. 1539.7 kJ
d. -1539.7 kJ

45. Use the given data at 298 K to calculate $\Delta G°$ for the reaction
H

$2Cl_2(g) + SO_2(g) \rightarrow SOCl_2(g) + Cl_2O(g)$

Substance	$Cl_2(g)$	$SO_2(g)$	$SOCl_2(g)$	$Cl_2O(g)$
ΔH_f° (kJ/mol)	0	-296.8	-212.5	80.3
$S°$(J/K·mol)	223.0	248.1	309.77	266.1

a. 129.3 kJ
b. 133.6 kJ
c. 196.0 kJ
d. 199.8 kJ

46. Hydrogen sulfide decomposes according to the following reaction
H

$$2H_2S(g) \rightarrow 2H_2(g) + S_2(g)$$

For this reaction at 298K $\Delta S° = 78.1$ J/K, $\Delta H° = 169.4$ kJ, and $\Delta G° = 146.1$ kJ. What is the value of $\Delta G°$ at 900 K?

a. 48.4 kJ
b. 99.1 kJ
c. 240 kJ
d. 441 kJ

47. Nitric oxide reacts with chlorine to form NOCl. The data refer to 298 K.
M

$$2NO(g) + Cl_2(g) \rightarrow 2NOCl(g)$$

Substance	$NO(g)$	$Cl_2(g)$	$NOCl(g)$
ΔH°_f (kJ/mol)	90.29	0	51.71
ΔG°_f (kJ/mol)	86.60	0	66.07
$S°$(J/K·mol)	210.65	223.0	261.6

What is the value of $\Delta G°$ for this reaction at 550 K?

a. -143.76 kJ
b. -78.78 kJ
c. -22.24 kJ
d. -10.56 kJ

48. Sulfuryl dichloride is formed when sulfur dioxide reacts with chlorine. The data refer to 298 K.
M

$$SO_2(g) + Cl_2(g) \rightarrow SO_2Cl_2(g)$$

Substance	$SO_2(g)$	$Cl_2(g)$	$SO_2Cl_2(g)$
ΔH°_f (kJ/mol)	-296.8	0	-364.0
ΔG°_f (kJ/mol)	-300.1	0	-320.0
$S°$(J/K·mol)	248.2	223.0	311.9

What is the value of $\Delta G°$ for this reaction at 600 K?

a. -162.8 kJ
b. -28.4 kJ
c. 28.4 kJ
d. 162.8 kJ

49. M The temperature at which the following process reaches equilibrium at 1.0 atm is the normal boiling point of hydrogen peroxide.

$H_2O_2(l) \rightleftarrows H_2O_2(g)$

Use the following thermodynamic information at 298 K to determine this temperature.

Substance	$H_2O_2(l)$	$H_2O_2(g)$
ΔH°_f (kJ/mol)	-187.7	-136.3
ΔG°_f (kJ/mol)	-120.4	-105.6
S° (J/K·mol)	109.6	232.7

a. 120°C
b. 144°C
c. 418°C
d. 585°C

50. M The temperature at which the following process reaches equilibrium at 1.0 atm is the normal melting point for phosphoric acid.

$H_3PO_4(s) \rightleftarrows H_3PO_4(l)$

Use the following thermodynamic data at 298 K to determine this temperature.

Substance	$H_3PO_4(s)$	$H_3PO_4(l)$
ΔH°_f (kJ/mol)	-1284.4	-1271.7
ΔG°_f (kJ/mol)	-1124.3	-1123.6
S°(J/K·mol)	110.5	150.8

a. 286 K
b. 305 K
c. 315 K
d. 347 K

51. M Consider the figure alongside which shows ΔG° for a chemical process plotted against absolute temperature. Which one of the following is an incorrect conclusion, based on the information in the diagram?

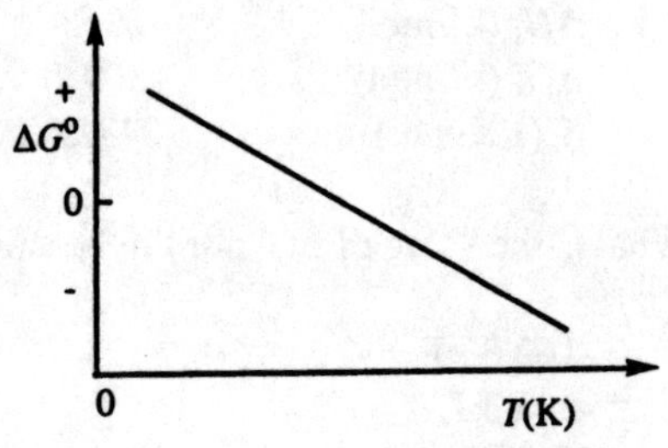

a. $\Delta H^\circ > 0$
b. $\Delta S^\circ > 0$
c. The reaction is spontaneous if the temperature is high enough
d. ΔS° increases with temperature while ΔH° remains constant
e. There exists a certain temperature at which $\Delta H^\circ = T\Delta S^\circ$

52. Consider the figure alongside which shows $\Delta G°$ for a chemical process plotted against absolute temperature. From this plot, it is reasonable to conclude that:
E

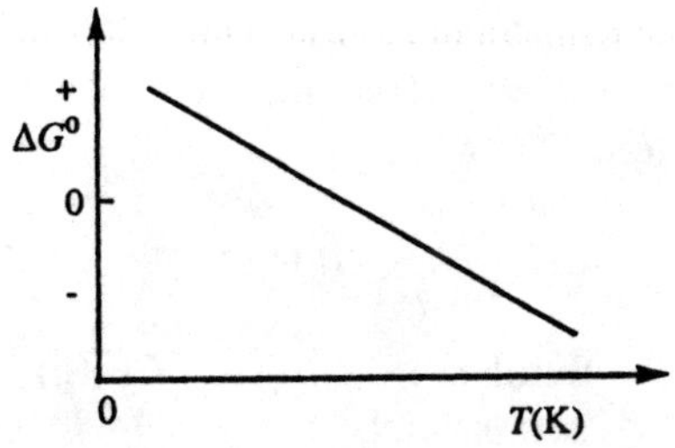

a. $\Delta H° > 0, \Delta S° > 0$
b. $\Delta H° > 0, \Delta S° < 0$
c. $\Delta H° < 0, \Delta S° > 0$
d. $\Delta H° < 0, \Delta S° < 0$
e. None of the above

53. Consider the figure alongside which shows $\Delta G°$ for a chemical process plotted against absolute temperature. From this plot, it is reasonable to conclude that:
E

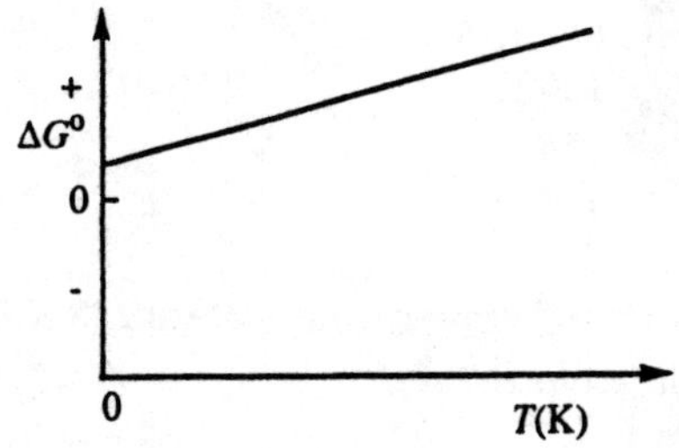

a. $\Delta H° > 0, \Delta S° > 0$
b. $\Delta H° > 0, \Delta S° < 0$
c. $\Delta H° < 0, \Delta S° > 0$
d. $\Delta H° < 0, \Delta S° < 0$
e. None of the above

54. Iron(III) oxide can be reduced by carbon monoxide.
H

$$Fe_2O_3(s) + 3CO(g) \rightleftarrows 2Fe(s) + 3CO_2(g)$$

Use the following thermodynamic data at 298 K to determine the equilibrium constant at this temperature.

Substance	$Fe_2O_3(s)$	$CO(g)$	$Fe(s)$	$CO_2(g)$
$\Delta H_f°$ (kJ/mol)	-824.2	-110.5	0	-393.5
$\Delta G_f°$ (kJ/mol)	-742.2	-137.2	0	-394.4
$S°$(J/K·mol)	87.4	197.7	27.78	213.7

a. 7.0×10^{-6}
b. 1.3×10^{-3}
c. 2.2×10^{4}
d. 1.4×10^{5}

55. Calculate the equilibrium constant at 25°C for the reaction of methane with water to form carbon dioxide and hydrogen. The data refer to 25°C.
H

$$CH_4(g) + 2H_2O(g) \rightleftarrows CO_2(g) + 4H_2(g)$$

Substance	$CH_4(g)$	$H_2O(g)$	$CO_2(g)$	$H_2(g)$
$\Delta H_f°$ (kJ/mol)	-74.87	-241.8	-393.5	0
$\Delta G_f°$ (kJ/mol)	-50.81	-228.6	-394.4	0
$S°$ (J/K·mol)	186.1	188.8	213.7	130.7

a. 0.58
b. 0.96
c. 1.2×10^{-20}
d. 8.2×10^{19}

56. M The reaction of methane with water to form carbon dioxide and hydrogen is non-spontaneous at 298 K. At what temperature will this system make the transition from non-spontaneous to spontaneous? The data refer to 298 K.

$$CH_4(g) + 2H_2O(g) \rightleftarrows CO_2(g) + 4H_2(g)$$

Substance	$CH_4(g)$	$H_2O(g)$	$CO_2(g)$	$H_2(g)$
ΔH°_f (kJ/mol)	-74.87	-241.8	-393.5	0
ΔG°_f (kJ/mol)	-50.81	-228.6	-394.4	0
S°(J/K·mol)	186.1	188.8	213.7	130.7

a. 658 K
b. 683 K
c. 955 K
d. 1229 K

57. H Use the thermodynamic data at 298 K below to determine the K_{sp} for barium carbonate, $BaCO_3$ at this temperature.

Substance	$Ba^{2+}(aq)$	$CO_3^{2-}(aq)$	$BaCO_3(s)$
ΔH°_f (kJ/mol)	-538.36	-676.26	-1219
ΔG°_f (kJ/mol)	-560.7	-528.1	-1139
S°(J/K·mol)	13	-53.1	112

a. 5.86
b. 6.30×10^8
c. 1.59×10^{-9}
d. 2.18×10^{-27}

58. M What is the free energy change, ΔG°, for the equilibrium between hydrogen iodide, hydrogen, and iodine at 453°C? $K_c = 0.020$

$$2HI(g) \rightleftarrows H_2(g) + I_2(g)$$

a. 6.4 kJ
b. 8.8 kJ
c. 15 kJ
d. 24 kJ

59. M The formation constant for the reaction

$$Ag^+(aq) + 2NH_3(aq) \rightleftarrows Ag(NH_3)_2^+(aq)$$

is $K_f = 1.7 \times 10^7$ at 25°C. What is ΔG° at this temperature?

a. -1.5 kJ
b. -3.5 kJ
c. -18 kJ
d. -41 kJ

Short Answer Questions

60. a. Explain what is meant by a spontaneous process.
E b. Is a spontaneous process necessarily a rapid one? Explain, and provide a real reaction as an example to illustrate your answer.

61. For each of the following pairs, predict which (A or B) will have the **greater** entropy, and in one sentence indicate your reasoning.
E

	A	B
a.	1 mole of $HI(g)$	1 mole of $HBr(g)$
b.	1 mole of $HI(g)$ at 20°C	1 mole of $HI(g)$ at 30°C
c.	3 moles of $H_2(g)$ + 1 mole of $N_2(g)$	2 moles of $NH_3(g)$
d.	1 mole of $H_2(g)$, pressure = 1 atm	1 mole of $H_2(g)$, pressure = 0.1 atm
e.	1 mole of $CO_2(g)$	1 mole of $CO_2(aq)$
f.	1 mole of $HCOOH(l)$	1 mole of $HCOOH(aq)$

62. State the second and third laws of thermodynamics.
E

63. In tables of thermodynamic data provided in chemistry books, one finds ΔH°_f, ΔG°_f and S° listed.
M Briefly, explain why the entropy data are supplied as S°, while the enthalpy and free energy data are in the form of ΔH°_f and ΔG°_f respectively.

64. Given: $C_2H_2(g) \rightarrow 2C(\text{graphite}) + H_2(g)$ $\quad\Delta G^\circ = -209$ kJ
E

A sample of gaseous C_2H_2 (acetylene, or ethyne) was stored for one year, yet at the end of this period the sample remained unchanged and no graphite or hydrogen gas had been formed. Briefly explain why there is no inconsistency between the sign of ΔG° and the apparent stability of the sample.

65. The complete combustion of liquid benzene is represented by the equation:
M

$$C_6H_6(l) + 7\tfrac{1}{2}O_2(g) \rightarrow 6CO_2(g) + 3H_2O(l)$$

Using the data below, calculate, for this reaction
a. ΔH° b. ΔS° c. ΔG° at 25°C.

Substance	$C_6H_6(l)$	$O_2(g)$	$CO_2(g)$	$H_2O(l)$
ΔH°_f (kJ/mol)	49	0	-394	-286
S°(J/mol·K)	173	205	214	70

66. For the reaction of xenon and fluorine gases to form solid XeF_4, $\Delta H^\circ = -251$ kJ and $\Delta G^\circ = -121$ kJ at
M 25°C. Calculate ΔS° for the reaction.

67. A chemical reaction has $\Delta G^\circ = 10.0$ kJ and $\Delta S^\circ = 50.0$ J/K
M a. Calculate ΔH° for this reaction at 25°C.
b. Could this reaction ever be spontaneous? Explain your answer.

68. Photosynthesis can be represented by the equation
M

$$6CO_2(g) + 6H_2O(l) \rightarrow C_6H_{12}O_6(s) + 6O_2(g)$$

a. Calculate ΔS° for this process, given the following data:

Substance	$CO_2(g)$	$H_2O(l)$	$C_6H_{12}O_6(s)$	$O_2(g)$
S° (J/(mol·K))	214	70	212	205

b. Given that ΔH° for the reaction is 2802 kJ, calculate ΔG° at 25°C.

69. A chemical reaction has $\Delta H° = 42.8$ kJ and $\Delta S° = 92.5$ J/K, at 25°C. Calculate the temperature at
M which $\Delta G° = 0$. State any approximation involved in your calculation.

70. Compare one mole of ice with one mole of liquid water, both at 1.0 atm and 0°C. The melting point
H of ice at 1.0 atm is 0°C. For the process

$$H_2O(s) \rightarrow H_2O(l)$$

under these conditions predict whether each of the following quantities will be greater than, less than, or equal to, zero (i.e. > 0, < 0 or = 0). Explain each prediction in one sentence.

a. $\Delta H°$ b. $\Delta S°$ c. $\Delta G°$

71. For what signs of ΔH and ΔS will a process
E
a. be spontaneous at high temperatures but not at low temperatures?
b. not be spontaneous at any temperatures?

72. A reaction has a positive value of $\Delta H°$ and a positive value of $\Delta S°$.
M Draw a neat, labeled schematic plot to show how $\Delta G°$ (y-axis) will depend on absolute temperature (x-axis)

73. The water-gas shift reaction plays an important role in the production of clean fuel from coal.
H

$$CO(g) + H_2O(g) \rightleftarrows CO_2(g) + H_2(g)$$

Use the following thermodynamic data to determine the equilibrium constant K_p at 700. K.

Substance	$CO(g)$	$H_2O(g)$	$CO_2(g)$	$H_2(g)$
$\Delta H_f°$ (kJ/mol)	-110.5	-241.8	-393.5	0
$S°$(J/mol·K)	197.7	188.8	213.7	130.7

True/False Questions

74. The higher the pressure of a gas sample, the greater is its entropy.
M

75. In a spontaneous process, the entropy of the system always increases.
E

76. In some spontaneous processes, the entropy of the surroundings decreases.
M

77. For a reaction at equilibrium, $\Delta S_{univ} = 0$.
M

78 The free energy of a perfect crystal at absolute zero, is zero.
M

79. For a given reaction, a change in the pressure may result in a change in the sign of ΔG.
M

80. For any reaction, if $\Delta G° > 0$, then $K < 1$.
E

81. As a chemical reaction proceeds toward equilibrium, the free energy of the system decreases.
M

Thermodynamics: Entropy, Free Energy, and the Direction of Chemical Reactions
Chapter 20
Answer Key

1.	c	26.	b	51.	d
2.	c	27.	e	52.	a
3.	d	28.	e	53.	b
4.	c	29.	b	54	d
5.	d	30.	a	55.	c
6.	b	31.	a	56.	c
7.	b	32.	d	57.	c
8.	a	33.	c	58.	d
9.	c	34.	a	59.	d
10.	d	35.	c		
11.	b	36.	b		
12.	b	37.	d		
13.	c	38.	b		
14.	b	39.	d		
15.	a	40.	e		
16.	c	41.	a		
17.	c	42.	c		
18.	a	43.	d		
19.	b	44.	d		
20.	c	45.	d		
21.	b	46.	b		
22.	b	47.	d		
23.	c	48.	c		
24.	d	49.	b		
25.	a	50.	c		

60.
a. A spontaneous process is one which will occur naturally, given enough time.
b. No, a spontaneous process may be immeasurably slow. An example is the conversion of diamond to graphite at room temperature and one atmosphere.

61.
a. A has greater entropy. HI and HBr are chemically similar, but HI has the higher molar mass.
b. B has greater entropy. At the higher temperature, the sample has greater energy and there are more ways to distribute this energy among the molecules in the sample.
c. A has greater entropy, as it has more moles of gas phase molecules.
d. B has greater entropy. At the lower pressure, the volume is larger and there is more positional disorder in the sample.
e. A has greater entropy. A substance has a greater entropy in the gas phase than in solution.
f. B has the greater entropy. When a solid or liquid dissolves, it has a greater volume available to it, and is thus more disordered.

62. All spontaneous processes are accompanied by an increase in the total entropy of the universe. The entropy of a perfect crystal at absolute zero, is zero.

63. The third law specifies the state of zero of entropy as being a perfect crystal at absolute zero. From this starting point, absolute entropies can be determined for other temperatures too. There is no corresponding absolute zero of enthalpy or free energy; their zero values are arbitrarily set as being the elements in their standard states at one atmosphere and a specified temperature. Standard enthalpies and free energies of formation are relative to this arbitrary zero.

64. Relative to graphite and hydrogen, acetylene is unstable. Its decomposition to form these products is spontaneous, as the negative sign of $\Delta G°$ suggests. However, the kinetics of the decomposition are immeasurably slow under normal conditions.

65. $\Delta H° = -3271$ kJ $\Delta S° = -217$ J/K $\Delta G° = -3206$ kJ

66. $\Delta S° = -436$ J/K

67. a. $\Delta H° = 24.9$ kJ
b. Yes. Assuming that $\Delta H°$ and $\Delta S°$ do not change much with temperature, at a sufficiently high temperature, $\Delta H° - T\Delta S°$ will be less than zero, and the reaction will be spontaneous.

68. a. $\Delta S° = -262$ J/K
b. $\Delta G° = 2880.$ kJ

69. $T = 463$ K. The calculation is based on the assumption that $\Delta H°$ and $\Delta S°$ do not change significantly with change in temperature.

70. a. $\Delta H° > 0$. The solid-to-liquid phase change is always endothermic
b. $\Delta S° > 0$. A liquid is always more disordered than the corresponding solid
c. $\Delta G° = 0$. At this temperature and pressure, the solid and liquid phases are in equilibrium

71. a. $\Delta H > 0$, $\Delta S > 0$
b. $\Delta H > 0$, $\Delta S < 0$

72.

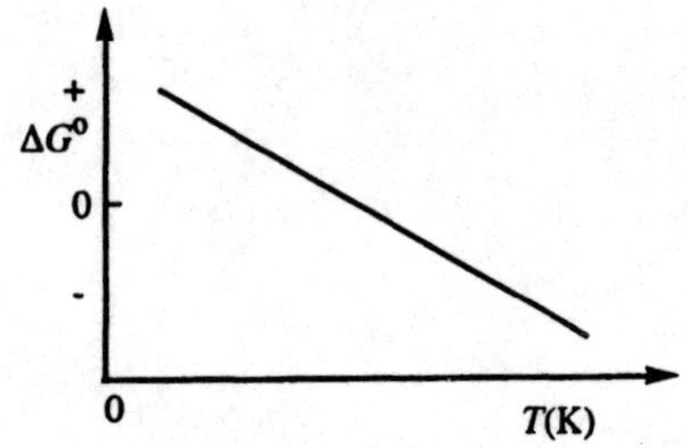

73. $K_p = 7.50$

74. F

75. F

76. T

77. T

78. F

79. T

80. T

81. T

Electrochemistry: Chemical Change and Electrical Work
Chapter 21

Multiple Choice Questions

1. Which one of the following is not a redox reaction?
E

a. $Al(OH)_4^-(aq) + 4H^+(aq) \rightarrow Al^{3+}(aq) + 4H_2O(l)$
b. $C_6H_{12}O_6(s) + 6O_2(g) \rightarrow 6CO_2(g) + 6H_2O(l)$
c. $Na_6FeCl_8(s) + 2Na(l) \rightarrow 8NaCl(s) + Fe(s)$
d. $2H_2O_2(aq) \rightarrow 2H_2O(l) + O_2(g)$
e. $CO_2(g) + H_2(g) \rightarrow CO(g) + H_2O(g)$

2. Consider the reaction
E

$$CuO(s) + H_2(g) \rightarrow Cu(s) + H_2O(l)$$

In this reaction, the oxidant and reductant are, respectively

a. CuO and H_2
b. H_2 and CuO
c. CuO and Cu
d. H_2O and H_2
e. none of the above

3. Consider the following balanced redox reaction
E

$$Mn^{2+}(aq) + S_2O_8^{2-}(aq) + 2H_2O(l) \rightarrow MnO_2(s) + 4H^+(aq) + 2SO_4^{2-}(aq)$$

Which of the following is true?

a. Mn^{2+} (*aq*) is the oxidizing agent and is reduced
b. Mn^{2+} (*aq*) is the oxidizing agent and is oxidized
c. Mn^{2+} (*aq*) is the reducing agent and is oxidized
d. Mn^{2+} (*aq*) is the reducing agent and is reduced

4. Consider the following balanced redox reaction
E

$$3CuO(s) + 2NH_3(aq) \rightarrow N_2(g) + 3H_2O(l) + 3Cu(s)$$

Which of the following is true?

a. CuO(*s*) is the oxidizing agent and Cu is reduced
b. CuO(*s*) is the oxidizing agent and Cu is oxidized
c. CuO(*s*) is the reducing agent and Cu is oxidized
d. CuO(*s*) is the reducing agent and Cu is reduced

5. H When the following redox equation is balanced with smallest whole number coefficients, the coefficient for nitrogen dioxide will be _____ .

$I_2(s) + HNO_3(aq) \rightarrow HIO_3(aq) + NO_2(g) + H_2O(l)$

a. 1
b. 2
c. 4
d. 10

6. H When the following redox equation is balanced with smallest whole number coefficients, the coefficient for the hydrogen sulfate ion will be ______ .

$Al(s) + HSO_4^-(aq) + OH^-(aq) \rightarrow Al_2O_3(s) + S^{2-}(aq) + H_2O(l)$

a. 1
b. 3
c. 4
d. 8

7. M When the following redox equation is balanced with smallest whole number coefficients, the coefficient for zinc will be _____.

$Zn(s) + ReO_4^-(aq) \rightarrow Re(s) + Zn^{2+}(aq)$ (acidic solution)

a. 2
b. 7
c. 8
d. 16

8. M When the following redox equation is balanced with smallest whole number coefficients, the coefficient for the iodide ion will be _____ .

$I^-(aq) + NO_3^-(aq) \rightarrow NO(g) + I_2(s)$ (acidic solution)

a. 2
b. 3
c. 6
d. 8

9. M When the following redox equation is balanced with smallest whole number coefficients, the coefficient for $Sn(OH)_3^-$ will be _____ .

$Bi(OH)_3(s) + Sn(OH)_3^-(aq) \rightarrow Sn(OH)_6^{2-}(aq) + Bi(s)$ (basic solution)

a. 1
b. 2
c. 3
d. 6

10. Consider the following redox equation
H

$$Mn(OH)_2(s) + MnO_4^-(aq) \rightarrow MnO_4^{2-}(aq) \quad \text{(basic solution)}$$

When the equation is balanced with smallest whole number coefficients, what is the coefficient for $OH^-(aq)$ and on which side of the equation is $OH^-(aq)$ present?

a. 4, reactant side
b. 4, product side
c. 6, reactant side
d. 6, product side

11. Two types of electrochemical cells which are based on thermodynamic principles can be described. Which of the following statements about these cells is correct?
M

a. The current in the external wire flows from cathode to anode in both cells.
b. Oxidation occurs at the cathode only in the voltaic cell.
c. The free energy change, ΔG, is negative for the electrolytic cell.
d. The cathode is positive in a voltaic cell but negative in an electrolytic cell.

12. Two types of electrochemical cells which are based on thermodynamic principles can be described. Which of the following statements about these cells is correct?
M

a. The anode will definitely gain weight in a voltaic cell.
b. Oxidation occurs at the cathode of both cells.
c. The free energy change, ΔG, is negative for the voltaic cell.
d. The current in the external wire flows from cathode to anode in an electrolytic cell.

13. Which one of the following pairs of substances could be used to construct a single electrode (i.e., they have an element in common, but in different oxidation states)?
M

a. HCl and Cl^-
b. H^+ and OH^-
c. H_2O and H^+
d. Fe^{3+} and Fe_2O_3
e. MnO_2 and Mn^{2+}

14. Which one of the following statements about electrochemical cells is correct?
M

a. In a salt bridge, current is carried by cations moving toward the anode, and anions toward the cathode.
b. In the external wire, electrons travel from cathode to anode.
c. The anode of a voltaic cell is labeled minus (-).
d. Oxidation occurs at the cathode, in an electrolytic cell.
e. None of the above statements is correct.

15. Which of the following solids would be considered an inactive electrode in an electrochemical cell?
E

a. zinc
b. graphite
c. copper
d. iron

16. Which component of the following cell notation is the anode?
E

$P \mid Q \parallel R \mid S$

a. P
b. Q
c. R
d. S

17. A voltaic cell is prepared using copper and silver. Its cell notation is shown below.
E

$Cu(s) \mid Cu^{2+}(aq) \parallel Ag^{+}(aq) \mid Ag(s)$

Which of the following half-reactions occurs at the cathode?

a. $Cu(s) \rightarrow Cu^{2+}(aq) + 2e^{-}$
b. $Cu^{2+}(aq) + 2e^{-} \rightarrow Cu(s)$
c. $Ag(s) \rightarrow Ag^{+}(aq) + e^{-}$
d. $Ag^{+}(aq) + e^{-} \rightarrow Ag(s)$

18. A voltaic cell prepared using aluminum and nickel has the following cell notation.
E

$Al(s) \mid Al^{3+}(aq) \parallel Ni^{2+}(aq) \mid Ni(s)$

Which of the following reactions occurs at the anode?

a. $Al(s) \rightarrow Al^{3+}(aq) + 3e^{-}$
b. $Al^{3+}(aq) + 3e \rightarrow Al(s)$
c. $Ni(s) \rightarrow Ni^{2+}(aq) + 2e^{-}$
d. $Ni^{2+}(aq) + 2e^{-} \rightarrow Ni(s)$

19. A voltaic cell prepared using aluminum and nickel has the following cell notation.
E

$Al(s) \mid Al^{3+}(aq) \parallel Ni^{2+}(aq) \mid Ni(s)$

Which of the following represents the correctly balanced spontaneous reaction equation for the cell?

a. $Ni^{2+}(aq) + Al(s) \rightarrow Al^{3+}(aq) + Ni(s)$
b. $3Ni^{2+}(aq) + 2Al(s) \rightarrow 2Al^{3+}(aq) + 3Ni(s)$
c. $Ni(s) + Al^{3+}(aq) \rightarrow Ni^{2+}(aq) + Al(s)$
d. $3Ni(s) + 2Al^{3+}(aq) \rightarrow 3Ni^{2+}(aq) + 2Al(s)$

20. A voltaic cell prepared using zinc and iodine has the following cell notation.
E

$Zn(s) \mid Zn^{2+}(aq) \parallel I^{-}(aq) \mid I_2(s) \mid$ graphite

Which of the following represents the correctly balanced spontaneous reaction equation for the cell?

a. $2I^{-}(aq) + Zn^{2+}(aq) \rightarrow I_2(s) + Zn(s)$
b. $I_2(s) + Zn(s) \rightarrow 2I^{-}(aq) + Zn^{2+}(aq)$
c. $2I^{-}(aq) + Zn(s) \rightarrow I_2(s) + Zn^{2+}(aq)$
d. $I_2(s) + Zn^{2+}(aq) \rightarrow 2I^{-}(aq) + Zn(s)$

21. E A cell can be prepared from copper and tin. What is the $E°_{cell}$ for the cell that forms from the following half reactions?

$Cu^{2+}(aq) + 2e^- \rightleftarrows Cu(s)$ $E° = 0.34$ V
$Sn^{4+}(aq) + 2e^- \rightleftarrows Sn^{2+}(aq)$ $E° = 0.13$ V

a. 0.47 V
b. 0.21 V
c. -0.21 V
d. -0.47 V

22. E What is the $E°_{cell}$ for the cell represented by the combination of the following half-reactions?

$2Hg^{2+}(aq) + 2e^- \rightleftarrows Hg_2^{2+}(aq)$ $E° = 0.92$ V
$Cr^{3+}(aq) + 3e^- \rightleftarrows Cr(s)$ $E° = -0.74$ V

a. -0.18 V
b. 0.18 V
c. 1.28 V
d. 1.66 V

23. E What is the $E°_{cell}$ for the cell represented by the combination of the following half-reactions?

$ClO_4^-(aq) + 8H^+(aq) + 8e^- \rightleftarrows Cl^-(aq) + 4H_2O(l)$ $E° = 1.389$ V
$VO_2^+(aq) + 2H^+(aq) + e^- \rightleftarrows VO^+(aq) + H_2O(l)$ $E° = 0.991$ V

a. -0.398 V
b. -2.380 V
c. 0.398 V
d. 2.380 V

24. M The redox reaction of peroxydisulfate with iodide has been used for many years as part of the iodine clock reaction which introduces students to kinetics. If $E°_{cell} = 1.587$ V and $E°_{iodine} = 0.536$ V, what is $E°_{peroxydisulfate}$

$S_2O_8^{2-}(aq) + 2H^+ + 2I^-(aq) \rightarrow 2HSO_4^-(aq) + I_2(aq)$

a. -1.051 V
b. -2.123 V
c. 1.051 V
d. 2.123 V

25. M The voltaic cell made up of cobalt and copper has $E°_{cell} = 0.62$ V. If $E°_{Cu} = 0.34$ V, what is $E°_{Co}$?

$Cu^{2+}(aq) + Co(s) \rightarrow Cu(s) + Co^{2+}(aq)$

a. -0.28 V
b. -0.96V
c. 0.28 V
d. 0.96 V

26. M Given that $E°$ for X + e^- → Y is greater than $E°$ for A + $2e^-$ → B, it is correct to say that, under standard conditions

a. X will oxidize A
b. Y will oxidize A
c. Y will reduce A
d. B will oxidize X
e. B will reduce X

27. E Examine the following half-reactions and select the strongest oxidizing agent among the substances.

$[PtCl_4]^{2-}(aq) + 2e^- \rightleftarrows Pt(s) + 4Cl^-(aq)$	$E° = 0.755$ V
$RuO_4(s) + 8H^+(aq) + 8e^- \rightleftarrows Ru(s) + 4H_2O(l)$	$E° = 1.038$ V
$FeO_4^{2-}(aq) + 8H^+(aq) + 3e^- \rightleftarrows Fe^{3+}(aq) + 4H_2O(l)$	$E° = 2.07$ V
$H_4XeO_6(aq) + 2H^+(aq) + 2e^- \rightleftarrows XeO_3(aq) + 3H_2O(l)$	$E° = 2.42$ V

a. $[PtCl_4]^{2-}(aq)$
b. $RuO_4(s)$
c. $HFeO_4^-(aq)$
d. $H_4XeO_6(aq)$

28. E Examine the following half-reactions and select the strongest oxidizing agent among the substances.

$Cr^{2+}(aq) + 2e^- \rightleftarrows Cr(s)$	$E° = -0.913$ V
$Co^{2+}(aq) + 2e^- \rightleftarrows Co(s)$	$E° = -0.28$ V
$Fe^{2+}(aq) + 2e^- \rightleftarrows Fe(s)$	$E° = -0.447$ V
$Sr^{2+}(aq) + 2e^- \rightleftarrows Sr(s)$	$E° = -2.89$ V

a. $Cr^{2+}(aq)$
b. $Co^{2+}(aq)$
c. $Fe^{2+}(aq)$
d. $Sr^{2+}(aq)$

29. E Examine the following half-reactions and select the weakest oxidizing agent among the substances.

$AuBr_4^-(aq) + 3e^- \rightleftarrows Au(s) + 4Br^-(aq)$	$E° = 0.854$ V
$Mn^{2+}(aq) + 2e^- \rightleftarrows Mn(s)$	$E° = -1.185$ V
$K^+(aq) + e^- \rightleftarrows K(s)$	$E° = -2.931$ V
$F_2O(aq) + 2H^+(aq) + 4e^- \rightleftarrows 2F^-(aq) + H_2O(l)$	$E° = 2.153$ V

a. $AuBr_4^-(aq)$
b. $Mn^{2+}(aq)$
c. $K^+(aq)$
d. $F_2O(aq)$

30. Examine the following half-reactions and select the strongest reducing agent among the substances.
E

$PbI_2(s) + 2e^- \rightleftarrows Pb(s) + 2I^-(aq)$	$E° = -0.365$ V
$Ca^{2+}(aq) + 2e^- \rightleftarrows Ca(s)$	$E° = -2.868$ V
$Pt^{2+}(aq) + 2e^- \rightleftarrows Pt(s)$	$E° = 1.18$ V
$Br_2(l) + 2e^- \rightleftarrows 2Br^-(aq)$	$E° = 1.066$ V

a. Pb(*s*)
b. Ca(*s*)
c. Pt(*s*)
d. $Br^-(aq)$

31. Examine the following half-reactions and select the strongest reducing agent among the substances.
E

$HgO(s) + H_2O(l) + 2e^- \rightleftarrows Hg(l) + 2OH^-(aq)$	$E° = 0.0977$ V
$Zn(OH)_2(s) + 2e^- \rightleftarrows Zn(s) + 2OH^-(aq)$	$E° = -1.25$ V
$Ag_2O(s) + H_2O(l) + 2e^- \rightleftarrows Ag(s) + 2OH^-(aq)$	$E° = 0.342$ V
$B(OH)_3(aq) + 7H^+(aq) + 8e^- \rightleftarrows BH_4^-(aq) + 3H_2O(l)$	$E° = -0.481$ V

a. Hg(*l*)
b. Zn(*s*)
c. Ag(*s*)
d. BH_4^- (*aq*)

32. Examine the following half-reactions and select the weakest reducing agent among the substances.
E

$Cr(OH)_3(s) + 3e^- \rightleftarrows Cr(s) + 3OH^-(aq)$	$E° = -1.48$ V
$SnO_2(s) + 2H_2O(l) + 4e^- \rightleftarrows Sn(s) + 4OH^-(aq)$	$E° = -0.945$ V
$MnO_2(s) + 4H^+(aq) + 2e^- \rightleftarrows Mn^{2+}(aq) + 2H_2O(l)$	$E° = 1.224$ V
$Hg_2SO_4(s) + 2e^- \rightleftarrows 2Hg(l) + SO_4^{2-}(aq)$	$E° = 0.613$ V

a. Cr(*s*)
b. Sn(*s*)
c. $Mn^{2+}(aq)$
d. Hg(*l*)

33. Calculate $E^°_{cell}$ and indicate whether the overall reaction shown is spontaneous or nonspontaneous.
M

$I_2(s) + 2e^- \rightleftarrows 2I^-(aq)$	$E° = 0.53$ V
$Cr^{3+}(aq) + 3e^- \rightleftarrows Cr(s)$	$E° = -0.74$ V

Overall reaction:

$$2Cr(s) + 3I_2(s) \rightarrow 2Cr^{3+}(aq) + (aq) + 6I^-(aq)$$

a. $E°_{cell} = -1.27$ V, spontaneous
b. $E°_{cell} = -1.27$ V, nonspontaneous
c. $E°_{cell} = 1.27$ V, spontaneous
d. $E°_{cell} = 1.27$ V, nonspontaneous

34. M

Calculate $E°_{cell}$ and indicate whether the overall reaction shown is spontaneous or nonspontaneous.

$Co^{3+}(aq)$ + e^- ⇄ $Co^{2+}(aq)$ — $E° = 1.82$ V
$MnO_4^-(aq)$ + $2H_2O(l)$ + $3e^-$ ⇄ $MnO_2(s)$ + $4OH^-(aq)$ — $E° = 0.59$ V

Overall reaction:

$MnO_4^-(aq)$ + $2H_2O(l)$ + $3Co^{2+}(aq)$ → $MnO_2(s)$ + $3Co^{3+}(aq)$ + $4OH^-(aq)$

a. $E°_{cell} = -1.23$ V, spontaneous
b. $E°_{cell} = -1.23$ V, nonspontaneous
c. $E°_{cell} = 1.23$ V, spontaneous
d. $E°_{cell} = 1.23$ V, nonspontaneous

35. M

Calculate $E°_{cell}$ and indicate whether the overall reaction shown is spontaneous or nonspontaneous.

$O_2(g)$ + $4H^+(aq)$ + $4e^-$ ⇄ $2H_2O(l)$ — $E° = 1.229$ V
$Al^{3+}(aq)$ + $3e^-$ ⇄ $Al(s)$ — $E° = -1.662$ V

Overall reaction:

$4Al(s)$ + $3O_2(g)$ + $12H^+(aq)$ → $4Al^{3+}(aq)$ + $6H_2O(l)$

a. $E°_{cell} = -2.891$ V, nonspontaneous
b. $E°_{cell} = -2.891$ V, spontaneous
c. $E°_{cell} = 2.891$ V, nonspontaneous
d. $E°_{cell} = 2.891$ V, spontaneous

36. M

Calculate $E°_{cell}$ and indicate whether the overall reaction shown is spontaneous or nonspontaneous.

$H_2O_2(aq)$ + $2H^+(aq)$ + $2e^-$ ⇄ $2H_2O(l)$ — $E° = 1.77$ V
$Fe^{3+}(aq)$ + e^- ⇄ $Fe^{2+}(aq)$ — $E° = 0.77$ V

Overall reaction:

$2Fe^{3+}(aq)$ + $2H_2O(l)$ → $H_2O_2(aq)$ + $2H^+(aq)$ + $2Fe^{2+}(aq)$

a. $E°_{cell} = -1.00$ V, nonspontaneous
b. $E°_{cell} = -1.00$ V, spontaneous
c. $E°_{cell} = 1.00$ V, nonspontaneous
d. $E°_{cell} = 1.00$ V, spontaneous

37. H

When metal A is placed in a solution of metal B salt, the surface of metal A changes its appearance. When metal B is placed in acid solution, gas bubbles form on its surface. When metal A is placed in a solution of metal C salt, no change is observed in the solution or on the surface of metal A. Which of the following reactions would not occur spontaneously?

a. $C(s)$ + $2H^+(aq)$ → $H_2(g)$ + $C^{2+}(aq)$
b. $C(s)$ + $A^{2+}(aq)$ → $A(s)$ + $C^{2+}(aq)$
c. $B(s)$ + $C^{2+}(aq)$ → $C(s)$ + $B^{2+}(aq)$
d. $A(s)$ + $2H^+(aq)$ → $H_2(g)$ + $A^{2+}(aq)$

38. Under which of the following conditions will a cell to do the most work?
M

a. $Q/K > 1$
b. $Q/K = 1$
c. $Q/K < 1$
d. $E° < 0$

39. A battery is considered "dead" when
E

a. $Q < 1$
b. $Q = 1$
c. $Q > 1$
d. $Q = K$

40. What is the value of the equilibrium constant for the reaction between Sn^{2+} and Fe(*s*) at 25°C? $E°_{cell} = 0.30$ V for the reaction
M

$$Sn^{2+}(aq) + Fe(s) \rightleftarrows Sn(s) + Fe^{2+}(aq)$$

a. 1.2×10^{5}
b. 1.4×10^{10}
c. 8.6×10^{-6}
d. 7.3×10^{-11}

41. What is the value of the equilibrium constant for the reaction of Cr(*s*) with Pb^{2+} (*aq*) at25°C? $E°_{cell} = 0.61$ V
M

$$2Cr(s) + 3Pb^{2+}(aq) \rightleftarrows 3Pb(s) + 2Cr^{3+}(aq)$$

a. 4.1×10^{20}
b. 8.2×10^{30}
c. 3.3×10^{51}
d. 7.4×10^{61}

42. The following half-reactions occur in the mercury battery used in calculators. If $E°_{cell} = 1.357$ V, calculate the equilibrium constant for the cell reaction at 25°C
M

$$HgO(s) + H_2O(l) + 2e^- \rightleftarrows Hg(l) + 2OH^-(aq)$$
$$ZnO(s) + H_2O(l) + 2e^- \rightleftarrows Zn(s) + 2OH^-(aq)$$

a. 9.4×10^{22}
b. 7.5×10^{45}
c. 6.4×10^{63}
d. 7.8×10^{91}

43. Consider the non-aqueous cell reaction
M

$$2Na(l) + FeCl_2(s) \rightarrow 2NaCl(s) + Fe(s)$$

for which $E°_{cell} = 2.35$ V at 200°C. The equilibrium constant K at this temperature is

a. 57.7
b. 115
c. 1.09×10^{25}
d. 1.19×10^{50}
e. $> 9.99 \times 10^{99}$

44. Consider the reaction in the lead-acid cell
M

$$Pb(s) + PbO_2(s) + 2H_2SO_4(aq) \rightarrow 2PbSO_4(aq) + 2H_2O(l)$$

for which $E^\circ_{cell} = 2.04$ V at 298 K. ΔG° for this reaction, in kJ, is

a. -3.94×10^5
b. -3.94×10^2
c. -1.97×10^5
d. -7.87×10^2
e. none of the above

45. The value of E°_{cell} for the reaction
M

$$2Cr^{3+}(aq) + 6Hg(l) \rightarrow 2Cr(s) + 3Hg_2^{2+}(aq)$$

is 1.59 V. Calculate ΔG° for the reaction.

a. -921 kJ
b. -767 kJ
c. -460 kJ
d. -307 kJ

46. Calculate ΔG° for the reaction of iron(II) ions with one mole of permanganate ions.
M

$$MnO_4^-(aq) + 8H^+(aq) + 5e^- \rightleftarrows Mn^{2+}(aq) + 4H_2O(l) \qquad E^\circ = 1.51 \text{ V}$$
$$Fe^{3+}(aq) + e^- \rightleftarrows Fe^{2+}(aq) \qquad E^\circ = 0.77 \text{ V}$$

a. -71.4 kJ
b. -286 kJ
c. -357 kJ
d. -428 kJ

47. Calculate ΔG° for the oxidation of 3 moles of copper by nitric acid.
M

$$Cu^{2+}(aq) + 2e^- \rightleftarrows Cu(s) \qquad E^\circ = 0.34 \text{ V}$$
$$NO_3^-(aq) + 4H^+(aq) + 3e^- \rightleftarrows NO(g) + 2H_2O(l) \qquad E^\circ = 0.957 \text{ V}$$

a. -120 kJ
b. -180 kJ
c. -240 kJ
d. -360 kJ

48. The value of the equilibrium constant for the reaction of nickel(II) ions with cadmium metal is 1.17×10^5.
M Calculate ΔG° for the reaction at 25°C.

a. -12.6 kJ
b. -28.9 kJ
c. 12.6 kJ
d. 28.9 kJ

49. Calculate E°_{cell} for the reaction of nickel(II) ions with cadmium metal at 25°C. $K = 1.17 \times 10^5$
M

$$Ni^{2+}(aq) + Cd(s) \rightarrow Cd^{2+}(aq) + Ni(s)$$

a. 0.075 V
b. 0.10 V
c. 0.15 V
d. 0.30 V

50. Consider the reaction of iodine with manganese dioxide
M

$$3I_2(s) + 2MnO_2(s) + 8OH^-(aq) \rightleftarrows 6I^-(aq) + 2MnO_4^-(aq) + 4H_2O(l)$$

The equilibrium constant for the overall reaction is 8.30×10^{-7}. Calculate ΔG° for the reaction at 25°C.

a. -15.1 kJ
b. -34.7 kJ
c. 15.1 kJ
d. 34.7 kJ

51. Consider the reaction of iodine with manganese dioxide
M

$$3I_2(s) + 2MnO_2(s) + 8OH^-(aq) \rightleftarrows 6I^-(aq) + 2MnO_4^-(aq) + 4H_2O(l)$$

The equilibrium constant for the overall reaction is 8.30×10^{-7}. Calculate E°_{cell} for the reaction at 25°C.

a. -0.36 V
b. -0.18 V
c. -0.12 V
d. -0.060 V

52. A voltaic cell consists of a Mn/Mn^{2+} electrode ($E^\circ = -1.18$ V) and a Fe/Fe^{2+} electrode ($E^\circ = -0.44$ V).
H Calculate $[Fe^{2+}]$ if $[Mn^{2+}] = 0.050\ M$ and $E_{cell} = 0.78$ V at 25°C.

a. 0.040 *M*
b. 0.24 *M*
c. 1.1 *M*
d. 1.8 *M*

53. A voltaic cell consists of an Au/Au^{3+} electrode ($E^\circ = 1.50$ V) and a Cu/Cu^{2+} electrode ($E^\circ = 0.34$ V).
H Calculate $[Au^{3+}]$ if $[Cu^{2+}] = 1.20\ M$ and $E_{cell} = 1.13$ V at 25°C.

a. 0.001 *M*
b. 0.002 *M*
c. 0.04 *M*
d. 0.2 *M*

54. A voltaic cell consists of a Hg/Hg_2^{2+} electrode ($E^\circ = 0.85$ V) and a Sn/Sn^{2+} electrode ($E^\circ = -0.14$ V).
H Calculate $[Sn^{2+}]$ if $[Hg_2^{2+}] = 0.24\ M$ and $E_{cell} = 0.94$ V at 25°C.

a. 0.0001 *M*
b. 0.0007 *M*
c. 0.005 *M*
d. 0.03 *M*

55. H A voltaic cell consists of a Cd/Cd^{2+} electrode ($E°$ = -0.40 V) and a Fe/Fe^{2+} electrode ($E°$ = -0.44 V) with the following initial molar concentrations: [Fe^{2+}] = 0.20 *M;* (Cd^{2+}) = 1.0 *M.* What is the equilibrium concentration of Cd^{2+}? (Assume the anode and cathode solutions are of equal volume, and a temperature of 25°C.)

a. 0.050 *M*
b. 0.21 *M*
c. 0.41 *M*
d. 1.2 *M*

56. H A voltaic cell consists of a Ag/Ag^{+} electrode ($E°$ = 0.80 V) and a Fe^{2+}/Fe^{3+} electrode ($E°$ = 0.77 V) with the following initial molar concentrations: [Fe^{2+}] = 0.30 *M;* [Fe^{3+}] = 0.10 *M;* [Ag^{+}] = 0.30 *M.* What is the equilibrium concentration of Fe^{3+}? (Assume the anode and cathode solutions are of equal volume, and a temperature of 25°C.)

a. 0.030 *M*
b. 0.11 *M*
c. 0.17 *M*
d. 0.21 *M*

57. M A concentration cell consists of two Zn/Zn^{2+} electrodes. The electrolyte in compartment A is 0.10 *M* $Zn(NO_3)_2$ and in compartment B is 0.60 *M* $Zn(NO_3)_2$. What is the voltage of the cell at 25°C?

a. 0.010 V
b. 0.020 V
c. 0.023 V
d. 0.046 V

58. M A concentration cell consists of two Al/Al^{3+}electrodes. The electrolyte in compartment A is 0.050 *M* $Al(NO_3)_3$ and in compartment B is 1.25 *M* $Al(NO_3)_3$. What is the voltage of the cell at 25°C?

a. 0.083 V
b. 0.062 V
c. 0.041V
d. 0.028 V

59. H Which one of the following statements relating to the glass electrode is correct?

a. The glass electrode detects hydrogen gas.
b. The glass of a glass electrode serves to conduct electrons.
c. When pH is measured, only a single electrode, the glass electrode, need be used.
d. The potential of the glass electrode varies linearly with the pH of the solution.
e. None of the above statements are correct

60. E A battery that cannot be recharged is a

a. fuel cell
b. primary battery
c. secondary battery
d. flow battery

61. Which of the following metals would be suitable for use as a sacrificial anode when used with iron pipe?
E $E°_{Fe}$ = -0.44 V

a. copper, Cu, $E°$ = 0.15 V
b. cobalt, Co, $E°$ = -0.28 V
c. chromium, Cr, $E°$ = -0.74 V
d. tin, Sn, $E°$ = -0.14 V

62. Which of the following metals would not be suitable for use as a sacrificial anode when used with iron pipe?
E $E°_{Fe}$ = -0.44 V

a. manganese, Mn, $E°$ = -1.18 V
b. cadmium, Cd, $E°$ = -0.40 V
c. magnesium, Mg, $E°$ = -2.37 V
d. zinc, Zn, $E°$ = -0.76 V

63. What product forms at the cathode during the electrolysis of molten lithium iodide?
M

a. $Li^+(l)$
b. $Li(l)$
c. $I^-(l)$
d. $I_2(g)$

64. What product forms at the anode during the electrolysis of molten NaBr?
M

a. $Na^+(l)$
b. $Na(l)$
c. $Br^-(l)$
d. $Br_2(g)$

65. Which of the following elements can be isolated by electrolysis of the aqueous salt shown?
M

a. sodium from $Na_3PO_4(aq)$
b. sulfur from $K_2SO_4(aq)$
c. silver from $AgNO_3(aq)$
d. potassium from $KCl(aq)$

66. Which of the following elements can be isolated by electrolysis of the aqueous salt shown?
M

a. phosphorus from $K_3PO_4(aq)$
b. iodine form $NaI(aq)$
c. aluminum from $AlCl_3(aq)$
d. fluorine from $KF(aq)$

67. What mass of silver will be formed when 15.0 A are passed through molten AgCl for
M 25.0 minutes?

a. 0.419 g
b. 0.557 g
c. 25.2 g
d. 33.4 g

68. M What mass of copper will be deposited when 18.2 A are passed through a $CuSO_4$ solution for 45.0 minutes?

a. 16.2 g
b. 33.4 g
c. 40.6 g
d. 81.3 g

69. M A solution is prepared by dissolving 32.0 g of $NiSO_4$ in water. What current would be needed to deposit all of the nickel in 5.0 hours?

a. 1.1 A
b. 2.2 A
c. 2.9 A
d. 5.8 A

70. M How many grams of oxygen gas will be produced in the electrolysis of water, for every gram of hydrogen gas formed?

Reaction: $2H_2O(l) \rightarrow 2H_2(g) + O_2(g)$

a. 31.7
b. 15.8
c. 7.94
d. 3.97
e. 1.98

71. M Two cells are connected in series, so that the same current flows through two electrodes where the following half-reactions occur

$Cu^{2+}(aq) + 2e^- \rightarrow Cu(s)$ and $Ag^+(aq) + e^- \rightarrow Ag(s)$

For every 1.00 g of copper produced in the first process, how many grams of silver will be produced in the second one?

a. 0.294
b. 0.588
c. 0.850
d. 1.70
e. 3.40

72. M A current of 250. A flows for 24.0 hours at an electrode where the reaction occurring is

$Mn^{2+}(aq) + 2H_2O(l) \rightarrow MnO_2(s) + 4H^+(aq) + 2e^-$

At the end of this time, the mass of solid MnO_2 deposited at the electrode, in kg, is

a. 19.5
b. 9.73
c. 4.87
d. 2.43
e. none of the above

Short Answer Questions

73. In one or two short sentences each, explain what is meant by the following terms
M

a. galvanic or voltaic cell
b. electrolytic cell
c. salt bridge
d. secondary battery or cell
e. primary battery or cell
f. glass electrode

74. A concentration cell is based on the aqueous reaction
M

$$Cu^{2+}(1.00\ M) \rightarrow Cu^{2+}(0.0100\ M)$$

The cell consists of copper electrodes dipping into solutions of Cu^{2+} ions. The anions present are sulfate ions. Draw a neat diagram to represent this cell, showing and labeling all necessary components including: anode, cathode, electron flow, cation flow and anion flow.

75. A concentration cell is based on the aqueous reaction
M

$$Cu^{2+}(1.00\ M) \rightarrow Cu^{2+}(0.0100\ M)$$

The cell consists of copper electrodes dipping into solutions of Cu^{2+} ions. The anions present are sulfate ions. Write the shorthand cell notation for this cell.

76. A much-studied cell in electrochemistry has the following cell notation:
H

$$Ag(s)\ |\ AgCl(s)\ |\ HCl(aq)\ |\ H_2(g)\ |\ Pt(s)$$

Bearing in mind that $HCl(aq)$ consists of $H^+(aq)$ and $Cl^-(aq)$, and that this solution is in contact with both electrodes (there is no salt bridge), write down balanced equations for

a. the anode half-reaction
b. the cathode half-reaction
c. the cell reaction

77. A concentration cell is based on the aqueous reaction
H

$$Cu^{2+}(1.00\ M) \rightarrow Cu^{2+}(0.0100\ M)$$

Calculate the potential of this cell if it operates at 25.0°C.

78. A galvanic cell is constructed using the two hypothetical half-reactions
M

	$A + e^- \rightarrow B$	$E° = 1.50$ V
and	$C + 2e^- \rightarrow D$	$E° = -0.50$ V

a. Write down the balanced equation representing the cell reaction
b. Calculate the standard potential of this cell, $E°_{cell}$
c. Calculate $\Delta G°$ for the cell reaction

79. Write down equations representing the anode half-reaction, the cathode half-reaction and the overall
H cell reaction for the lead-acid battery.

80. M Explain what is meant by a fuel cell. Provide a balanced equation to represent the reaction in any fuel cell of your choice.

81. M A current of 1000. A flows for exactly 1 hour, through a cell in which the following reaction occurs at one of the electrodes.

$$Mg^{2+} + 2e^- \rightarrow Mg$$

a. Calculate the charge, in coulombs, which passes through the circuit in this time.
b. Calculate the theoretical mass of Mg (magnesium metal) which is produced in this time.

82. M Manganese dioxide (MnO_2) for use in dry cells is made by electrolyzing solutions of Mn^{2+}, where the following reaction occurs at the anode:

$$Mn^{2+}(aq) + 2H_2O(l) \rightarrow MnO_2(s) + 4H^+(aq) + 2e^-$$

Using a current of 100. amperes (A), how many hours will it take to produce 5.00 kg of MnO_2, according to the above reaction?

83. H
a. Write a balanced equation to represent the overall reaction you would expect to occur in the electrolysis of molten KCl.
b. Write a balanced equation to represent the overall reaction you would expect to occur in the electrolysis of aqueous KCl.
c. Clearly explain why the products of the two processes are not the same.

True/False Questions

84. E Electrolytic cells utilize electrical energy to drive non-spontaneous redox reactions.

85. E For the reaction occurring in a voltaic (galvanic) cell, $\Delta G > 0$.

86. M Oxidation occurs at the cathode of a galvanic cell, but at the anode of an electrolytic cell.

87. M Electrons are produced at the cathode of a voltaic cell.

88. M If the electrodes of a voltaic cell are connected with an external wire, electrons will flow in this wire from the cathode to the anode.

89. M In the electrolyte of an electrochemical cell, current is carried by anions moving toward the anode and cations moving in the opposite direction.

90. E In the electrolyte of an electrochemical cell, current is carried by electrons moving from the anode to the cathode.

91. E A salt bridge provides a path for electrons to move between the anode and cathode compartments of a voltaic cell.

92. E In the shorthand notation for cells, a double vertical line is used to separate the reduced and oxidized forms of a redox couple.

93. In the shorthand notation for cells, a single vertical line represents a salt bridge.
E

94. A secondary cell (battery) can operate either as a galvanic or an electrolytic cell.
M

95. A primary battery is one which can be recharged.
E

96. The lead-acid battery is an example of a secondary battery.
E

97. In a fuel cell, an external source of electrical power is used to drive a non-spontaneous reaction in which a fuel is produced.
M

Electrochemistry: Chemical Change and Electrical Work
Chapter 21
Answer Key

1.	a	26.	e	51.	d
2.	a	27.	d	52.	c
3.	c	28.	b	53.	c
4.	a	29.	c	54.	c
5.	d	30.	b	55.	a
6.	b	31.	b	56.	c
7.	b	32.	c	57.	c
8.	c	33.	c	58.	d
9.	c	34.	b	59.	d
10.	c	35.	d	60.	b
11.	d	36.	a	61.	c
12.	c	37.	c	62.	b
13.	e	38.	c	63.	b
14.	c	39.	d	64.	d
15.	b	40.	b	65.	c
16.	a	41.	d	66.	b
17.	d	42.	b	67.	c
18.	a	43.	d	68.	a
19.	b	44.	b	69.	b
20.	b	45.	a	70.	c
21.	b	46.	c	71.	e
22.	d	47.	d	72.	b
23.	c	48.	b		
24.	d	49.	c		
25.	a	50.	d		

73.

a. A galvanic (voltaic) cell is one in which a spontaneous cell reaction is harnessed to produce electrical energy.

b. An electrolytic cell is one in which an external voltage source is used to drive the cell reaction. The cell reaction is normally a non-spontaneous one.

c. A salt bridge is a conducting solution used to place the anode and cathode compartments of a cell in electrical contact. The solution is often in gel form and contains a salt such as KCl for conductivity.

d. A secondary battery or cell is a voltaic cell which can be recharged. After being discharged, it is recharged using an external voltage source to reverse the current and reverse the cell reaction.

e. A primary battery or cell is a voltaic cell which cannot be recharged. After being discharged, the cell is discarded.

f. The glass electrode is an electrode used to measure pH. It consists of an Ag/AgCl electrode in aqueous HCl, encased in a special glass which responds to the hydrogen ion concentration of a solution.

74.

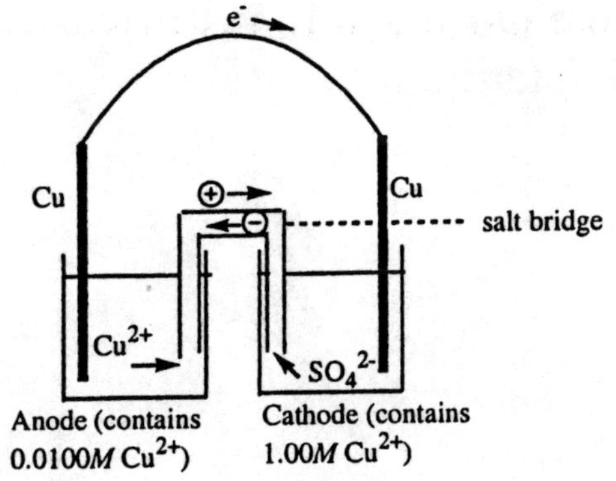

75. $Cu(s) \mid Cu^{2+}(aq,\ 0.0100\ M) \parallel Cu^{2+}(aq,\ 1.00\ M) \mid Cu(s)$

76. a. $Ag(s) + Cl^-(aq) \rightarrow AgCl(s) + e^-$
b. $2H^+(aq) + 2e^- \rightarrow H_2(g)$
c. $2Ag(s) + 2Cl^-(aq) + 2H^+(aq) \rightarrow 2AgCl(s) + H_2(g)$

77. 0.0591 V

78. a. $2A + D \rightarrow 2B + C$
b. 2.00 V
c. -386 kJ

79. Anode half-reaction: $Pb(s) + SO_4^{2-}(aq) \rightarrow PbSO_4(s) + 2e^-$
Cathode half-reaction: $PbO_2(s) + SO_4^{2-}(aq) + 4H^+(aq) + 2e^- \rightarrow PbSO_4(s) + 2H_2O(l)$
Cell reaction: $Pb(s) + PbO_2(s) + 2H_2SO_4(aq) \rightarrow 2PbSO_4(s) + 2H_2O(l)$

80. A fuel cell is a voltaic cell in which the reactants are supplied from an external reservoir and products are removed, both in a continuous fashion, so that the cell never discharges.
e.g. $2H_2(g) + O_2(g) \rightarrow 2H_2O(l)$

81. a. 3.60×10^6 C
b. 453 g

82. 30.8 h

83. a. $2KCl(l) \rightarrow 2K(l) + Cl_2(g)$
b. $2KCl(aq) + 2H_2O(l)) \rightarrow 2KOH(aq) + Cl_2(g) + H_2(g)$
c. At the cathode in the aqueous system water is reduced to hydrogen gas in preference to the reduction of potassium ions to the metal, since the former has the more favorable (higher) reduction potential. There is no water present in the molten salt system.

84. T
85. F
86. F
87. F
88. F
89. T
90. F
91. F
92. F
93. F
94. T
95. F
96. T
97. F

The Elements in Nature and Industry
Chapter 22

Multiple Choice Questions

1. The Earth's core consists mainly of
E

a. Ni
b. O
c. Al
d. Si
e. Fe

2. The main effect of the biosphere on the chemistry of the earth's crust has been to
E

a. create a reducing atmosphere
b. create an oxidizing atmosphere
c. increase the relative humidity
d. decrease the relative humidity
e. increase the level of atmospheric carbon dioxide

3. The most abundant element in the Earth's crust is
M

a. H
b. O
c. Al
d. Si
e. Fe

4. Transition elements from the left side of the periodic table are generally found as
M

a. sulfides because they tend to give up bonding electrons and form crystals with high lattice energies with sulfur.
b. sulfides because their electronegativities favor formation of covalent bonds with the polarizable sulfide anion.
c. oxides because they tend to give up bonding electrons and form crystals with high lattice energies with oxygen.
d. oxides because oxygen, with its high electronegativity, forms strong covalent bonds with them.

5. Transition elements from the right side of the periodic table are generally found as
M

a. sulfides because they tend to give up bonding electrons and form crystals with high lattice energies with sulfur.
b. sulfides because their electronegativities favor formation of covalent bonds with the polarizable sulfide anion.
c. oxides because they tend to give up bonding electrons and form crystals with high lattice energies with oxygen.
d. oxides because oxygen, with its high electronegativity, forms strong covalent bonds with them.

6. E The process in which a gaseous substance is converted into a condensed, more usable form is

a. differentiation
b. leaching
c. fixation
d. smelting

7. M The atmosphere contains about 700. billion metric tons of carbon in the form of carbon dioxide. In the carbon cycle, about 200. billion metric tons of carbon (as carbon dioxide) enter the atmosphere each year. Assuming that the cycle is in a steady-state situation, how many years does the average carbon atom spend in the atmosphere during each cycle?

a. 0.29
b. 3.5
c. 500.
d. 140,000
e. need more information in order to calculate the answer

8. M Nitrogen fixation occurs through atmospheric, industrial, and biological processes. Which of these fixes the most nitrogen?

a. atmospheric
b. industrial
c. biological
d. industrial ≈ biological

9. E The main mineral sources of phosphorus consist of

a. phosphides
b. phosphites
c. phosphors
d. phosphates
e. elemental phosphorus

10. M Plants extract phosphate from the soil

a. by converting it to phosphoric acid
b. by converting it to the dihydrogen phosphate ion by addition of acid to the soil near roots
c. by osmosis
d. by leaching

11. E The most common source for commercial production of aluminum is

a. aluminite
b. bauxite
c. galena
d. cinnabar

12. E The most common source for commercial production of sodium is

a. sodalite
b. limestone
c. halite
d. galena

13. The process that selectively extracts a metal from its ore, by dissolving it, is called
E

a. roasting
b. leaching
c. smelting
d. flotation

14. The process of converting metal sulfides to metal oxides is called
E

a. smelting
b. flotation
c. roasting
d. leaching

15. The debris attached to a mineral is
M

a. slag
b. gangue
c. ore
d. halite

16. The process which generates silicon with a purity of more than 99.999999% is
M

a. electrorefining
b. distillation
c. zone refining
d. alloying

17. Pyrometallurgy uses ________________ to separate a metal from its ore.
M

a. solid phase chemical properties
b. electrical processes
c. thermal processes
d. aqueous chemical properties

18. Electrometallurgy uses ________________ to separate a metal from its ore.
E

a. solid phase chemical properties
b. electrical processes
c. thermal processes
d. aqueous chemical properties

19. Hydrometallurgy uses ________________ to separate a metal from its ore.
E

a. solid phase chemical properties
b. electrical processes
c. thermal processes
d. aqueous chemical properties

20. Alloying a metal is done to
E

a. make its extraction from its ore easier
b. convert the metal to an oxide
c. enhance properties like conductivity
d. prepare ultrapure metal samples

21. Stainless steel is an alloy of
M

a. Fe and C
b. Fe and Mn
c. Fe and Ni
d. Fe and Ni
e. Fe, Cr and Ni

22. The final step of the purification of copper involves electrorefining in which copper is separated from nickel
M and iron by being reduced at the cathode of a cell. Why are nickel and iron not reduced?

a. Their reduction potentials are more positive than copper's.
b. Their reduction potentials are more negative than copper's.
c. They cannot be deposited on a copper electrode.
d. Their reduction potentials are more negative than water's.

23. The ______________ process uses a boiling 30% sodium hydroxide solution to extract aluminum from
M its ore.

a. Bayer
b. Hall-Heroult
c. Dow
d. Frasch

24. The Hall-Heroult process refers to
M

a. the production of aluminum by electrolysis
b. the recovery of sulfur from underground deposits
c. the manufacture of sulfuric acid
d. the production of ammonia from nitrogen and hydrogen gases
e. the isolation of Al_2O_3 from bauxite

25. Cryolite, Na_3AlF_6, is used in the electrolysis of aluminum oxide because
M

a. it is a good source of fluoride ions
b. it reduces the energy requirement of the process, due to its low melting point
c. it provides a source of fluorine, an oxidizing agent
d. it provides a source of sodium, a reducing agent

26. The kinetic isotope effect is the basis for
M

a. the radioactivity of tritium
b. isolation of pure D_2O
c. tritium's high reactivity
d. the increase in boiling point when comparing hydrogen, deuterium and tritium

27. Which of the following reactions is a good small-scale laboratory method for preparation of hydrogen?
M

a. $CH_4(g) + H_2O(g) \rightarrow CO(g) + 3H_2(g)$
b. $3FeCl_2(s) + 4H_2O(g) \rightarrow Fe_3O_4(s) + 6HCl(g) + H_2(g)$
c. $NaH(s) + H_2O(l) \rightarrow NaOH(aq) + H_2(g)$
d. $CO(g) + H_2O(g) \rightarrow CO_2(g) + H_2(g)$

28. The Downs cell is used in the production of
E

a. copper
b. sodium
c. iron
d. magnesium

29. Calcium oxide is added to molten iron in the production of carbon steel
M

a. to serve as a scrubber to remove sulfur dioxide from the gases leaving the furnace
b. to remove any traces of acid- which could weaken the steel
c. to convert silicon and phosphorus oxides to slag which can be decanted from the molten steel
d. to add a small amount of oxygen to the steel to prevent corrosion and increase its strength

30. The alkali metals are isolated from non-aqueous systems. Why is this necessary?
E

a. The electrolysis of aqueous solutions of the alkali metals requires more energy than electrolysis of the molten salts.
b. The dissolved alkali earth halides are too reactive to be electrolyzed.
c. The aqueous metal ions are more difficult to reduce than water.
d. The reduction potentials of the alkali metals are more positive than the reduction potential of water.

31. Electrolysis is used as the last step in isolating pure ________________.
E

a. iron
b. boron
c. aluminum
d. selenium

32. Roasting in oxygen is the first step in isolating ________________ from its ore.
M

a. phosphorus
b. antimony
c. manganese
d. chromium

33. Which of the following elements is used in xerography?
M

a. titanium
b. germanium
c. chromium
d. selenium

34. Which of the following elements combines with uranium to form a compound used to enrich nuclear fuel?
M

a. sulfur
b. iodine
c. fluorine
d. astatine

35. Which of the following elements is used in high-speed semiconductors?
M

a. zinc
b. gallium
c. tellurium
d. antimony

36. Which of the following elements is not extracted from sea water or sea water residues?
M

a. potassium
b. bromine
c. iodine
d. magnesium

37. The Frasch process is used to
M

a. convert aluminum oxide to metallic aluminum
b. convert copper ore to copper sulfide
c. mine magnesium
d. mine elemental sulfur

38. Sulfur trioxide is the anhydride of sulfuric acid. However, SO_3 is not added directly to water during the
M synthesis of sulfuric acid because

a. hydration of SO_3 is very exothermic and difficult to control.
b. hydration of SO_3 is very endothermic and proceeds too slowly to be profitable.
c. at high temperatures water vapor catalyzes polymerization of SO_3.
d. vapor phase sulfuric acid is corrosive and limits the useful life of the reactor.

39. The chlor-alkali process produces chlorine, $Cl_2(g)$, in large quantities. What other industrially important
M substances are produced in this process?

a. $Na(s)$, $H_2(g)$
b. $H_2(g)$, $O_2(g)$
c. $NaOH(aq)$, $H_2(g)$
d. $Na(s)$, $O_2(g)$

Short Answer Questions

40. Describe, and give reasons for, the fundamental bonding differences between transition metal
M sulfides and transition metal oxides.

41. Give two important reasons why there is so much more Na^+ in the oceans than K^+, despite their
M similar abundances in the earth's crust.

42. The mass of the atmosphere has been estimated to be 5.2×10^{21} g, while the average molar mass of
H atmospheric gases is 28.9 g/mol. If the concentration of carbon dioxide in the atmosphere is 360 ppm by volume, calculate the mass of this gas in the atmosphere, in billions of metric tons, bmt. (1 bmt = 10^{12} kg)

43. Briefly describe the three main pathways for nitrogen fixation.
M

44. Lightning, fires and other atmospheric processes result in the formation of nitric acid, HNO_3, from
H atmospheric N_2, O_2 and H_2O. Write balanced chemical equations to represent the three reactions involved in the formation of HNO_3 by this route.

45. The leaching of gold can be represented by the equation
E

$$4Au(s) + O_2(s) + 8CN^-(aq) + 2H_2O(l) \rightarrow 4Au(CN)_2^-(aq) + 4OH^-(aq)$$

In this reaction, identify the oxidant, the reductant and the oxidation numbers of the elements which change.

46. What gas is produced at the anode in the Downs cell, in which molten NaCl is electrolyzed?
E

47. Potassium is produced industrially by the reaction
M

$$Na(l) + K^+(l) \rightleftarrows Na^+(l) + K(g)$$

The reaction is carried out at 850°C, but the equilibrium K_c constant strongly favors the reactants. Explain how the conditions are manipulated to ensure a good yield of potassium despite the small value of K_c.

48. In the commercial production of aluminum metal, the final stage involves the reaction
H

$$2Al_2O_3(\text{in molten } Na_3AlF_6) + 3C(\text{graphite}) \rightarrow 4Al(l) + 3CO_2(g)$$

Briefly outline the method and equipment with which the reaction is accomplished.

49. What gas is produced at the anode in the Hall-Heroult process for the electrolytic production of
M aluminum?

50. Aluminum and magnesium form are light metals which are used in many structural applications.
M Although they are highly active metals (as their electrode potentials suggest), they can often be used without the use of paint or other applied finishes to protect them against corrosion. How is this possible?

51. Compared to the asbestos diaphragm-cell used in the chlor-alkali process, list
H
a. one main advantage and one disadvantage of mercury cells
b. two important advantages of polymer membrane-cells

True/False Questions

52. Sulfide ores are frequently treated by flotation in order to concentrate them.
E

53. Carbon atoms in the carbon cycle spend most of their time in the oceans.
E

54. The industrial production of ammonia from nitrogen and hydrogen accounts for a greater amount of
M nitrogen fixation than either the atmospheric or the biological pathways.

55. The purpose of anodizing aluminum is to remove the oxide layer from the metal's surface.
E

56. In the electrolysis of water, the gas evolved at the cathode is enriched in deuterium, ${}^{2}_{1}H$, compared
M to the more common isotope, ${}^{1}_{1}H$.

57. In the industrial electrolysis of aqueous NaCl (the chlor-alkali process), the modern trend is toward
M the use of cells incorporating polymer membranes to separate the anode and cathode solutions.

The Elements in Nature and Industry
Chapter 22
Answer Key

1.	e	14.	c	27.	c
2.	b	15.	b	28.	b
3.	b	16.	c	29.	c
4.	c	17.	c	30.	c
5.	b	18.	b	31.	c
6.	c	19.	d	32.	b
7.	b	20.	c	33.	d
8.	c	21.	e	34.	c
9.	d	22.	b	35.	b
10.	b	23.	a	36.	c
11.	b	24.	a	37.	d
12.	c	25.	b	38.	c
13.	b	26.	b	39.	c

40. Transition metal oxides are ionic in nature. They tend to form if the transition metal has a low electronegativity, which is the case for elements in the early part of the transition series. The big electronegativity difference between metal and oxygen, and the high lattice energy associated with the small oxide ion, favor ionic bonding. Transition metals with higher electronegativities tend to form sulfides. This occurs for elements in the later (right hand) part of the transition series. The electronegativity difference between metal and sulfur is small, and the bond is largely covalent.

41. Clays bind K^+ more effectively than Na^+, so the former is preferentially retained in soils. Plants also utilize K^+, reducing the amount reaching the oceans.

42. 2.9×10^3 bmt of CO_2.

43. 1. Atmospheric. N_2 and O_2 gases combine directly to form NO in lightning, fires and other high temperature situations. The NO is further oxidized by atmospheric O_2 to NO_2, which dissolves to form nitric acid (HNO_3) which reaches the ground in rain.
2. Industrial. N_2 and H_2 combine to form ammonia (NH_3) in the presence of a catalyst. The ammonia is converted to nitric acid and nitrate fertilizers.
3. Marine algae and nitrogen-fixing bacteria associated with legumes reduce atmospheric nitrogen to ammonia. Other bacteria convert this to nitrate.

44. $N_2(g) + O_2(g) \rightarrow 2NO(g)$
$2NO(g) + O_2(g) \rightarrow 2NO_2(g)$
$3NO_2(g) + H_2O(l) \rightarrow 2HNO_3(aq) + NO(g)$

45. The oxidant is O_2, the reductant is Au. The oxidation number of O changes from 0 to -2; the oxidation number of Au changes from 0 to +1.

46. Cl_2

47. The gaseous potassium product is constantly removed and pumped away. The system is not allowed to reach equilibrium and more products are continually being produced.

48. This is done electrolytically, using a molten salt as the electrolyte. Al^{3+} is reduced to Al at the cathode; the carbon of the anode is oxidized and combines with oxygen (from the Al_2O_3) to form CO_2.

49. CO_2

50. Both Al and Mg have strong, adherent, relatively insoluble oxides. These form naturally on the metal surface, in the presence of atmospheric oxygen, and serve to protect the underlying metal from further oxidation/corrosion.

51. a. purer NaOH product is formed, but toxic mercury is released into the environment.
b. purer NaOH product is formed, and the electrical costs are lower.

52. T

53. F

54. F

55. F

56. F

57. T

The Transition Elements and Their Coordination Compounds
Chapter 23

Multiple Choice Questions

1. The most common oxidation state for ions of the transition elements is
M

a. +2
b. +3
c. +5
d. +7

2. The most common oxidation state for ions of the inner transition elements is
H

a. +2
b. +3
c. +5
d. +7

3. The ground state electronic configuration of Cr^{2+} is
E

a. $[Ar]4s^1 3d^5$
b. $[Ar]4s^2 3d^4$
c. $[Ar]3d^4$
d. $[Ar]4s^1 3d^3$
e. $[Ar]4s^2 3d^2$

4. The ground state electronic configuration of Zn^{2+} is:
E

a. $[Ar]4s^2 3d^8$
b. $[Ar]4s^2 3d^{10}$
c. $[Ar]4s^1 3d^9$
d. $[Ar]3d^{10}$
e. $[Ar]3d^8$

5. Which of the following atoms has the biggest radius?
M

a. Ti
b. Cr
c. Fe
d. Ni
e. Zn

6. A certain transition element has the stable oxidation states of +2, +4, +5, and +6. In which state will the
M element be most likely to form a covalent bond with chlorine?

a. +2
b. +4
c. +5
d. +6

7. A certain transition element has the stable oxidation states of +2, +3, +5, and +6. In which state will the
M element be most likely to form an ionic bond with chlorine?

a. +2
b. +4
c. +5
d. +6

8. Which of the following transition elements will form an ion with the largest oxidation number?
M

a. chromium, Cr, Group 6B(6)
b. manganese, Mn, Group 7B(7)
c. iron, Fe, Group 8B(8)
d. cobalt, Co, Group 8B(9)

9. If M represents a transition element, which of the following oxides should be the
M least basic?

a. MO
b. M_2O_3
c. MO_2
d. MO_3

10. Which of the following ions is most likely to form colored compounds?
E

a. Sc^{3+}
b. Cu^{+}
c. Zn^{2+}
d. Cr^{3+}

11. Which of the following ions is least likely to form colored compounds?
E

a. Mn^{2+}
b. Cr^{5+}
c. Sc^{3+}
d. Fe^{3+}

12. Which of the following will be diamagnetic?
M

a. Ni^{2+}
b. Ti^{4+}
c. Mn^{2+}
d. Co^{3+}

13. Which of the following will be paramagnetic?
M

a. V^{5+}
b. Ni^{2+}
c. Mn^{7+}
d. Ti^{4+}

14. What is the highest possible oxidation state for palladium, Pd?
H

a. +2
b. +4
c. +8
d. +10

15. What is the highest possible oxidation state for molybdenum, Mo?
M

a. +2
b. +4
c. +6
d. +8

16. Which of the following should be the strongest reducing agent?
H

a. Fe
b. Ru
c. Os
d. Re

17. Which of the following will be the strongest oxidizing agent?
E

a. Cr
b. Cr(II)
c. Cr(III)
d. Cr(VI)

18. Which of the oxidation states of chromium has the largest valence-state electronegativity?
M

a. chromium(II)
b. chromium(III)
c. chromium(IV)
d. chromium(VI)

19. Chromium and manganese are among the transitions elements that form several different oxides. Which of
H the following statements characterize these oxides?

a. As the oxidation number on the metal increases, the valence-state electronegativity increases and the oxides change from acidic to basic.
b. As the oxidation number on the metal increases, the valence-state electronegativity increases and the oxides change from basic to acidic.
c. As the oxidation number on the metal increases, the valence-state electronegativity decreases and the oxides change from acidic to basic.
d. As the oxidation number on the metal increases, the valence-state electronegativity decreases and the oxides change from basic to acidic.

20. Aluminum reacts with oxygen in the air to form a protective oxide coating. Silver also reacts with compounds
M in air to form a black coating. What substance is formed?

a. silver oxide
b. silver chloride
c. silver sulfide
d. silver carbonate

21. Which of the following is not a property of silver that is important in black and white photography?
M

a. silver halides are not soluble in water
b. silver metal is easily oxidized
c. silver halides undergo a redox reaction when exposed to light
d. silver ions will form stable, water-soluble complex ions

22. Mercury(II) compounds are assimilated into the food chain because
M

a. they are very soluble in water and easily ingested
b. they are nonpolar substances that are concentrated in fatty tissues as they move up the food chain
c. they are ionic substances that are concentrated in cell tissue
d. they complex with compounds in the blood which carries them to the muscle tissue of organisms

23. A certain transition metal complex has the formula MX_4^{2+}. If the metal ion has a d^8 electron configuration, what is the shape of the complex?
H

a. octahedral
b. square planar
c. tetrahedral
d. trigonal pyramid

24. Which of the following is considered a bidentate ligand?
E

a. cyanide, CN^-
b. thiocyanate, SCN^-
c. oxalate, $C_2O_4^{2-}$
d. nitrite, NO_2^-

25. A characteristic of ligands is that
E

a. they are Lewis acids
b. they are Lewis bases
c. they are ions
d. they are electron pair acceptors

26. 10.0 mL of a 0.100 mol/L solution of a metal ion M^{2+} is mixed with 10.0 mL of a 0.100 mol/L solution of a ligand L. A reaction occurs in which the product is ML_3^{2+}. Approximately, what is the maximum concentration of ML_3^{2+}, in mol/L, which could result from this reaction?
H

a. 0.100
b. 0.050
c. 0.033
d. 0.025
e. 0.017

27. What is the coordination number of cobalt in the complex ion $[Co(en)Cl_4]^-$? (en = ethylene diamine)
M

a. 1
b. 2
c. 4
d. 6
e. 8

28. The oxidation and coordination numbers of cobalt in the compound $[Co(NH_3)_5Cl]Cl_2$ are,
M respectively

a. 2 and 6
b. 2 and 8
c. 3 and 6
d. 3 and 8
e. none of the above

29. In the compound $K[Co(C_2O_4)_2(H_2O)_2]$ (where $C_2O_4^{2-}$ = oxalate) the oxidation number and
H coordination number of cobalt are, respectively:

a. -1 and 4
b. -1 and 6
c. 3 and 4
d. 3 and 6
e. 1 and 6

30. In the compound $[Ni(en)_2(H_2O)_2]SO_4$ (where en = ethylene diamine) the oxidation number and
M coordination number of nickel are, respectively:

a. 2 and 6
b. 4 and 6
c. 6 and 6
d. 2 and 4
e. 4 and 4

31. Give the systematic name for $[Cu(NH_3)_4]Cl_2$.
E

a. dichlorotetraamminecuprate(II)
b. tetraamminecopper(II) chloride
c. copper(II)ammonium chloride
d. tetraaminocopper(II) chloride

32. The compound $K_3[Fe(CN)_6]$ is used in calico printing and wool dyeing. Give its systematic name.
E

a. potassium iron(III)hexacyanate
b. tripotassium iron(III)hexacyanate
c. potassium hexacyanoferrate(III)
d. potassium hexacyanideferrate

33. Give the systematic name for $Cr(CO)_3(NH_3)_3$.
E

a. chromiumtriaminotricarbonyl
b. triamminechromium carbonate
c. triamminetricarbonylchromate(0)
d. triamminetricarbonylchromium(0)

34. Give the systematic name for $[CoCl_3(H_2O)]^-$.
M

a. cobalt(II) chloride monohydrate
b. aquatrichlorocobalt(II)
c. aquatrichlorocobaltate(II)
d. aquatrichlorocobaltite(I)

35. Write the formula for pentaamminechlorocobalt(III) chloride.
E

a. $[Co(NH_3)_5Cl]Cl$
b. $[Co(NH_3)_5Cl]Cl_2$
c. $[Co(NH_3)_5Cl]Cl_3$
d. $[Co(NH_3)_5Cl]Cl_4$

36. Write the formula for diamminedichloroethylenediaminecobalt(III) bromide.
E

a. $[CoCl_2(en)(NH_3)_2]Br$
b. $[CoCl_2(en)(NH_3)_2]Br_2$
c. $[CoCl_2(en)_2(NH_3)_2]Br$
d. $[CoCl_2(en)_2(NH_3)_2]Br_2$

37. Write the formula for sodium tetracyanonickelate(II).
M

a. $Na[Ni(CN)_4]$
b. $Na[Ni(CN)_4]_2$
c. $Na_2[Ni(CN)_4]$
d. $Na_4[Ni(CN)_4]$

38. Which of the following could participate in linkage isomerism?
M

a. NH_3
b. H_2O
c. NO_2^-
d. OH^-

39. Which of the following species could exist as isomers?
E

a. $[Co(H_2O)_4Cl_2]^+$
b. $[Pt(NH_3)Br_3]^-$
c. $[Pt(en)Cl_2]$
d. $[Pt(NH_3)_3Cl]^+$

40. Consider the following octahedral complex structures, each involving ethylene diamine and two different, unidentate ligands X and Y.
M

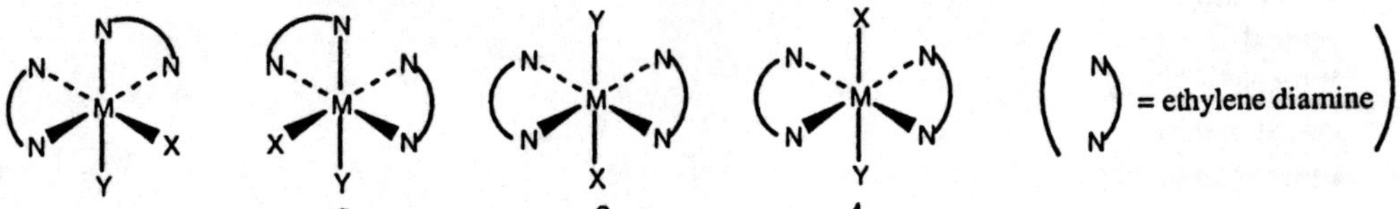

1 2 3 4

Which, if any, of the following pairs are optical isomers?

a. 1 and 2
b. 1 and 3
c. 1 and 4
d. 3 and 4
e. none of the above

41.
H

Consider the following octahedral complex structures, each involving ethylene diamine and two different, unidentate ligands X and Y.

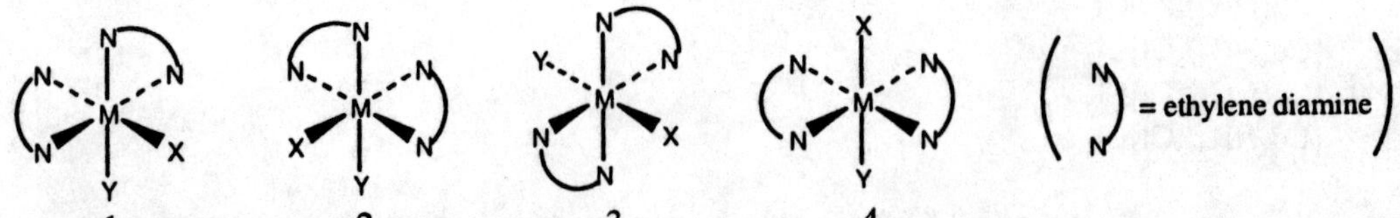

Which one of the following statements about these structures is incorrect?

a. Structures 1 and 2 are optical isomers
b. Structures 1 and 3 are optical isomers
c. Structures 1 and 3 are different complexes
d. Structures 1 and 4 are geometrical isomers
e. Structures 3 and 4 are the same complex

42.
H

Consider the following structures (1 and 2 are octahedral; 3 and 4 are square planar).

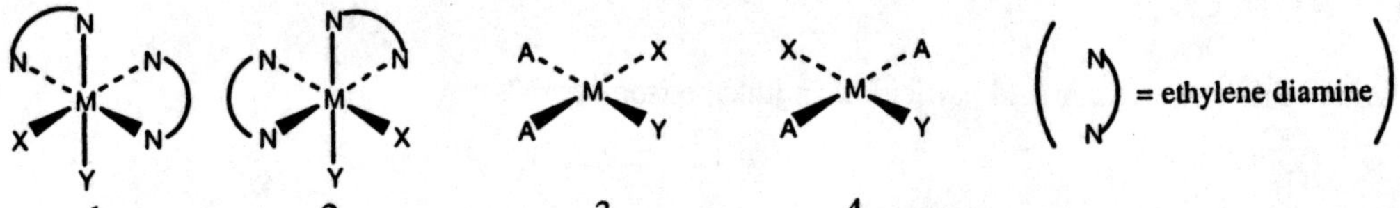

Which one of the following statements about the above structures is correct?

a. 1 and 2 are superimposable
b. 1 and 2 are geometric isomers
c. 3 and 4 are structural isomers
d. 3 and 4 are optical isomers
e. 3 and 4 are geometric isomers

43.
M

In a coordination compound involving a complex ion of square planar geometry, which of the following types of isomerism is/are never possible?

a. geometric
b. optical
c. linkage
d. coordination
e. more than one of the above

44.
E

According to valence bond theory, what would be the set of hybrid orbitals used when a Period 4 transition metal forms a square planar complex?

a. d^2sp
b. dsp^2
c. dsp^3
d. sp^3

45. According to valence bond theory, what would be the set of hybrid orbitals used when a Period 4 transition metal forms a tetrahedral complex?
E

a. d^2sp
b. dsp^2
c. dsp^3
d. sp^3

46. According to Valence Bond theory, in the square planar $Ni(CN)_4^{2-}$ complex ion, the orbital hybridization pattern is
M

a. sp^3
b. dsp^2
c. d^2sp
d. d^2sp^3
e. none of the above

47. The crystal field splitting energy, Δ,
M

a. is larger for tetrahedral complexes than for octahedral complexes
b. depends on the metal but not on the ligand
c. determines the color of a complex
d. is larger for ionic ligands like chloride than for molecular ligands like carbon monoxide, CO

48. In the spectrochemical series, which one of the following ligands has the strongest field?
M

a. H_2O
b. CN^-
c. NH_3
d. OH^-
e. Cl^-

49. Which of the following ions could exist in either the high-spin or low-spin state in an octahedral complex?
E

a. Sc^{3+}
b. Ni^{2+}
c. Mn^{2+}
d. Ti^{4+}

50. Which of the following ions could exist in only the high-spin state in an octahedral complex?
E

a. Cr^{2+}
b. Mn^{4+}
c. Fe^{3+}
d. Co^{3+}

51. In the presence of a strong octahedral ligand field, the number of unpaired electrons in Co(III) will be
M

a. 0
b. 2
c. 4
d. 6
e. none of the above

52. Which of the following octahedral complexes should have the largest crystal field splitting energy, Δ?
M

a. $[Cr(H_2O)_6]^{3+}$
b. $[Cr(SCN)_6]^{3-}$
c. $[Cr(NH_3)_6]^{3+}$
d. $[Cr(CN)_6]^{3-}$

53. Which of the following ligands is most likely to form a low-spin octahedral complex with iron(III)?
M

a. Cl^-
b. H_2O
c. CO
d. OH^-

54. Which of the following ligands is most likely to form a high spin octahedral complex with cobalt(II)?
M

a. CN^-
b. I^-
c. NO_2^-
d. CO

55. Iron(III) forms an octahedral complex with the ligand CN^-. How many unpaired electrons are in the d-orbitals of iron?
M

a. 1
b. 3
c. 5
d. 7

56. In a high-spin octahedral complex, the number of unpaired electrons in Fe(II) will be
H

a. 0
b. 2
c. 4
d. 6
e. none of the above

57. If a solution absorbs green light, what is its likely color?
E

a. red
b. violet
c. orange
d. yellow
e. blue

Short Answer Questions

58. Why is the +2 oxidation state so common among transition elements?
E

59. Give one important use of the following metals or their compounds.
E
a. chromium
b. manganese
c. mercury
d. silver

60. What is the difference between a coordination compound and a complex ion?
E

61. a. How can the formation of a complex ion be described in terms of a theory of acids and bases?
M
b. What is the essential requirement for a molecule or ion to act as a ligand?

62. What geometry is particularly common for complexes of d^{10} metal ions?
H

63. The compound $Rh(CO)(H)(PH_3)_2$ forms *cis* and *trans* isomers. Use this information to predict the geometry of this complex, and draw the geometric isomers.
M

64. a. State the requirement for two molecules to be optical isomers.
M
b. A complex ion $MABCD^{2+}$ (where A, B, C and D are different unidentate ligands) rotates the plane of polarized light. Deduce the geometry of the complex and draw the optical isomers of this ionic formula.

65. Apply the valence bond theory to predict the electronic structure and hybridization pattern of chromium in the complex ion $Cr(NH_3)_6^{3+}$.
H

66. a. The d_{xy} and the $d_{x^2-y^2}$ orbitals both lie in the *xy* plane, yet for a metal ion in an octahedral complex the energy of the d_{xy} orbital is lower than that of the $d_{x^2-y^2}$ orbital. Explain this using the arguments of crystal field theory.
H

67. a. How many unpaired 3*d* electrons will there be in (i) high and (ii) low-spin complexes of Co(II)?
M
b. How can high and low-spin complexes be recognized and distinguished experimentally?

68. a. Explain how the crystal field theory can use the magnitude of the splitting energy Δ to provide an explanation of the color and magnetic properties of octahedral complexes.
H
b. In promoting an electron from the t_{2g} set of orbitals to the e_g set, an octahedral complex absorbs a photon with a wavelength λ of 523 nm. Calculate the value of Δ in the complex, in kJ/mol.

True/False Questions

69. All atoms of the first transition series of elements have the ground state electronic configuration $[Ar]4s^2 3d^x$, where *x* is an integer from 1 to 10.
M

70. The M^{2+} ions of the first transition series of elements all have the general electronic configuration $[Ar]4s^2 3d^x$, where *x* is an integer from 1 to 8.
E

71. Of the 3*d* transition series of elements, scandium has the greatest atomic radius.
M

72. Of the 3*d* transition series of elements, zinc has the greatest atomic radius.
M

73. The maximum oxidation state of an element in the first transition series never exceeds its group number.
E

74. The inner transition series of elements arise from the filling of *f* orbitals.
E

75. All the actinide series of transition elements are radioactive.
E

76. The conversion of the chromate ion (CrO_4^{2-}) to the dichromate ion ($Cr_2O_7^{2-}$) is a redox process.
M

77. The permanganate ion (MnO_4^-) is a powerful reducing agent.
E

78. In complexes of transition metals, the maximum coordination number of the metal is equal to its number of d electrons.
M

79. Octahedral complexes can exhibit geometric, optical and linkage isomerism.
M

80. Tetrahedral complexes can exhibit both optical and linkage isomerism.
M

81. Square planar complexes can exhibit both geometric and optical isomerism.
M

82. Valence Bond theory rationalizes octahedral geometry by assuming a d^2sp^3 hybridization pattern.
M

The Transition Elements and Their Coordination Compounds
Chapter 23
Answer Key

1. a
2. b
3. c
4. d
5. a
6. d
7. a
8. b
9. d
10. d
11. c
12. b
13. b
14. b
15. c
16. a
17. d
18. d
19. b
20. c
21. b
22. b
23. b
24. c
25. b
26. e
27. d
28. c
29. d
30. a
31. b
32. c
33. d
34. c
35. b
36. a
37. c
38. c
39. a
40. a
41. b
42. e
43. b
44. b
45. d
46. b
47. c
48. b
49. c
50. b
51. a
52. d
53. c
54. b
55. a
56. c
57. a

58. The outermost (ns^2) electrons are easily lost, producing the +2 oxidation state.

59.
a. protective/decorative chromium plating; manufacture of stainless steel
b. component of some steels; MnO_2 is the oxidant in dry cells
c. thermometers; barometers; electrical switches; chlor-alkali process
d. photography

60. A coordination compound generally consists of a complex ion and a counter ion; a complex ion has a central metal and surrounding ligands.

61.
a. According to Lewis, in an acid-base reaction, the acid accepts an electron pair from a base. Thus, in the formation of a metal-ligand complex, the metal is an acid and the ligand is a base.
b. A ligand must have a lone pair of electrons available to donate in forming a bond with the metal.

62. tetrahedral

63. The complex has four ligands and there are geometric isomers; therefore it is square planar rather than tetrahedral.

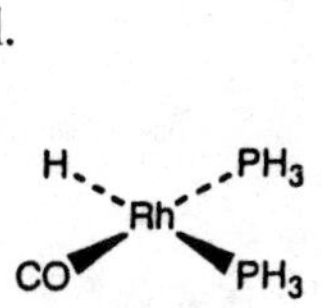

cis

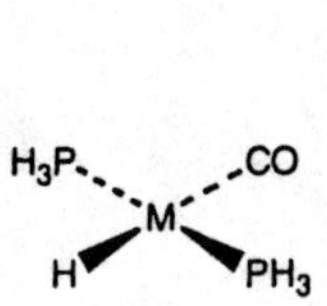

trans

64. (a) The molecules must be non-superimposable mirror images of each other. (b) The complex has four unidentate ligands and in order to display optical isomerism it must be tetrahedral in shape.

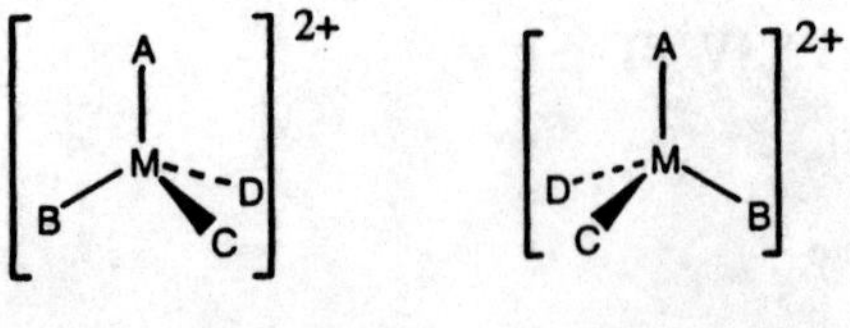

65. The three *d* electrons of the Cr(III) will occupy three 3*d* orbitals, singly. The remaining two 3*d* orbitals will be hybridized with the 4*s* and 4*p* orbitals, yielding six d^2sp^3 hybrid orbitals, which are used to accommodate the six lone pairs of the ammonia ligands.

66. The d_{xy} orbital lobes are directed between the *x* and *y* axes, whereas the $d_{x^2-y^2}$ orbital lobes point along these axes. Thus, four ligands with their lone pairs, approaching the central atom along the + and - directions of the *x* and *y* axes will interact more strongly (repulsively) with electrons in the $d_{x^2-y^2}$ orbital, raising it in energy above that of the d_{xy} orbital.

67. a. (i) high-spin, 3 unpaired electrons (ii) low-spin, 1 unpaired electron
b. unpaired electrons interact strongly with a magnetic field, and this property is used to determine the number of such electrons in a complex.

68. a. The energy needed for a photon to excite an electron from the lower to the higher set of the split *d* orbitals, will increase as Δ increases. This will mean the complex absorbs at shorter wavelengths, influencing its color. Also, a large splitting energy Δ will tend to produce low-spin complexes.
b. 229 kJ/mol

69. F

70. F

71. T

72. F

73. T

74. T

75. T

76. F

77. F

78. F

79. T

80. T

81. F

82. T

Nuclear Reactions and Their Applications
Chapter 24

Multiple Choice Questions

1. Who discovered radioactivity?
E

a. Geiger
b. Curie
c. Roentgen
d. Becquerel
e. Rutherford

2. Which one of the following is an incorrect representation of the indicated particle or nucleus?
E

a. positron: ${}^{0}_{1}\beta$
b. neutron: ${}^{1}_{0}n$
c. helium-3: ${}^{5}_{2}He$
d. alpha particle: ${}^{4}_{2}He$
e. proton: ${}^{1}_{1}p$

3. ${}^{108}_{49}In \rightarrow {}^{108}_{48}Cd + ?$
E

In the equation above, what particle or type of radiation needs to be included on the right hand side in order to balance it?

a. alpha
b. beta
c. gamma
d. positron
e. proton

4. Which one of the following equations correctly represents alpha decay of ${}^{222}_{86}Rn$?
M

a. ${}^{222}_{86}Rn \rightarrow {}^{218}_{84}Po + {}^{4}_{2}He$
b. ${}^{222}_{86}Rn \rightarrow {}^{220}_{82}Pb + {}^{2}_{4}He$
c. ${}^{222}_{86}Rn \rightarrow {}^{222}_{87}Fr + {}^{0}_{-1}\beta$
d. ${}^{222}_{86}Rn + {}^{4}_{2}He \rightarrow {}^{226}_{88}Ra$
e. ${}^{222}_{86}Rn \rightarrow {}^{222}_{85}At + {}^{0}_{1}\beta$

5. Which one of the following equations correctly represents positron decay of ${}^{49}_{19}K$?
M

a. ${}^{40}_{19}K \rightarrow {}^{36}_{17}Cl + {}^{4}_{2}He$
b. ${}^{40}_{19}K + {}^{0}_{-1}e \rightarrow {}^{40}_{18}Ar$
c. ${}^{40}_{19}K + {}^{0}_{1}e \rightarrow {}^{40}_{20}Ca$
d. ${}^{40}_{19}K \rightarrow {}^{40}_{20}Ca + {}^{0}_{-1}\beta$
e. ${}^{40}_{19}K \rightarrow {}^{40}_{18}Ar + {}^{0}_{1}\beta$

6. Which one of the following equations correctly represents electron capture by the $^{204}_{84}Po$ nucleus?
M

a. $^{204}_{84}Po \rightarrow ^{204}_{85}At + ^{0}_{-1}\beta$

b. $^{204}_{84}Po \rightarrow ^{204}_{83}Bi + ^{0}_{1}\beta$

c. $^{204}_{84}Po + ^{0}_{-1}e \rightarrow ^{204}_{83}Bi$

d. $^{204}_{84}Po + ^{0}_{-1}e \rightarrow ^{204}_{85}At$

e. $^{204}_{84}Po + ^{0}_{1}\beta \rightarrow ^{204}_{85}At$

7. Select the nuclide that completes the following nuclear reaction.
E

$^{147}_{62}Sm \rightarrow ? + ^{4}_{2}He$

a. $^{151}_{64}Gd$
b. $^{149}_{64}Gd$
c. $^{145}_{60}Nd$
d. $^{143}_{60}Nd$

8. Select the nuclide that completes the following nuclear reaction.
E

$^{216}_{84}Po \rightarrow ? + ^{4}_{2}He$

a. $^{212}_{82}Pb$
b. $^{214}_{82}Pb$
c. $^{220}_{86}Rn$
d. $^{218}_{86}Rn$

9. Select the nuclide that completes the following nuclear reaction.
E

$^{123}_{52}Te + ^{0}_{-1}e \rightarrow ?$

a. $^{122}_{51}Sb$
b. $^{122}_{52}Te$
c. $^{123}_{51}Sb$
d. $^{123}_{53}I$

10. Select the nuclide that completes the following nuclear reaction.
E

$^{50}_{23}V + ^{0}_{-1}e \rightarrow ?$

a. $^{50}_{22}Ti$
b. $^{50}_{24}Cr$
c. $^{51}_{23}V$
d. $^{49}_{23}V$

11. Select the nuclide that completes the following nuclear reaction.
E

$${}^{115}_{49}\text{In} \rightarrow {}^{0}_{-1}\beta + ?$$

a. ${}^{114}_{49}\text{In}$
b. ${}^{115}_{50}\text{Sn}$
c. ${}^{115}_{48}\text{Cd}$
d. ${}^{116}_{50}\text{Sn}$

12. Select the nuclide that completes the following nuclear reaction.
E

$${}^{40}_{19}\text{K} \rightarrow {}^{0}_{-1}\beta + ?$$

a. ${}^{39}_{19}\text{K}$
b. ${}^{39}_{20}\text{Ca}$
c. ${}^{40}_{18}\text{Ar}$
d. ${}^{40}_{20}\text{Ca}$

13. Select the nuclide that completes the following nuclear reaction.
E

$${}^{26}_{13}\text{Al} \rightarrow {}^{0}_{1}\beta + ?$$

a. ${}^{25}_{12}\text{Mg}$
b. ${}^{25}_{13}\text{Al}$
c. ${}^{26}_{12}\text{Mg}$
d. ${}^{26}_{14}\text{Si}$

14. Select the nuclide that completes the following nuclear reaction.
E

$${}^{58}_{27}\text{Co} \rightarrow {}^{0}_{1}\beta + ?$$

a. ${}^{58}_{26}\text{Fe}$
b. ${}^{58}_{28}\text{Ni}$
c. ${}^{57}_{26}\text{Fe}$
d. ${}^{57}_{27}\text{Co}$

15. An isotope with a high value of *N/Z* will tend to decay through
E

a. α decay
b. β decay
c. positron decay
d. electron capture

16. The radioisotope ${}^{90}_{37}\text{Rb}$ will decay through
M

a. α decay
b. β decay
c. positron decay
d. electron capture

17. An isotope with a low value of *N/Z* will generally decay through

E

a. α decay
b. β decay
c. γ decay
d. electron capture

18. The radioisotope $^{82}_{38}Sr$ will decay through

M

a. α decay
b. β decay
c. γ decay
d. electron capture

19. An isotope with $Z > 83$, which lies close to the band of stability, will generally decay through

E

a. α decay
b. β decay
c. γ decay
d. positron decay

20. The radioisotope $^{222}_{86}Rn$ will decay through

M

a. α-decay
b. β-decay
c. γ-decay
d. positron decay

21. The isotopes $^{20}_{10}Ne$, $^{21}_{10}Ne$ and $^{22}_{10}Ne$ are all stable, while $^{23}_{10}Ne$ is radioactive. The mode of decay for $^{23}_{10}Ne$ is most likely to be

M

a. positron decay
b. alpha decay
c. beta decay
d. electron capture
e. fission

22. The isotopes $^{28}_{14}Si$, $^{29}_{14}Si$ and $^{30}_{14}Si$ are all stable, while $^{27}_{14}Si$ is radioactive. The mode of decay for $^{27}_{14}Si$ is most likely to be

M

a. positron decay
b. alpha decay
c. beta decay
d. gamma decay
e. fission

23. The isotope $^{14}_{6}C$ is unstable because

M

a. $N/Z \neq 1$
b. N/Z is relatively low and $Z < 20$
c. N/Z is relatively large and $Z < 20$
d. Z is small

24. The isotope ${}^{42}_{21}Sc$ is unstable because
M

a. The number of neutrons is too large.
b. The number of neutrons equals the number of protons and the atomic number is greater than 20.
c. The atomic number is too large.
d. The mass number is too large.

25. Select the isotope that should be unstable.
M

a. ${}^{20}_{10}Ne$
b. ${}^{72}_{37}Rb$
c. ${}^{16}_{8}O$
d. ${}^{11}_{5}B$

26. Which of the following isotopes should be unstable?
M

a. ${}^{19}_{9}F$
b. ${}^{39}_{19}K$
c. ${}^{58}_{24}Cr$
d. ${}^{66}_{30}Zn$

27. Which of the following isotopes should be unstable?
M

a. ${}^{23}_{12}Mg$
b. ${}^{28}_{14}Si$
c. ${}^{38}_{18}Ar$
d. ${}^{46}_{22}Ti$

28. Which of the following isotopes should be unstable?
M

a. ${}^{150}_{62}Sm$
b. ${}^{197}_{79}Au$
c. ${}^{191}_{77}Ir$
d. ${}^{204}_{84}Po$

29. The isotopes of promethium, ${}^{144}_{59}Pr$ and ${}^{134}_{59}Pr$, are unstable. Which of the following combinations represents
H the type of decay for each isotope?

a. promethium-144, β decay; promethium-134, positron decay
b. promethium-144, positron decay; promethium-134, β decay
c. promethium-144, positron decay; promethium-134, electron capture
d. promethium-144, electron capture; promethium-134, positron decay

30. Detection of radiation by a Geiger-Müller counter depends on
M

a. the ability of a particle to excite an atom
b. the ability of a particle to ionize a gas
c. emission of a photon of light by the radioactive particle
d. the ability of a photomultiplier tube to amplify the electrical signal from a phosphor

31. A scintillation counter

M

a. measures the signal coming from an ionized gas.
b. measures light emissions from excited atoms.
c. depends on an avalanche of electrons generated as a particle moves through a tube of argon gas.
d. detects high energy radiation better than low energy radiation.

32. What is the specific activity (in Ci/g) of an isotope if 3.56 mg emits 4.26×10^8 β particles per second?

M

a. 0.003232
b. 0.0115
c. 0.309
d. 3.23

33. A certain isotope has a specific activity of 7.29×10^{-4} Ci/g. How many α particles will a 75.0 mg sample emit in one hour?

H

a. 9.99×10^4
b. 2.02×10^6
c. 7.28×10^9
d. 1.29×10^{12}

34. The radiochemist, Will I. Glow, studied thorium-232 and found that 2.82×10^{-7} moles emitted 8.42×10^6 α particles in one year. What is the decay constant for thorium-232?

M

a. 3.35×10^{-14} yr^{-1}
b. 4.96×10^{-11} yr^{-1}
c. 1.40×10^{10} yr^{-1}
d. 2.99×10^{13} yr^{-1}

35. A 7.85×10^{-5} mol sample of copper-61 emits 1.47×10^{19} positrons in 90.0 minutes. What is the decay constant for copper-61?

M

a. 0.00230 h^{-1}
b. 0.00346 h^{-1}
c. 0.207 h^{-1}
d. 0.311 h^{-1}

36. The isotope $^{28}_{12}Mg$ has a half-life of 21 hours. If a sample initially contains exactly 10000 atoms of $^{28}_{12}Mg$, approximately how many of these atoms will remain after one week?

E

a. 1250
b. 78
c. 39
d. 0
e. None of the above

37. The isotope $^{179}_{79}Au$ has a half-life of 7.5 seconds. If a sample contains 144 atoms of $^{179}_{79}Au$, approximately how many such atoms were there present 30 seconds earlier?

E

a. 576
b. 1152
c. 2304
d. 4320
e. 4.30×10^8

38. A 9.52×10^{-5} mol sample of rubidium-86 emits 8.87×10^{16} β particles in one hour. What is the half-life of rubidium-86?
H

a. 2.23×10^{-3} h
b. 1.55×10^{-3} h
c. 448 h
d. 645 h

39. Iodine-131, $t_{1/2} = 8.0$ days, is used in diagnosis and treatment of thyroid gland diseases. If a laboratory sample of iodine-131 initially emits 9.95×10^{18} β particles per day, how long will it take for the activity to drop to 6.22×10^{17} β particles per day?
M

a. 2.0 days
b. 16 days
c. 32 days
d. 128 days

40. Cesium-134 is a β emitter with a half-life of 2.0 years. How much of a 2.50-g sample of cesium-134 will remain after 10 years?
M

a. 0.0024 g
b. 0.078 g
c. 0.25 g
d. 0.50 g

41. Palladium-107 undergoes β decay ($t_{1/2} = 6.5 \times 10^5$ yr) to form silver-107. How long will it take for 0.150 mol of silver-107 to form from 1.25 mol of palladium-107?
H

a. 8.3×10^5 y
b. 1.2×10^6 y
c. 1.4×10^7 y
d. 2.0×10^7 y

42. A pure sample of tritium, ^3H, was prepared and stored for a number of years. Tritium undergoes β decay with a half-life of 12.32 years. How long has the container been sealed if analysis of the contents shows there are 5.25 mol of ^3H and 6.35 mol of ^3He?
E

a. 2.34 y
b. 3.38 y
c. 9.77 y
d. 14.1 y

43. Identify the missing species in the following nuclear transmutation.
E

$$^{45}_{21}Sc(n, ?)^{42}_{19}K$$

a. $^{3}_{2}He$
b. $^{4}_{2}He$
c. $2\ ^{2}_{1}H$
d. $3\ ^{1}_{1}H$

44. Identify the missing species in the following nuclear transmutation.
E

$^{16}_{8}O(n, ?)^{1}_{1}H$

a. $^{17}_{8}O$
b. $^{15}_{7}N$
c. $^{16}_{7}N$
d. $^{15}_{9}F$

45. Identify the missing species in the following nuclear transmutation.
E

$^{246}_{96}Cm + ^{12}_{6}C \rightarrow 4\ ^{1}_{0}n + ?$

a. $^{254}_{102}No$
b. $^{258}_{102}No$
c. $^{238}_{98}Cf$
d. $^{238}_{90}Th$

46. Identify the missing species in the following nuclear transmutation.
E

$^{238}_{92}U + ? \rightarrow 5\ ^{1}_{0}n + ^{249}_{100}Fm$

a. $^{11}_{8}O$
b. $^{16}_{8}O$
c. $^{16}_{13}Al$
d. $^{11}_{3}Li$

47. An 85-kg person exposed to barium-141 receives 2.5×10^{5} β particles, each with an energy of 5.2×10^{-13} J.
H How many rads does the person receive?

a. 2.4×10^{-20}
b. 1.5×10^{-7}
c. 1.8×10^{-16}
d. 6.1×10^{-15}

48. A 55-kg person exposed to thorium-234 receives 7.5×10^{4} β particles, each with an energy of 1.6×10^{-14} J.
H How many rads does the person receive?

a. 2.1×10^{-19}
b. 1.2×10^{-17}
c. 2.2×10^{-9}
d. 1.2×10^{-9}

49. A 30.0-kg child receives 2.65×10^{7} β particles, each with an energy of 4.60×10^{-13} J. If the RBE = 0.78, how
H many millirem did the child receive?

a. 3.2×10^{-7}
b. 5.2×10^{-7}
c. 5.2×10^{-4}
d. 3.2×10^{-2}

50. H A patient's thyroid gland is to be exposed to an average of 5.5 μCi for 16 days as an ingested sample of iodine-131 decays. If the energy of the β radiation is 9.7×10^{-14} J and the mass of the thyroid is 32.0 g, what is the dose received by the patient?

a. 0.027 rad
b. 1.2 rads
c. 37 rads
d. 85 rads

51. M Exposure to 10 nCi for 10 minutes is more hazardous for a child than for an adult because

a. the child's cells are dividing more rapidly than the adult's and are, therefore, more susceptible to the radiation.
b. the child's smaller body size makes the effective dose larger for the child than for the adult.
c. the child's immune system is not developed well enough to resist damage.
d. the child's skin is not as thick as an adult's and cannot block as much radiation.

52. M Carbon-14 will emit a β particle with an energy of 0.1565 MeV. What is this energy in joules?

a. 1.0×10^{-24}
b. 2.5×10^{-20}
c. 1.0×10^{-18}
d. 2.5×10^{-14}

53. M Sodium-21 will emit positrons each having an energy of 4.0×10^{-13} J. What is this energy in MeV?

a. 4.0×10^{-7}
b. 2.5
c. 40
d. 2.5×10^{6}

54. H Calcium-39 undergoes positron decay. Each positron carries 5.49 MeV of energy. How much energy will be emitted when 0.0025 mol of calcium-39 decays?

a. 13.2 kJ
b. 1.32×10^{4} kJ
c. 1.32×10^{6} kJ
d. 1.32×10^{9} kJ

55. M Which of the following materials is put into a nuclear reactor to slow the chain reaction?

a. heavy water
b. moderators
c. control rods
d. reflectors

56. It is believed that two carbon-12 nuclei can react in the core of a supergiant star to form sodium-23 and
H hydrogen-1. Calculate the energy released from this reaction for each mole of hydrogen formed. The masses of carbon-12, sodium-23, and hydrogen-1 are 12.0000 amu, 22.989767 amu, and 1.007825, respectively.

$$^{12}_{6}C + ^{12}_{6}C \rightarrow ^{23}_{11}Na + ^{1}_{1}H$$

a. 2.16×10^{14} kJ
b. 2.16×10^{11} kJ
c. 2.16×10^{8} kJ
d. 2.16×10^{5} kJ

57. Which one of the following elements is formed largely in supernova explosions?
M

a. H
b. He
c. Mg
d. Fe
e. U

Short Answer Questions

58. Fill in missing sub- and superscripts for all particles to complete the following equation for alpha decay
E

$$^{257}_{100}Fm \rightarrow Cf + He$$

59. Fill in missing sub- and superscripts for all particles to complete the following equation for beta decay
E

$$^{35}_{16}S \rightarrow Cl + \beta$$

60. Fill in missing sub- and superscripts for all particles to complete the following equation for positron decay
E

$$^{64}_{29}Cu \rightarrow Ni + \beta$$

61. Write a complete, balanced equation to represent the alpha decay of radon-210.
M

62. Write a complete, balanced equation to represent the beta decay of thallium-207.
M

63. Write a complete, balanced equation to represent the electron capture decay of argon-37.
M

64. Write a complete, balanced equation to represent the formation of manganese-55 by the beta decay
H of another nuclide.

65. Explain how the number of protons and neutrons in a radioactive nucleus can be used to predict its
M probable mode of decay. Illustrate your answer with a schematic graph, properly labeled, showing stable nuclides (nuclei) in relation to number of protons and neutrons.

66. A bottle of vintage red wine has lost its label. The concentration of tritium ($^{3}_{1}H$) in the wine is 0.34
M times that found in freshly bottled wines. If the half-life of tritium is 12.3 years, estimate the time elapsed since the wine was bottled.

67. Bombardment of uranium-238 nuclei by carbon-12 nuclei produces californium-246 and neutrons.
H Write a complete, balanced equation for this nuclear process.

68. Briefly, explain the relationship between the rad and the rem as units of radiation dosage.
H

69. When an electron and its anti-particle, a positron, collide, they annihilate each other. Calculate the
M energy released in this process, in J. (The positron mass is the same as the electron mass, namely 9.11×10^{-31} kg.)

70. Calculate to four significant figures
H
a. the mass defect in kg and
b. the energy released in kJ/mol, when a neutron decays to produce a proton and an electron. The neutron, proton and electron masses are 1.67493×10^{-27} kg, 1.67262×10^{-27} kg and 9.10939×10^{-31} kg, respectively.

71. The masses of a potassium-40 atom, a proton and a neutron are 39.963999 amu, 1.007825 amu and
H 1.008665 amu, respectively. Calculate to four significant figures
a. the mass defect in amu and
b. the energy released in MeV/nucleon, in the formation of ${}^{40}_{19}K$ from the appropriate number of protons and neutrons.

72. What is the mechanism by which control rods slow down the fission rate in a nuclear reactor?
M

73. Of the naturally-occurring elements on earth today, identify by their chemical symbols:
H

a. two which would have resulted directly from the "big bang"
b. one which can only be formed in supernova explosions
c. two which are formed during the normal life of first generation stars

74. What features do the r- and s-processes for element formation have in common? How do they differ?
H

True/False Questions

75. Gamma rays are high energy electrons.
E

76. Gamma rays are not deflected by an electric field.
E

77. Positron decay and electron capture have the same net effect on the Z and N values of a nucleus.
M

78. No alpha decay is observed for isotopes of elements with $Z < 83$.
M

79. Radioactive decay follows zero-order kinetics.
M

80. After 4 half-lives, the fraction of a radioactive isotope which still remains is approximately one eighth.
M

81. Most foodstuffs contain natural, radioactive isotopes.
M

82. The (negative) binding energy per nucleon reaches a maximum for the isotope $^{12}_{6}C$.
E

83. The r-process occurs during supernova explosions.
M

84. The s-process involves a slow succession of neutron absorption and beta decay processes during the normal life of a star.
M

Nuclear Reactions and Their Applications
Chapter 24
Answer Key

1.	d	20.	a	39.	c
2.	c	21.	c	40.	b
3.	d	22.	a	41.	b
4.	a	23.	c	42.	d
5.	e	24.	b	43.	b
6.	c	25.	b	44.	c
7.	d	26.	c	45.	a
8.	a	27.	a	46.	b
9.	c	28.	d	47.	b
10.	a	29.	a	48.	c
11.	b	30.	b	49.	d
12.	d	31.	b	50.	d
13.	c	32.	d	51.	b
14.	a	33.	c	52.	d
15.	b	34.	b	53.	b
16.	b	35.	c	54.	c
17.	d	36.	c	55.	c
18.	d	37.	c	56.	c
19.	a	38.	c	57.	e

58. $^{257}_{100}\text{Fm} \rightarrow {}^{253}_{98}\text{Cf} + {}^{4}_{2}\text{He}$

59. $^{35}_{16}\text{S} \rightarrow {}^{35}_{17}\text{Cl} + {}^{0}_{-1}\beta$

60. $^{64}_{29}\text{Cu} \rightarrow {}^{64}_{28}\text{Ni} + {}^{0}_{1}\beta$

61. $^{210}_{86}\text{Rn} \rightarrow {}^{206}_{84}\text{Po} + {}^{4}_{2}\text{He}$

62. $^{207}_{81}\text{Ti} \rightarrow {}^{207}_{82}\text{Pb} + {}^{0}_{-1}\beta$

63. $^{37}_{18}\text{Ar} + {}^{0}_{-1}\beta \rightarrow {}^{37}_{17}\text{Cl}$

64. $^{55}_{24}\text{Cr} \rightarrow {}^{55}_{25}\text{Mn} + {}^{0}_{-1}\beta$

65. The "line of stability" on the accompanying diagram shows the region in which the combination of the number of neutrons (*N*) and the number of protons (atomic number, *Z*) tends to result in stable nuclides. Nuclides lying on either side of the line, or beyond its end, are radioactive. As a general rule, the nuclide formed in a radioactive decay process tends to lie closer to the line of stability than the one from which it is formed. Thus, nuclides with relatively large *N/Z* tend to undergo beta decay, those with low *N/Z* undergo electron capture or positron decay, and those with $Z > 83$ tend to undergo alpha decay.

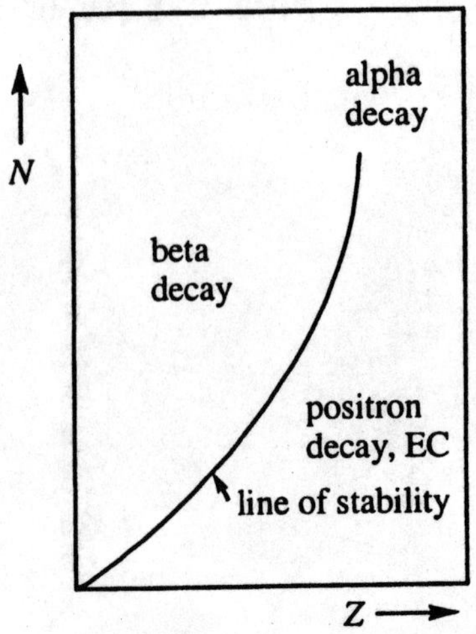

66. 19 years

67. $^{238}_{92}U + ^{12}_{6}C \rightarrow ^{246}_{98}Cf + 4^{1}_{0}n$

68. The rad (radiation absorbed dose) measures the energy of the absorbed radiation per unit mass of absorbing tissue. The rad does not take into account the relative effects of different kinds of radiation, the exposure time and the type of tissue, which the rem (roentgen equivalent for man) does. 1 rem = 1 rad × RBE. The RBE (relative biological effectiveness) depends on these factors, making the rem a more useful unit than the rad in assessing exposure.

69. 1.64×10^{-13} J

70. a. 1.399×10^{-30} kg
b. 7.572×10^{7} kJ/mol

71. a. 0.3666 amu
b. 8.538 MeV/nucleon

72 Control rods absorb neutrons which would otherwise be absorbed by nuclei in the fuel rods, inducing fission and carrying on the chain reaction. Absorption of some neutrons slows down the rate of the reaction.

73. a. H, He
b. U
c. C, Fe, O, Mg and many others

74. Both involve neutron absorption by a nucleus, followed by beta decay, thus producing a nucleus of greater atomic number *Z*. The s-process is very gradual, occurring over the lifetime of a star. The r-process is very rapid, involving multiple neutron absorption and beta decay events during the short time of a supernova.

75. F
76. T
77. T
78. F
79. F
80. F
81. T
82. F
83. T
84. T